Environmental History

Volume 4

Series editor

Mauro Agnoletti, Florence, Italy

More information about this series at http://www.springer.com/series/10168

Mauro Agnoletti · Simone Neri Serneri
Editors

The Basic Environmental History

Springer

Editors
Mauro Agnoletti
DEISTAF
University of Florence
Florence
Italy

Simone Neri Serneri
Political and International Sciences
University of Siena
Siena
Italy

ISSN 2211-9019 ISSN 2211-9027 (electronic)
ISBN 978-3-319-35188-9 ISBN 978-3-319-09180-8 (eBook)
DOI 10.1007/978-3-319-09180-8

Springer Cham Heidelberg New York Dordrecht London

Printed on acid-free paper

Springer is part of Springer Science+Business Media (www.springer.com)

Environmental History and other Histories. A Foreword

Environmental history has by now acquired a history of its own. The theme has been treated by generations of scholars who have produced a great number of research studies and compared notes and findings in numerous conferences, associations and academic journals. The fields of interest are many and varied, as are the methods of survey, which have often matured at the crossroads between arts and humanities, social and natural sciences.

What is Environmental History?

The recurring debate on "what is environmental history?" has received numerous and basically converging responses. One of the most concise considers that its purpose is the study of "man and the rest of nature". A decidedly controversial definition in respect of the distinction, when not contraposition, between the human world and the natural world, underlying dominant cultural and scientific tradition, not only in historical studies, in the modern world. With regard to the object and to the end proposed by studies in environmental history, it would, however, appear more incisive to speak of a discipline that has the purpose of studying the relationships between man and the environment in their historical dynamics.

The definition presents various original heuristic implications, but ultimately it is probably more suitable and tends to suggest a holistic approach to the history of man and nature. An approach which, moreover, is widespread among environmental historians, largely derived from studies in natural history, historical ecology, forest history, historical geography and concerned primarily with delineating the numerous changes in the natural environment—from the history of climate change, to changes in landscape or forest cover, from the history of natural disasters to that of epidemics or the variation in animal species, which have been induced by or, on the contrary, condition man's social life. Furthermore, the above-mentioned disciplines remind us that the history of relationships between man and nature did not begin with studies in environmental history, nor with the work by John Perkins

Marsh, but had already been put forward in the early eighteenth century in Germany with the work of Friedrich Stisser. That definition and that approach, however, risk depicting the relationship between human societies and the natural world in excessively naturalistic terms, thus overshadowing the tension between the two areas or considering it as solved. The natural world and human societies are more easily understandable when they are considered as two systemic and complex realities, fully interactive with each other. The dynamics of the natural world, or, better, of the ecosystems and the dynamics of anthropic societies are the most strongly interactive with each other because they rest on the same material, physical, chemical and biological base. But for this very reason, an irreducible state of tension is created which sometimes opens the way to widespread conflict.

In history the tension between anthropic dynamics and ecological dynamics has always been an evident reality, albeit with different modes, intensities and outcomes. It was during the twentieth century, however, that it developed and expressed its explosive power. The main cause for this marked discontinuity was technological development which basically reversed the relationship of dependency between the environmental context and the anthropic context; since then, at least in the short term, human societies have been more successful in adapting ecosystems to their needs rather than the reverse, as occurred previously.

The enormous and, at times, threatening consequences of this change in reciprocal adaptability remind us that—as Donald Worster noted—men are more than ever simultaneously agents and victims of environmental history. But they also induce us not to stop at considering only the most sensational changes in landscape, extinction of animal species or the most conspicuous forms of pollution and to perceive behind these phenomenons the emergence of the most critical forms of tension intrinsic in the constant interaction between the reproductive dynamics of anthropic and environmental systems. These reproductive dynamics proceed through a partial, yet continuous, reciprocal incorporation between the two systems. In turn, this incorporation occurs with processes and intensities which are mediated and progressively redefined by available technology. The outcomes are the consequence of the interaction between reproductive mechanisms and therefore reflect the capacity of anthropic and environmental systems to reproduce through a succession of equilibrium and disequilibrium phases. Increasingly over the last century and latter decades, the negative effects of the dynamics between man and nature have become more and more evident. As a consequence of the rapid change in environmental structures, the sustainability of the reproduction processes of anthropic systems—those that permit the satisfaction of basic needs and the more complex manifestations of social life—has become more and more uncertain. Moreover, the very concept of sustainability, however widespread in political spheres, is subject to growing criticism in scientific circles. The idea of the sustainability of development based on the conservation of a determined quota of systems defined as "natural", is largely a cultural construction given that, strictly speaking, systems that are really natural are now very limited on a planetary scale. More often it is naturalness on the rebound after previous anthropic impacts, or semi-naturalness, whereas the sustainability necessary for the life of man refers to environmental parameters, rather than to quotas of naturalness for the conservation of various animal and vegetable species. The return to nature proposed by much of environmental literature, as a remedy for the disequilibria referred to above, at least from the nineteenth century onwards, is in effect largely the result of the cultural hegemony of currents of thought in Northern Europe and North America which have imposed the value of natural landscapes on that of cultural landscapes which for four or five centuries have represented the template, as described in the Grand Tour literature.

The aim of environmental history is, therefore, to rebuild the relationships and interactions between anthropic and environmental systems, as they were historically set up. Environmental history moves from its awareness of the relative autonomy that characterises the reproductive dynamics of both. It is gradually freeing itself of the merely conservationist perspective that has characterised and still characterises most of its approaches, because its object of study is strictly the changing transformative equilibrium that is set up between social systems and ecosystems. In fact, the relationships between them have anything but a static nature, but rather processual, because it stretches over time and is therefore eminently historical. In other words, historicity is an intrinsic quality in relationships between anthropic and environmental systems precisely because they interact during their respective reproduction processes which, far from reproducing their initial conditions—have a developmental and transformative nature. It also follows that historicity is manifold, if we consider the different levels over which it spreads—"historical times, biological times" wrote Enzo Tiezzi over 30 years ago—but profoundly unitary because anthropic and environmental systems are ultimately part of the same context: the former are, however, an expression of one of the most specialised of the innumerable biological forms that populate the latter.

In conclusion, environmental history is, by definition, a field of tension. Not only, as referred above, because attention can be calibrated to the relationship between man and the rest of nature, privileging either its unitary profile or internal dualism. But—and this is the aspect that most interests us—because, while it develops as a distinct disciplinary area, at the same time it proposes to be a means of critical comparison with more consolidated areas of historical research: economic history, urban history, the history of technology, the history of ideas and cultural history, the history of public policies and, last but not least, social history. On the other hand, it is no coincidence that many scholars from the above-recalled fields of research have become animators of environmental history, bringing with them debatable issues fuelled by the motivating force, sensitivity and knowledge of environmentalist mobilisation which in the 1970s spread throughout Europe, the United States and more widely in Asia, Africa and the American continent. And indeed they have impregnated environmental history research with traditions and cultural and social experiences from their various areas of origin.

Another Point of View: Themes and Suggestions

The essays in this volume mainly reflect this acceptation of environmental history and aim to compare, stimulate and even contest widely consolidated knowledge and compartmentation of predominant historiography. Altogether, the collection of essays make the book first and foremost an introductory instrument to the main themes of environmental history, illustrating its development over time, methodological implications, results achieved and those still under discussion. However, the problem is not that of proposing environmental history as another, distinct and, as such, delimited disciplinary area in search of legitimacy in its own right. Or to offer an overview of the main research studies and consequently the potentialities of environmental history. Quite the opposite, for the overriding aspiration is to show that the doubts, methods and knowledge elaborated by environmental history have a heuristic value that is far from negligible precisely in its attitude to the most consolidated major historiography. For this reason, this book gives an overview of the main themes of environmental history as it is an essential component of the basic knowledge of global history. But, at the same time, it introduces specific aspects which are useful both for anyone wanting to deepen his/her studies of environmental historiography and for those interested in one of the many disciplinary areas—from rural history to urban history, from the history of technology to the history of public health, etc.—with which environmental history, often with some difficulty, develops a dialogue.

The choice of themes, therefore, is not encyclopaedic, but intentionally selective. The expositive approach does not consider environmental history from within, as a primary disciplinary area, nor does it illustrate the making of this historiography. On the contrary, it endeavours to place environmental issues within a much wider field of research and its manifold thematic stratifications. Least of all, the book intends to denounce the gravity of environmental issues—not because they are not serious or worthy of denunciation—but because its concern is primarily with promoting knowledge of the past rather than recounting the present-day crisis.

Circumscribed, but nonetheless challenging, tasks. We hope to succeed in our undertaking. Nor is it the task of the book, let alone of this introduction, to identify dominating lines in the environmental history of the planet, or of any other continent or other thematic area. We do not propose to give a brief outline of the environmental history of the planet or part of it. Many already exist, albeit frequently characterised by limits and typical of attempts to reduce to a global-scale processes that are decidedly more complex which can only be studied on a local scale. We shall merely summarise introductory knowledge, but also—while making no claim to sufficiency or exclusivity—propose methods and analytical and interpretative concepts, the fruit of long and qualified experience acquired by the authors of the essays in their respective areas of research and, more in general, of their in-depth knowledge of European and global environmental historiography.

Various essays have different approaches. All share a comprehensive overview of their own theme and develop a narration that necessarily leaves in the background the history of policies and practices and environmental conflicts. But the choice of the central theme and expositive style responds to different criteria, because the preference is given to descriptive and interpretative efficacy rather than to analytical orderliness. In some cases, a certain environmental medium has been used as barycentre: soil, air and water. In others, a process, such as growth, has been taken as the main theme, and a certain factor, like energy or the interaction between a multitude of factors has been considered. Or, again, production and reproduction processes have been used as a reference, to examine, in one case, waste and residues and, in another, the most acute and serious critical manifestations, chiefly those caused by inappropriate, and therefore risky technologies. Lastly, in another case, the chief observation point is the urban structure that organizes media, resources and processes. Without prejudice to these distinctions, echoes of each of these different approaches can easily be perceived in all the essays.

Likewise, various asymmetries are also seen in the capacity of each essay to communicate critically with the other historical disciplines: a capacity that is unquestionably evident and incisive in the case of urban history or, for example, of economic growth problems or the role of energy, but—on the contrary—forcedly more restrained in the case of environmental history of the soil, an area of investigation still in its infancy. Each essay deals with numerous distinct themes and those that generally circulate, return and in various ways aggregate all together in the essays. Particularly worthy of attention is the vast theme of growth, in the sense of material and, consequently, economic growth, because it deals with the connection between nature and social development, growth being none other than the use of natural resources to the advantage of human society. So to study growth from the viewpoint of environmental history means not only proposing responses to many aporias or highlighting choices, paths, crises, etc., but—as Tello and Javier recount in their essay—explaining how economic growth takes place. On the other hand, precisely the theme of growth shows how the nature/society connection has an intrinsic historicity, because its processuality not only determines different ways of realization—depending on the various factors available—but determines its cyclicity, since the availability of resources depends on their characteristics and therefore is a constitutive rather than a marginal growth factor.

The other theme that is closely linked and, to a large extent, recurrent since it is crucial in mediating between nature and society, is technological development. Technology is the means by which portions of nature become available resources for the productive and reproductive processes of anthropic societies: it is the instrument of what the economists call their valorisation, in other words, of their utilisation for economic and social development. So technology—with its specific modes of action—largely determines the methods, intensities and outcomes of the incorporation of part of the ecosystems in anthropic processes. Also for this reason, the technological question largely characterises and supports many essays in the book. It applies to the use of soil, especially after agricultural practices underwent great innovation with the advance of industrialization. But of similar relevance is the story of water, air or waste or, evidently, risks, accidents and disasters caused by the use of technology in industrial society. It is understandably at the centre of the environmental history of urban systems which, by definition, are the outcome of the functional integration of numerous technologies aimed at diversifying and articulating the social life of a multitude of people and, at the same time, making it less dependent upon nature's reproduction cycles.

In other words, it is evident that the themes dealt with, the approaches and methods of research and interpretative proposals—far from being self-referential and determined by ideological and militant impulses—establish a close, albeit critical, dialogue with the questions and results of consolidated major historiography. Environmental history has the merit of broadening the view of historical reflection. Because, metaphorically speaking, it forces taking into consideration other points of view, other methods of knowledge and other disciplinary competences. But also in a real sense, because environmental history has an intrinsic spatial dimension that is difficult to define, since it continually calls upon the cohesion or concatenation of ecosystems and always refers to the direct connections that unite the local context to the global context.

Even a brief overall consideration confirms that the essays in this book have several common and peculiar traits which deserve to be stressed because they highlight the richness of the environmental historical approach. Only apparently more extrinsic is the question of periodisation, the conceptual barycentre of every historical reflection. In a formal consideration, the periodisation adopted varies in the different essays: in one respect it is easy to perceive the tendency to stretch backwards in respect of the present in search of anchors to account for the body of changes, but also their different ways of gathering together. In another respect there is a common second tendency to concentrate narration in the centuries that are closest to us. This arrangement is partly for practical reasons—to respond to present-day doubts—but above all derives from environmental history's historiographic solicitations: over the last two centuries anthropic societies have succeeded in making an unparalleled and exceptional impact on the natural world leading to an undoubted acceleration in the history of environmental changes. Those changes have always occurred, sometimes with important, indeed catastrophic, consequences in local and regional and even continental contexts—suffice it to recall the so-called "Columbian exchange" which followed the mass arrival of Europeans on the American continent—but from the end of the eighteenth century, they acquired an unprecedented rhythm, intensity and extension on global scale.

Generally speaking, the essays do not, however, treat their respective themes in a systematically global dimension, aimed at embracing the entire planet as a whole. They do, however, endeavour, with inevitably diverse possibilities and results—to assume a worldwide perspective that takes into account the plurality of the planet's experiences, their connections in history and in the present. Within these coordinates it is easy to perceive first that environmental history is simultaneously the history of relationships between anthropic systems and ecosystems and the history of man's knowledge of nature, as well as the history of the policies and practices that have consequently been implemented. So, for example, the environmental history of soil tells us about technical knowledge, agricultural practices, the culture of agricultural societies, which have characterised much of human history. But it looks at anthropic practices (the use of forests, livestock breeding, cultivations, irrigation, etc.), hinging on the ecosystem in which they are immersed and which they influence, in the awareness that those practices are within that ecosystem; they are the mode of constructing man's ecological niche. So they do not alter a given equilibrium in itself, but introduce themselves into transformative dynamics that are wide-ranging and more complex. An analytical perspective reminds us that the natural, environmental dimension is a constituent of anthropic practices, not only preliminary to them.

This observation should, in turn, be placed in relation to another which, as various essays suggest, attributes to technology—insofar as it is a crucial instrument of mediation between nature and society—a key role in determining the periodisation of environmental history, marked by the transition between successive states of equilibrium between social structures and ecosystems. Not because technological innovations shape periodisation deterministically, maybe after the hypothetical formation of environmental bottlenecks caused by the obsolescence of a technology and a corresponding depletion of a primary resource. But rather because the transition to different ways of relating between society and environment—for example in the epochal transition to the large-scale exploitation of fossil fuels, the treatment of urban waste, the change in use of agricultural land, etc.—hinges on technological innovations which at times are seen to be comparatively more remunerative as much in terms of cost as in use value, in the exploitation of one natural resource or another which they allow to be incorporated in social reproduction processes. So even in this regard anthropic dynamics—those relating to the economic profitability of a certain technology—and ecosystem dynamics—deriving from its environmental impact—are inextricably intertwined.

The integration between social factors and ecological factors is in fact at the centre of the analytical and interpretative models proposed by environmental history. Whether the approach is "socio-metabolic", borrowed from ecological economy, "urban metabolism" or "ecological heritage", the essays in this book prove their originality and fecundity, compared with traditional approaches which consider development and social changes determined almost exclusively by intrinsic cultural or institutional factors. To consider the capacity, or lack of it, to introduce portions of ecosystems into anthropic systems and the methods for realizing it, as decisive explicative factors of the dynamics of social development is however an extremely innovative and promising approach. Mainly for two reasons: First because it calls for greater attention to the quantity and quality of the overall patrimony of available resources—in the various contexts—for social development. Second, and more in general, because it prompts the abandonment of a solipsistic, accumulative and linear vision of social development and invites consideration of the fact that the circulation of resources (between ecosystems and anthropic systems, but to a likewise significant extent also within these) fuels close interaction between the various systems.

That interaction, and the flows and exchanges that fuel it—even more so following the epochal changes induced by the advent of urban-industrial society—frustrate all investigations that consider social development separately, territory by territory and country by country. But they impose the repositioning of development processes in a multiplicity of spatial, local, regional and global contexts that accounts for the procurement of the resources that fuel them, the dislocation of residue from anthropic processes and above all of the interaction and accumulation phenomenons consequent to those flows. The result is a conception of development as a composite and plural process, of variable intensity, with a helical trend and partially reversible. The only one that makes it possible to explain the otherwise misleadingly defined "aporias" of development and to fully assess the sustainability of present social and ecosystem structures, if not of future ones. Because, even in the case of environmental history, although knowledge of the past does not place us in a position to foresee the future, it undoubtedly gives us a better understanding of the times in which we live.

Mauro Agnoletti
Simone Neri Serneri

Contents

Editors and Contributors

About the Editors

Mauro Agnoletti is an Associate Professor at the University of Florence, where he teaches landscape planning and environmental history at the Faculty of Agriculture. He has an abilitation to full-time Professor in Landscape Planning and Economic History. Most of his studies and activities have been dedicated to forest and landscape history and to transfer research findings into policies. He chairs the unit on landscape policies at the Italian Ministry of Agriculture, Food and Forestry. He is a scientific expert for UNESCO, CBD, European Landscape Convention, FAO. He is a codirector of the scientific "Journal Global Environment" and member of the board of the International Association of Environmental History Organizations. He has produced more than 120 scientific articles and 20 books. www.landscape.unifi.it.

Simone Neri Serneri completed his Ph.D. in History at the University of Pisa (Italy). He is a full-time Professor of Contemporary History at the Department of Political and International Sciences at the University of Siena (Italy) and Director of the Istituto Storico della Resistenza in Toscana (Florence). He is a member of the editorial board "Global environment" and "Contemporanea. Rivista di storia dell'800 e del 900". He has been a member of the Board and Italian Regional Representative of the European Society for Environmental History. In the field of environmental history, he researched mainly about urban and industrial development, water resources and pollution and environmental policies in Italy from the late nineteenth century to the present. He is author of *Incorporare la natura. Storie ambientali del Novecento* [Rome, 2005] and many articles in collective books and co-edited the books *Industria, ambiente e territorio. Per una storia ambientale delle aree industriali in Italia* [Bologna, 2009]; *Storia e ambiente. Città, risorse e territori nell'Italia contemporanea* [Rome, 2007] and the on line *World environmental history* by Eolls.

About the Contributors

Sabine Barles is a Professor at the University Paris 1 Panthéon-Sorbonne and a member of the laboratory Géographie-Cités (University Paris 1, University Paris 7 and French National Research Council). She is a Civil Engineer (1988) and obtained a Master's degree in Urbanism (1989) and a Master's degree in History of Technology (1990) and later a Ph.D. in Urbanism (1993). The focus of her research is the history of technology and of the urban environment and the interactions between societies and nature (eighteenth–twentieth centuries) through urban metabolism and urban and territorial ecology. She has published *La ville délétère. Médecins et ingénieurs dans l'espace urbain, XVIIIe–XIXe siècles* (Seyssel, Champ Vallon, 1999) and *L'invention des déchets urbains, France*, 1790–1970 (Seyssel, Champ Vallon, 2005), and articles and chapters of books about the urban environment, mostly about Paris (see for instance Barles "The Seine and Parisian Metabolism: Growth of Capital Dependencies in the nineteenth and twentieth Centuries", *in*: Castonguay, S., Evenden, M.D. (eds.), *Urban Waters: Rivers, Cities and the Production of Space in Europe and North America,* Pittsburgh: Pittsburgh University Press, 2012, p. 94–112; Billen, G., Garnier, J., Barles (eds.), Special issue "History of the urban environmental imprint", *Regional Environmental Change* 12(2), 2012).

Stéphane Frioux is an Assistant Professor of History at the Université Lyon 2, France, where he teaches European modern history and urban history, and research in urban environmental history at the Laboratoire de recherche historique Rhône-Alpes (UMR CNRS 5190 LARHRA). He published *Les batailles de l'hygiène. Villes et environnement de Pasteur aux Trente Glorieuses* (Paris, PUF, 2013), in which he examines the municipal policies of environmental sanitation and the implementation of water and waste treatment facilities in French cities in the first half of the twentieth century. Among his articles, "At a green crossroads: recent theses in urban environmental history in Europe and North America", *Urban History*, vol. 39/3, 2012: 529–539, and "Pour une histoire politique de l'environnement au 20e siècle", *Vingtième siècle. Revue d'histoire*, 113, 2012/1: 3–12. He is currently working on environmental protection policies in twentieth century France, shifting his focus from water pollution to air pollution.

Gabriel Jover-Avellà is an Associate Professor of the Department of Economics at University of Girona (http://www.udg.edu/personal/tabid/8656/Default.aspx?ID=52454) and collaborator researcher of the international project *Sustainable Farm Systems: Long-Term Socio-Ecological Metabolism of Western Agriculture* funded from 2012 to 2017 by the Social Sciences and Humanities Research Council of Canada, together with the Ministry of Economy and Competitiveness in Spain. His research is focused on agrarian history from the sixteenth to the eighteenth centuries. He uses farm accounts from the Majorca Island to analyse the changes in Mediterranean organic agro-systems. He has published recently the first results in Gabriel Jover and Jerònia Pons (2012) *Possessions, renda de la terra i treball* *assalariat. L'illa de Mallorca*, 1400–1660, Documenta Universitària-Biblioteca d'Història Rural, Girona [Farms, rents and wage-labour. Majorca Island, 1400–1660]. En Enric Saguer, Gabriel Jover i Helena Benito (2013) *Comptes de senyor, comptes de pages. Les comptabilitats en la història rural*. Documenta Universitaria-Biblioteca d'Història Rural [*Landlor accounts, Peasant accounts. Accounting in Rural History*]. He has published in journals like *Revista de Historia Económica-Journal of Iberian and Latin American Economic History, Histoire & Mesure,* among others.

Paolo Malanima is a Professor of Economic History and Economics (University «Magna Graecia» in Catanzaro). He received his education at the Scuola Normale Superiore (Pisa) and University of Pisa. Malanima is Co-President of the European School for Training in Economic and Social Historical Research (ESTER) (University of Leiden) and a member of the editorial board of the journals *Società e Storia* and *Rivista di Storia Economica*, corresponding editor of the *International Review of Social History*, member of the Consejo of *Investigaciones de Historia Economica*, and *Revista de Istorie A Moldovei,* member of the editorial board of the *Economic History Review* and *Scandinavian Review of Economic History*. His research is long-term economic history and the history of energy. His book *Pre-Modern European Economy. One Thousand Years (tenth–nineteenth Centuries),* Brill: Leiden-Boston, 2009; german translation as *Europäische Wirtschaftsgeschichte* 10–19. *Jarhundert*. Wien: Böhlau, 2010, refers to both these areas of research. He is the author of *Le energie degli italiani. Due secoli di storia*, Milano, B. Mondadori, 2013, and coauthor of A. Kander, P. Malanima, P. Warde, *Power to the people. Energy in Europe over the last five centuries*, Princeton, Princeton University Press, 2013.

Stephen Mosley completed both his MA and Ph.D. in History at Lancaster University. He is now a Senior Lecturer in History in the School of Cultural Studies at Leeds Metropolitan University. Mosley's research interests are in environmental history, particularly the history of environmental pollution and associated socio-economic and health issues. His publications include: *Common Ground: Integrating the Social and Environmental in History* (2011, with Geneviève Massard-Guilbaud) which opens up a dialogue between the two disciplines; *The Chimney of the World: A History of Smoke Pollution in Victorian and Edwardian Manchester* (2008 edn.), which examines the human and environmental costs of smoke pollution in the world's first industrial city; and *The Environment in World History* (2010), which offers a fresh environmental perspective on familiar world history narratives of imperialism and colonialism, trade and commerce, technological progress and the advance of civilisation. He has been an Editor of the journal *Environment and History* since 2010.

Dieter Schott studied History, Political Science and English at the University of Konstanz and the Free University of Berlin. He gained his Ph.D. with a thesis on history of the city of Konstanz in the interwar-period. His habilitation thesis *Die Vernetzung der Stadt* (=*Networking the City*) (Darmstadt University of Technology

1996, published 1999) analyses urban electrification processes in three German cities in the context of wider processes of urban development in the period 1880–1918. From 2000 to 2004, he taught as Professor for the History of Urban Planning at the Centre for Urban History, University of Leicester, UK. Since 2004, he teaches Modern History at Darmstadt University of Technology. He was particularly involved with promoting international exchange on urban environmental history, co-editing for example the proceedings of a 2002 conference at Leicester on *Resources of the City* (Aldershot 2005). He has published widely in the fields of urban and environmental history of the nineteenth and twentieth century, on natural disasters, energy and infrastructures, rivers and cities. His most recent book is a text book on European Urbanization with a particular focus on city-environment-relations (*Die Urbanisierung Europas. Eine umweltgeschichtliche Einführung*, to be released 3/2014). He is the president of the German Society for Urban History and a member of the International Council of the European Association of Urban History (EAUH).

Gianni Silei is an Aggregate Professor of Social History in the Dipartimento di Scienze Politiche e Internazionali and coordinator of the Observatory on Risks and Natural and Technological Events and Disasters (Osservatorio Rischi e Eventi Naturali e Tecnologici, Orent) under the Centro Interuniversitario per la Storia del Cambiamento Sociale e dell' Innovazione (Ciscam) at the University of Siena. He is a member of the European Society for Environmental History (Eseh). His research interests include welfare state and social protection policies history, contemporary fear culture and natural and man-made disasters history. Among his recent publications: *Le radici dell'incertezza. Storia della paura tra Otto e Novecento* (2008); *Ambiente, rischio sismico e prevenzione nella Storia d'Italia* (2011); *Volontariato e mutua solidarietà. 150 anni di previdenza in Italia* (2011); *Espansione e crisi: le politiche di welfare in Italia tra gli anni Settanta e Ottanta*, in *Momenti del welfare in Italia. Storiografia e percorsi di ricerca* (2012); *Breve storia dello Stato sociale* (2013).

Enric Tello-Aragay is a full-time Professor of the Department of Economic History and Institutions at the University of Barcelona (http://www.ub.edu/histeco/eng/inici.htm) and a co-researcher of the international project *Sustainable Farm Systems: Long-Term Socio-Ecological Metabolism of Western Agriculture* funded from 2012 to 2017 by the Social Sciences and Humanities Research Council of Canada, together with the Ministry of Economy and Competitiveness in Spain, that assembles seven universities in six countries. He publishes on environmental as well as economic history of Catalonia and Spain, using socio-metabolic approaches to energy and material balances of agricultural systems, as well as landscape ecology analysis of land cover land-use changes. Some of his last publications are: (with Parcerisas, L.; Marull, J.; Pino, J.; Coll, F. and Basnou, C., 2012): Land use changes, landscape ecology and their socioeconomic driving forces in the Spanish Mediterranean coast (El Maresme County, 1850–2005), *Environmental Science & Policy* 23:120–132; (with Garrabou, R.; Cussó, X.; Olarieta, J.R. and Galán, E.

(2012): Fertilizing methods and nutrient balance at the end of traditional organic agriculture in the Mediterranean bioregion: Catalonia (Spain) in the 1860s, *Human Ecology* 40(3):369–383; or (with Ostos, J.R., 2012): Water consumption in Barcelona and its regional environmental imprint: a long-term history (1717–2008), *Regional Environmental Change* 12(2):347–361.

Verena Winiwarter was first trained as a Chemical Engineer. After years of working in atmospheric research, she earned her Ph.D. in Environmental History at the University of Vienna in 1998. She was granted the venia legendi in Human Ecology in 2003. From 2003 to 2006, she held a postdoctoral fellowship in environmental history (APART fellowship) awarded by the Austrian Academy of Sciences at the Institute for Soil Research, University of Natural Resources and Applied Life Sciences, Vienna and at the Faculty for Interdisciplinary Research of Alpen-Adria-Universität Klagenfurt, where she holds the first chair in Environmental History in Austria since 3/2007. Since 2010, she also serves as Dean of the faculty for interdisciplinary studies there. Her main research interests comprise the history of landscapes, in particular rivers, waste, images and the environmental history of soils. She has been among the founding members of ESEH, the European Society for Environmental History. From 2001 to 2005, she served as President of ESEH. A corresponding member of the Austrian Academy of Sciences, she has published numerous articles and edited several books. Her CV can be downloaded at: http://www.uni-klu.ac.at/socec/downloads/CV_VW_01-10_13_new.pdf.

Chapter 1
Energy in History

Paolo Malanima

Abstract The topic of energy is of central interest today. Although a long-term view can be useful in order to clarify contemporary trends and future perspectives, scholarly literature provides little information on the consumption of energy sources by past societies, before the beginning of the 20th century. In the following analysis, the topic of energy will be discussed from the viewpoint of economics, with a long-term historical perspective. After a brief introduction in Sects. 1.1 and 1.2 will examine some definitions and concepts, useful when dealing with energy and the role of energy within the economy. Section 1.3 will focus on the relationship between humans and energy in pre-modern societies. Section 1.4 will discuss the energy transition, that is changes in energy and environment from the early modern age to the present day. In the Conclusion (Sect. 1.5) general estimates will be proposed of past energy consumption on the whole.

1.1 Introduction

Scholars disagree about the role of energy within the economy. An optimistic view is shared by many economists. Their opinion is that raw materials played virtually no role in the modern development of the economy, as growth depended and continues to depend on knowledge, technical progress and capital. The contribution of natural resources to past and present growth is almost non-existent; and energy is a natural resource. After all energy represents today—they say—something less than 10 % of aggregate demand in the advanced economies.

Scholars with interest in environmental changes support the opposite view on the role of material goods and nature in the economy. Environment and natural materials played an important function in the development of human societies and in history on the whole. Energy in particular is of central importance in economic

P. Malanima (✉)
Università Magna Graecia, Catanzaro, Italy
e-mail: mala1950@hotmail.it

M. Agnoletti and S. Neri Serneri (eds.), *The Basic Environmental History*,
Environmental History 4, DOI 10.1007/978-3-319-09180-8_1

life and is also a central concern, given the heavy impact of energy consumption on the environment, especially in the last two centuries. Material underpinnings to economic success are not to be underrated, in their opinion.[1]

1.2 Definitions and Concepts

1.2.1 An Economic Definition

In daily life we have direct contact with matter, but not with energy. Matter can be touched, its form described and it is to be found underfoot as well as around us. With energy it is different. Its indirect effects are only perceived deriving from changes either in the *structure,* that is the molecular or atomic *composition* of matter, or in its *location* in space, such as in the case of a stream of water or wind, whose potential energy we can exploit. In both cases effects such as movement, heat or light reveal the presence of what we call energy from about 200 years.

In physics energy is defined as the ability of bodies to perform work.[2] Since work is the result of force by distance, then energy includes any movement of some material body in space together with the potential energy deriving from its position. Heat as well is the result of the movement of the components of matter. When dealing with the economy and then with the interrelationship between humans and the environment, our definition must be a little different. We could define energy in economic terms as *the capacity of performing work, useful for human beings, thanks to changes introduced with some cost or effort in the structure of the matter or its location in space*. Solar heat is of primary importance for the existence of life. The definition of energy in physics includes it. Since it is a free source of energy, it is not included in our economic definition; whereas the capture of solar rays by means of some mechanism in order to heat water or produce electric power is included. In the first case solar heat is not an economic resource, while it is in the second. The formation of biomass in a forest is a transformation of the Sun's energy by the plants through photosynthesis and is not included in this definition either. On the other hand, firewood is included, which is a part of forest biomass used by human beings for heating, cooking and melting metals. Food is a source of energy in economic terms, since its consumption enables the performance of useful work and its production implies some cost. Food for animals is only exploitable, and then it is an economic resource, when metabolised by those animals utilized by humans for agricultural work. It is their fuel, and, since the power of the working animals is exploited by the people, its calories have to be divided among the consumers (such as the fuel of our cars today is divided among the population and is part of their per capita consumption). When consumed by wild animals in a forest, however, these

---

[1] On these topics see the first two chapters of Kander et al. (2013) chaps. 1 and 2.

[2] Useful the discussion of the definitions of energy in Kostic (2004, 527–538) (2007).

calories are not a source of mechanical power for humans and then are not included in our calculation of past energy consumption. Both fossil fuels used today and uranium are also energy carriers. They were not until a quite recent epoch, since they were not utilized in order to produce economic goods and services.

Although the definition of energy in physics is much wider than in economics, the definition here proposed is much wider than the ordinary meaning of the term energy. Many people immediately think of modern sources, when speaking of energy, and do not include daily food consumption. It is well known that working animals played a central role in pre-modern agricultural economies, but their feed is not considered as a main source of energy for humans. The lack of a clear definition, common to most contributions devoted to the history of energy, prevents from the possibility of calculating energy consumption in past societies.

1.2.2 Energy and Production

In the long history of technology, main developments consisted in the increasing knowledge about the possibility of "extracting" energy from the input of natural resources. The production process and the role of energy can be represented by the following diagram (Fig. 1.1).

The diagram can be seen as an illustration of the ordinary production function:

$$Y = AF(L, R, K).$$

Labour (*L*) and capital (*K*), the factors of any productive process of useful goods and services (*Y*), can be better defined, from the viewpoint of energy, as *converters* able to extract energy from resources (*R*) in order to transform materials into commodities. Y is in fact a function (*F*) of the converters. The progress of technical knowledge embodied in *A*, plays a central role in the production function. In one

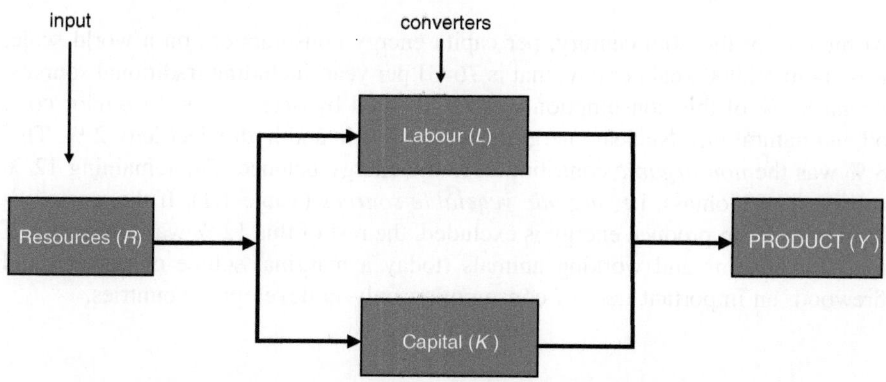

Fig. 1.1 Natural resources, converters of energy, product

sense, energy is the main input; that is to say, the main input is that part of matter (resources *R*) transformed by the converters, that is by workers (*L*), who metabolize food, and capital (*K*), which transforms some materials such as firewood, coal, oil, gas and electricity into mechanical work, heat and light.

The increase in productivity of energy, as a consequence both of discoveries of new sources and technologies (*macro-inventions*) or improvements in the exploitation of those already existing (*micro-inventions*)[3] can be represented by the following ratio:

$$\pi = \frac{Y}{E}$$

where *Y* is output (in value) and *E* is the total input of energy in physical terms (in Calories or joules or any other energy measure). The formula represents the productivity of energy, that is the product generated by the unit of energy. It is the reciprocal of the better known energy intensity (*i*), or the energy we need to produce an unit of GDP:

$$i = \frac{E}{Y}$$

In the previous diagram, energy productivity is the result of the ratio between the final product (in money) and the input of matter (food, coal, oil…) transformed into energy by the converters (in kcal, joules…). It is a measure of the efficiency of the energy converters from a technical viewpoint. The result is also conditioned by changes in the structure of the product. The increasing importance of less energy intensive sectors can result in an increase in energy productivity (or decline in energy intensity) even without any technical change.

1.2.3 Energy and History

At the end of the 20th century, per capita energy consumption, on a world scale, was about 50,000 kcal per day; that is 76 GJ per year, including traditional sources. About 80 % of this consumption was represented by *organic fossil sources*; coal, oil and natural gas. Nuclear energy represented 6 % and hydroelectricity 2 %. This 8 % was the *non organic* contribution to the energy balance. The remaining 12 % consisted of biomass, i.e. *organic vegetable sources* (Table 1.1). If the waste utilized in order to produce energy is excluded, the rest of this 12 % was composed of food for humans and working animals (today a marginal source of power), and firewood, an important item of consumption only in developing countries.

3 For the terms "micro-" and "macro-inventions" see Mokyr (1990).

Table 1.1 Daily and yearly per capita consumption of energy worldwide around 2000 (kcal, Toe and %)

	Sources	kcal per capita per day	Toe per capita per year	(%)
3	Non organic	4,000	0.15	8
2	Organic fossil	40,000	1.47	80
1	Organic vegetable	6,000	0.22	12
		50,000	1.84	100

Source IEA, *World Energy Outlook 2010*, OECD/IEA, Annex A, Tables for Scenario Projections
Note Organic Vegetable food, firewood and feed for working animals; *Organic Fossil* coal, oil, natural gas; *Non organic* nuclear, wind, hydro, photovoltaic. Toe = ton oil equivalent = 10 million kcal

This composition of the energy balance reveals the strata of a long history of technical conquests.[4] The history of energy technology is nothing else than the chronological analysis of our present energy balance, in order to single out the various ways of extracting energy from matter to produce heat, movement, light, work etc. Following Table 1.1, we will track the history of energy consumption from the most remote layer (1) that is *Organic vegetable sources*, to the development of *Organic fossil sources*, the intermediate stratum (2), and subsequently to the progressing *Non organic sources* (3), which will be the basis of our future energy systems.[5]

From the viewpoint of energy, the long history of mankind could be divided into two main epochs (corresponding to the first two lines of Table 1.1):

- *First epoch* the about 5–7 million years from the birth of the human species until the early modern age, that is about 5 centuries ago, and
- *Second epoch* the recent history of the last 500 years, which has witnessed a fast acceleration in the pace of energy consumption.

In the first long epoch, energy sources were represented by *food for humans*, *fodder for animals* and *firewood*, that is biomass, with a small addition of *water* and *wind power*. The second epoch witnesses the rapid partial replacement of the old sources by *fossil carriers*, which became and still are the main energy sources. While in the first epoch energy was scarce, expensive and environmental changes heavily influenced its availability, during recent history energy has been plentiful, its price relatively low and the influence of the energy consumption on the environment considerable.

[4] Still important on the big changes in the history of energy is the book by Cipolla (1962).

[5] "Organic economies" is the expression used by Wrigley (1988). With reference to the history of energy, the same term of "organic" had been used before by Cottrell (2009), See also Wrigley (2010).

Here is a synthetic view of the sources characterizing these two main epochs:

First epoch	Second epoch
Food	Coal
Firewood	Oil
Fodder (for working animals)	Primary electricity
Water power	Natural gas
Wind power	Nuclear power

Although the energy system prevailing today is apparently different from the simple digestion of food (the first energy source), or from the burning of firewood by our primitive ancestors, it is based on the same principle, which is the oxidation of Carbon compounds by breaking their chemical ties. Since Carbon compounds are defined in chemistry as organic compounds and organic chemistry is the chemistry of organic compounds, we could define all the energy systems which have existed until today as organic and the economies based on those organic sources as *organic economies*. Coal, oil and natural gas, the basic sources oxidized today in order to bring about organized, that is mechanical, work, heating or light are carbon compounds such as bread or firewood. The difference between pre-modern and modern energy systems depends on the fact that, until the recent energy transition, organic vegetable sources were exploited, whilst from then on organic fossil energy sources became the basis of our economy. Since organic vegetable sources of energy were transformed into work by biological converters (animals) and fossil sources are transformed by mechanical converters (machines), we are able to distinguish past economies according to the system of energy they employed and the prevailing kind of converters in:

1. *organic vegetable economies* or *biological economies*;
2. *organic fossil economies* or *mechanical economies*.[6]

Given the importance of energy in human history, changes in the use of this main input mark the evolution of humans in relation to their environment much more than changes in the use of those materials, such as stone and metals, ordinarily utilized by the historians to distinguish the main epochs of human history.

[6] In chemistry "organic" refers to Carbon compounds. The term has been used by F. Cottrell and A. Wrigley (see the previous footnote) to distinguish past agricultural economies (whose base was an organic energy system) from modern economies (based on mineral fossil sources). However, fossil fuels are also organic compounds. To avoid misunderstandings I think it useful to distinguish "Past agricultural organic vegetable economies" from "Modern organic fossil economies".

1.3 Pre-modern Organic Vegetable Economies

At the end of the 18th century three were the main economic sources of energy; corresponding to three different kinds of biomass. According to the age of the discovery and exploitation of these three sources, three ages can be distinguished in the distant past (that is in the First epoch identified in Sect. 1.2.3). The original source was *food*, the second was *firewood* and the third was *fodder for working animals*. A relatively small contribution came from two other carriers: *falling water*, the potential energy of which was exploited by watermills; and *wind*, utilized both by sailboats, and, much later, mills.

1.3.1 The First Age: Food

Since the birth of the human species some 5–7 million years ago, and then for some 85–90 % of human history, food was the only source of energy. In this long period, the only transformation of matter in order to engender movement and heat was the metabolism of organic material either produced spontaneously by plants and vegetation or converted into meat by some other animal consumed by humans as food. Although nothing certain can be said about energy consumption per head at that time, given the stature and physical structure of these early humans, consumption per day of about 2,000 Cal could be plausible. Their own body was the early machine used by humans. An animal body is not very efficient in the conversion of energy. Only 15–20 % of the input of energy, that is 300–400 Cal, is transformed into work, while the rest is utilized in order to support the metabolism and dispersed in the environment as heat and waste. The economic output of these far ancestors consisted in collecting, transporting and consuming this original input of energy.

1.3.2 The Second Age: Fire

The use of fire has been the main conquest in the history of energy.[7] The first evidence of fire being used by humans refers to several different regions of the world and can be dated between 1 million and 500,000 years ago. Fire was a conquest of independent groups of humans in several parts of the world and the main source of energy for several millennia. Its use spread slowly. In this case, as in the case of food, an estimate of the level of energy consumption by our distant ancestors can only be speculative. As far as is known for much more recent ages, the level of firewood consumption in different regions in pre-modern times may have varied from 1 kg per head per day to 10 in cold climates, that is between

[7] On the discovery of fire see particularly Perlès (1977) and Goudsblom (1992).

3,000–4,000 and 30,000–40,000 Cal. A daily consumption of about 1 kg per capita could be assumed for the humans living in relatively warm climates. In northern regions firewood consumption was considerably higher. Fire could be used for heating, cooking, lighting, and for protection against wild animals. Although, with fire, Calories per head drastically increased from 2,000 to 3,000–4,000 per day or more, that is 5–6 GJ per year, the efficiency in its use was very low. The useful energy exploited by the population did not exceed 5 % of its Calories, the rest being lost in the air.

1.3.3 The Third Age: Agriculture

During the Mesolithic, the end of glaciations and the rise in temperature enabled humans to increase the cultivation of vegetables and particularly cereals. The overall availability of energy in the form of food increased dramatically and supported the growth of population. In per capita terms, the perspective is different. Since population increased rapidly in the agricultural regions of the World, availability of food per head did not increase. A diet based on cereals represented a deterioration, as is witnessed by the decrease in stature following the spread of agriculture. Agriculture, as the main human activity, progressed quite slowly, if we compare the diffusion of this technological conquest to the following ones. From the Near East, where primarily developed 10,000 years ago, agriculture progressed towards Europe at the speed of 1 km per year. Within 3,000 years, agriculture reached northern Europe. At the same time, the new economic system was spreading from northern China and central America, the regions of the world where agriculture independently developed at the same time or a little later than in the Near East.

A new development in the agricultural transition took place during a second phase: from about 5,000 years until 3000 BCE. The period can be considered as a true revolution. The fundamental change was represented by the taming of animals, (oxen, donkeys, horses and camels), and their utilization in agriculture and transportation. Humans' energy endowment was rising. If we consider a working animal as a machine and divide his daily input of energy as food—about 20,000 Cal—among the humans who employed him, consumption per head may have increased by 20–50 % or more, according to the ratio between working animals and human beings; which is not easy to define for these distant epochs. Only about 15 % of this input represented, however, useful energy, that is energy converted into work.

During this age, several innovations allowed a more efficient utilization of humans' power, fuels and animals; e.g. the wheel, the working of metals, pottery, the plough, and the sail. The sail was previously used, but it only spread widely during this revolutionary epoch. The use of wind was the first example of the utilization of a non-organic source of energy, not generated by the photosynthesis of vegetables. Labour productivity rose markedly. Even though some changes in the agricultural energy system also took place in the following centuries, technical progress was modest on the whole. Water and windmills, invented respectively 3 centuries BCE (as recent research suggests) and in the 7th century CE, were the main innovations in the energy basis of the agrarian civilisations. Although important from a technological viewpoint, these changes added very little in terms of energy availability: ordinarily no more than 1–2 %.[8]

1.3.4 Main Features of the Organic Vegetable Economies

Although several important differences exist among the three ages of our organic vegetable past, there are also some analogies; especially when dealing with the relationship between humans and environment. The dependence of this energy system on soil implies several constraints to the possibilities of economic development.

1. *Reproducible sources* Vegetable energy carriers are reproducible. They are based on solar radiation and since the Sun has existed for 4.5 billion years and will continue to exist for 5 billion years, vegetable materials may be considered as an endless source of energy. Organic vegetable economies have been sustainable since solar energy allowed a continuous flow of exploitable biomass. However, only a negligible part of solar radiation reaching the Earth, less than 1 %, is transformed into phytomass by the vegetable species. Of this 1 %, only an insignificant part is utilized by humans and working animals. On the other hand, increase in the exploitation of phytomass was far from easy. The availability of more vegetable sources implied extension of the arables and pastures and the gathering of firewood, which was difficult to transport over long distances. The ways of utilizing the phytomass were also in conflict, since more arables implied less pastures and woods. Thus, while the availability of these carriers was endless, their exploitation was hard and time consuming. The production of phytomass was, furthermore, subject to climatic changes both in the short and long run and heavily influenced by temperature changes and weather variations. Long-term climatic changes could also raise or diminish the extent of cultivation and wood productivity. Past organic vegetable economies, based on reproducible sources of energy, were the economies of poverty and famine.
2. *Climate and energy* Given that, in pre-modern organic vegetable energy systems, transformation of the Sun's radiation into biomass by means of photosynthesis was fundamental and since the heat of the Sun is not constant on Earth, the energy basis—phytomass—of any human activity was subject to changes. Climatic phases have thus marked the history of mankind. The availability of phytomass deeply varied and strongly influenced human economies. Glaciations caused a decline in available energy and therefore in the

[8] On the quantification of water and wind power see Malanima (1996).

number of humans and the evolution of their settlements. The end of the glaciations provoked changes in the main human activities; from hunting and gathering to agriculture. Agricultural civilizations were also deeply influenced by climatic variations. While warm periods were favourable to the spread of cultivations and the multiplication of mankind, cold epochs corresponded to demographic declines. Roman civilisation flourished in a warm period and was accompanied by population rise, while the early Middle Ages, characterized by a cold climate, was an epoch of demographic decline. The so-called warm Medieval Climatic Optimum coincided with worldwide population increase, between 900 and about 1270, while the following Little Ice Age, from 1270 until 1820, was again a period of economic hardship and population stability or slow increase. While present day energy systems heavily influence the environment and climate, until a few centuries ago the opposite was true.

3. *Efficiency and energy intensity* Only a part of energy input is actually transformed into useful energy (or energy services, that is mechanical work, light and useful heat). How great this share is depends on the efficiency of the converters of energy, that is labour (*L*) and capital goods (*K*). The thermodynamic efficiency (η) of the system of energy can be represented through the following ratio between the energy services (*Eu*) and the total input of energy (*Ei*):

$$\eta = \frac{Eu}{Ei}$$

Today, in our developed economies, this ratio is about 0.35; that is 35 % of the input of energy becomes actual mechanical work, light or useful heat. In past agricultural civilizations, the efficiency was much lower. A plausible calculation is easier for the past, when biological converters prevailed, than for the present. Today, in fact, the variety of machines, with diverse yields, make hard any estimate. The ratio between useful mechanical work and input of energy into biological converters, such as humans and working animals, is around 15–20 %.[9] Part of the intake of energy in the form of food is not digested and is expelled as waste, whilst the main part is utilized as metabolic energy in order to repair the cells, digest and preserve body heat. A human being or animal consumes even when inactive. The use of firewood is even less efficient. The greater part of the heat is dispersed without any benefit for those who burn the wood. Its yield is about 5–10 %. Overall, the efficiency of a vegetable energy system based on biological converters, such as that of ancient civilizations, was around 15 % at the most: that is 1,000–1,500 kcal. were transformed into useful mechanical work or useful heat; the rest was lost. Thermal machines are much more efficient than biological converters such as animals and humans.

4 *Low Power* Power is defined as the maximum of energy liberated in a second by a biological or technical engine. In the economies of the past another

[9] See the useful Herman (2007).

consequence of the usage of biomass converted into work was the low level of power attainable. The power of a man using a tool is about 0.05 horsepower (HP). That of a horse or donkey can be 10 times higher. A watermill can provide 3–5 HP, while a windmill can reach 8–10 HP. As a comparison, a steam engine could attain 8,000–12,000 HP around 1900, while a nuclear plant can reach 2 million HP. The conquest of power meant an incredible advance in the possibility of harnessing the forces and materials of the environment. To clarify this central point about the differences between past and modern energy systems, we must remember that the power of an average car (80 kW) is today equal to the power of 2,000 people and that the power of a large power station generating electricity (800 mW) is the same as that of 20 million people. The electric power of a medium sized nation of 40–60 million inhabitants, some 80,000 mW, equals the power of 2 billion people. Today, a nuclear plant or a nuclear bomb can concentrate millions of HP, or the work of many generations of humans and draft animals, into a small space and a fraction of time. This *concentration of work* allows humans to accomplish tasks that were barely imaginable just a few lifetimes ago.

1.4 Modern Organic Fossil Economies

At the start of modern growth around 1800, on the world scale, energy consumption was about 8,000–9,000 kcal per capita per day, that is 13 GJ per year.[10] The main sources were those already seen, that is different kinds of biomass (food, firewood and fodder). Water and wind were the only non organic sources. In 1800, throughout western Europe, the energy balance per head was 20 GJ per year, that is 13,000 Cal per day, excluding coal, which was then widely used only in England. On the continent, many differences existed in the levels of energy consumption. While in Mediterranean countries it was about 15 GJ per year (10,000 Cal per day), in Scandinavia it was 45 (30,000 Cal per day). In pre-modern Europe, the main energy carrier was firewood. It represented 50 % in the south and more than 70 % in the northern regions, followed by fodder for working animals and food for the population.[11]

In Europe, energy consumption was higher than in other agricultural civilisations, both in Asia and southern America, for two reasons:

1. the European civilisation was the most northern agrarian civilisation and, since temperature was a main determinant of energy consumption, wood consumption was higher than in coeval agrarian economies;

[10] On the relationship Modern Growth—Energy see: Ayres and Warr (2009).

[11] See the estimates by Kander (2002) and Malanima (2006).

2. in the dry European agriculture, the utilisation of animals in agriculture and transportation was more widespread than elsewhere. In both China and southern America, the presence of animals in agriculture was far more modest. In pre-modern centuries, probably only in India was animal power exploited to the same extent as in Europe.

1.4.1 The Start of the Energy Transition

Modern growth, from about 1820 until today, has marked a sharp rise both in the sources utilized and in the efficiency of their utilization.[12] We could define this change as an *energy transition.* It was an important support to the growth in the capacity to produce. Although not *sufficient condition* of modern growth, energy transition was a *necessary condition.*[13] Without this transition, modern growth could not occur. As has been seen, although some other deep changes occurred in the use of energy before the modern era, this last transition is often represented, for its rapidity and intensity, as the "transition" par excellence or the period that marked a break between past and present.

Fossil sources, coal, oil, natural gas, were also products of photosynthetic processes, such as food and firewood. Their formation had taken place in the Carboniferous era, some 300–350 million years ago. This underground forest had been mineralized or transformed into liquid fuel and gas in the course of several millennia.[14] In various parts of the world and in England and other northern European regions, coal was easily extracted. If by the start of the epoch of fossil fuels we refer to the period when they began to develop, the second half of the 16th century could be defined as the starting point. It was then that they began to be employed on a large scale by English manufacturers and for domestic use. If, instead, we want to single out the epoch when they began to play an important role on the European and non European economy, this age is the first half of the 19th century.

The existence of fossil fuels had been known in Europe since the times of ancient Rome. During the late Middle Ages, in those northern European regions where coal was easily available, its consumption spread, as its price was far lower than that of firewood. In China coal was also widely used in metallurgy during the late Middle Ages. From the second half of the 16th century, the use of coal increased in England, above all. The rising population and particularly that of London represented a strong stimulus towards the consumption of a much less expensive fuel than firewood. In the whole of England the production of coal increased 7–8 times between 1530 and 1630, thanks to the greater depth of the shafts and better drainage

---

[12] On this phase in the history of energy see the still useful article by Bairoch (1983) and particularly Kander et al. (2013). A brief, useful reconstruction is that provided by Grübler (2004).

[13] Malanima (2012).

[14] On the transition to fossil sources of energy, it is useful Sieferle (2001).

Table 1.2 Share of coal production in England and the rest of Europe 1800–1870 (%)

	England	Rest of Europe
1800	96	4
1830	79	21
1840	73	27
1850	73	27
1860	65	35
1870	58	42

Source Etemad and Luciani (1991, 256)

of the mines and by the 1620s it had become more important than wood as a provider of thermal energy. For a long period, England was by far the main producer of coal. Only at the end of the 19th century, was the rest of Europe able to compete with England (Table 1.2).

The share of coal on total energy consumed in England was 12 % in 1560, 20 % in 1600, and 50 % in 1700. Coal consumption from 1560 until 1900 shows an almost stable rate of growth (Fig. 1.2). In The Netherlands another fossil fuel, peat, began to be used on a wide scale from the 17th century onwards. It was an important support of the Dutch Golden Age, but did not cause such fundamental changes in the economy as coal in England.

One of the reasons for the transition to a new source of energy was the growth in population throughout the continent from the last decades of the 17th century onwards. While in 1650 the European population numbered 112 million, in 1800 it was already 189 million and in 1850 it was 288 million. The main converter of the organic vegetable energy system, land, was becoming scarcer. Energy consumption of traditional sources was diminishing in per capita terms, whereas food, fodder and above all firewood were becoming more expensive. The price of these sources increased across the whole continent from the second half of the 18th century

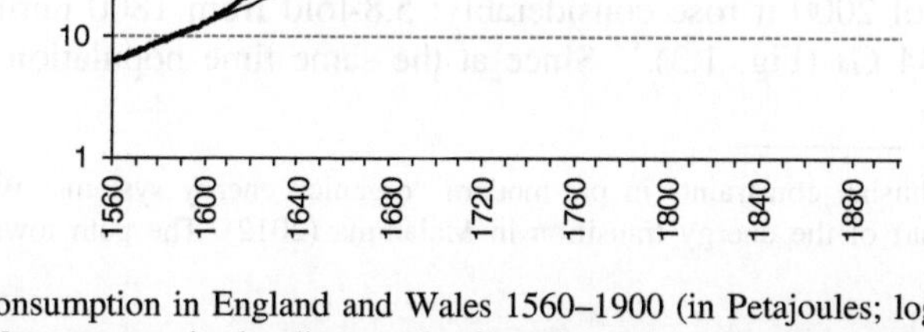

Fig. 1.2 Coal consumption in England and Wales 1560–1900 (in Petajoules; log scale). *Source* Warde (2007). *Note* you see in the diagram the formula of the exponential interpolating curve. 1 PJ = 1,000 billion KJ

onwards. Land per capita outside Europe was also diminishing.[15] The European population growth was part of the demographic transition taking place worldwide. World population rose from 600 million in 1650 to 1 billion in 1820.

The shift to new fuels represented one aspect of the energy transition then in act. It was not, however, the most important. The main technological change was the new utilisation of fuels, that is, the techniques designed to employ in a different way the heat of these organic sources. For about one million year, fuels had been utilized for heating, lighting and melting metals, while work, in economic terms, that is organized movement in order to produce commodities and services, was only provided by humans and animals; apart from wind and water (whose mechanical work, in any case, was not the conversion of a fuel). The only engines able to provide work were biological machines. The introduction of machines in order to convert heat into mechanical power was the main change in the energy system, comparable in importance to the discovery of fire. It was only during the 18th century, with the invention of the steam engine by Thomas Newcomen and James Watt, that the *Age of the Machines* really began. The fundamental technological obstacle that had for millennia limited the capacity of the economic systems to perform work, was only then overcome. In 1824, the French physicist Sadi Carnot clearly pointed out the great novelty represented by what he called the "machines à feu", the thermal machines.[16] In his opinion they would have replaced soon both the force of animals and that of water and wind. This is precisely what happened over the last two centuries. The age of machinery began with the steam engine and such energy transition resulted in great changes in:

- the volume and trend of energy consumption;
- the process of substitution of energy carriers;
- the geography of energy production;
- the price of energy;
- the relationship energy-economy;
- the relationship energy-environment.

The following sections are devoted to these changes.

1.4.2 The Volume and Trend of Energy Consumption

Energy consumption per head diminished in Europe during the 18th century, whilst from 1800 until 2000 it rose considerably: 5.8-fold from 1800 until 2000, that is from 23 to 134 GJ (Fig. 1.3).[17] Since at the same time population increased 3.5

[15] On the Malthusian constraints in pre-modern "organic" energy systems: Wrigley (1989). I examined the start of the energy transition in Malanima (2012). The path towards the modern economy.

[16] Carnot (1824).

[17] On energy consumption in Europe, see Bartoletto (2012).

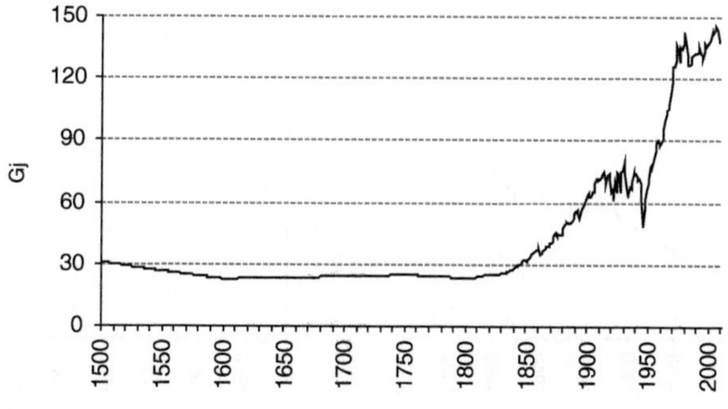

Fig. 1.3 Per capita energy consumption of traditional and modern carriers in western Europe 1500–2005 (GJ). *Source* Kander et al. (2013). *Note* 1 GJ = 1 million KJ = 0.0239 Toe

times, total energy consumption registered a 20-fold increase (Table 1.3). The global crisis of the first decade of the 21st century resulted in a fall of energy consumption.

Until about 1840, energy consumption per head did not increase in Europe, since the input of fossil fuels rose at the same rate as the population. From 1840 onwards until the First World War, growth was instead remarkable. After a period of stability between the two World Wars, a significant increase took place from the 1950s until the 1970s, followed by a slower rise. In the long run the growth witnesses an almost constant rate with brief deviations due to wars or epochs of fast economic rise (Fig. 1.4).

On the World scale, the rise of per capita consumption has been 5.7 times between 1850 and 2000. Since population growth was 5.8-fold, the aggregate rise

Table 1.3 Energy consumption in western Europe from 1800 until 2000 in kcal per capita per day, in Toe per capita per year, population and total energy consumption in Mtoe

	kcal per capita per day	Toe per capita per year	Traditional sources (%)	Rate of growth (%)	Population (000)	Total Mtoe
1800	15,300	0.56	77		96,950	54
1830	16,700	0.61	62	0.29	118,800	72
1900	42,000	1.53	20	1.32	194,800	299
1950	46,500	1.70	13	0.20	254,500	432
1970	82,200	3.00	7	2.85	293,700	880
1990	86,800	3.17	7	0.27	316,900	1,004
2000	90,700	3.31	8	0.44	327,400	1,084
2010	88,000	3.21	8	−0.30	336,000	1,079

Source Kander et al. (2013)
Note data refer to western Europe: Sweden, The Netherlands, Germany, France, Spain, Portugal, Italy. 1 Megatoe = 1 million Toe

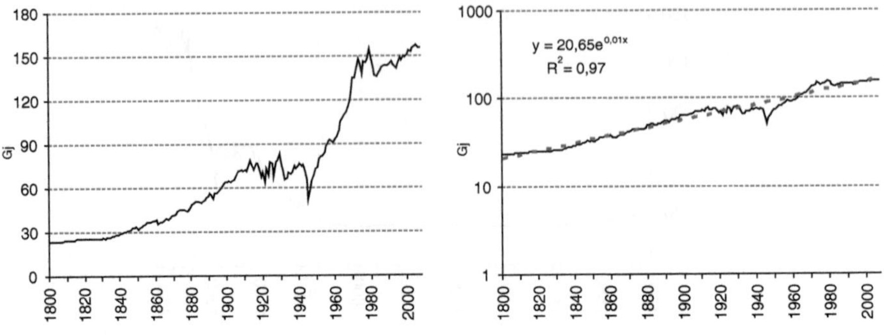

Fig. 1.4 Per capita energy consumption in western Europe 1800–2007 (GJ) (on the *right* log *vertical axis*, trend and the equation of the trend). *Source* Kander et al. (2013)

was 33 times (Table 1.4). We see that modern or commercial sources overcame traditional sources, or the phytomass, around 1900, or the epoch of the second industrial revolution.

1.4.3 The Process of Substitution

In organic vegetable economies any discovery of a new source was an addition to the balance of energy and not a substitution. With fossil sources it was different. Fossil sources replaced a large part of the traditional carriers, which lost their importance in relative and sometimes in absolute terms. While food consumption rose in aggregate and per capita terms, the power of working animals diminished and, in developed economies, totally disappeared. Firewood continued to represent an important share of energy consumption only in relatively backward areas. On the world scale, traditional sources of energy diminished from 98 % in 1800 to 50 in 1900 and only 14 in 2000–2010. In Europe the decline was still higher. England was the only important consumer of coal at the beginning of the 19th century. Traditional sources then represented the greater majority throughout the continent, that is almost 90 % of the overall consumption (when England is excluded). Their share decreased to 25 % in 1900 and was only 5 % in 2000 (always excluding England).

For several millennia changes in the energy system had been very slow. From 1800 transitions and substitutions began to dominate the picture. If we look at the fuels utilized in Europe from 1800 until 2000, we see that, in terms of Calories, firewood still dominated in 1800, while coal represented about 30 %. Wood consumption was, in relative terms, already relatively modest in 1900, while coal equalled about 80 %. Oil began to be used during the last decades of the 19th century and only in the 1960s exceeded coal. Natural gas spread from the 1970s on a large scale and only in the 1990s did it overtake coal; although its share was less than half that of oil. While coal dominated for a long period in the last half century,

Table 1.4 World energy consumption from 1800 until 2010 in kcal per capita per day, in Toe per year, world population and total in Mtoe

	kcal per capita per day	Toe per capita per year	Traditional sources (%)	Rate of growth (%)	World population (000,000)	Total Mtoe
1800	8,500	0.31	98		950	295
1850	9,800	0.36	88	0.30	1,180	425
1880	13,000	0.47	65	0.89	1,365	642
1900	18,400	0.67	50	1.77	1,560	1,045
1950	28,200	1.00	33	0.80	2,527	2,527
1970	45,900	1.67	20	2.56	3,691	6,164
1985	48,100	1.76	16	0.35	4,838	8,515
2000	49,000	1.79	14	0.11	6,077	10,878
2010	55,700	2.03	14	1.26	6,850	13,906

Sources on the World scale energy consumption and production can be assumed to be equal. Data on the production of modern sources of energy are from Etemad and Luciani (1991). The consumption of traditional energy carriers is based on plausible figures on the relative share of the modern sources (United Nations 1956) and Fernandes et al. (2007) and also the auxiliary material in http://onlinelibrary.wiley.com/doi/10.1029/2006GB002836/suppinfo on consumption of biofuels
Note 1 Megatoe = 1 million Toe

and although oil holds a central position, the picture is more varied and variety is ever increasing with the rising exploitation of solar power, wind, biomass and nuclear power as sources of primary electricity (Fig. 1.5).

Electricity is in any case a secondary energy source, a transformation, that is, of other sources. Even when electricity is generated by a water turbine, the primary source of power is represented by falling water, that is, by the change in its potential

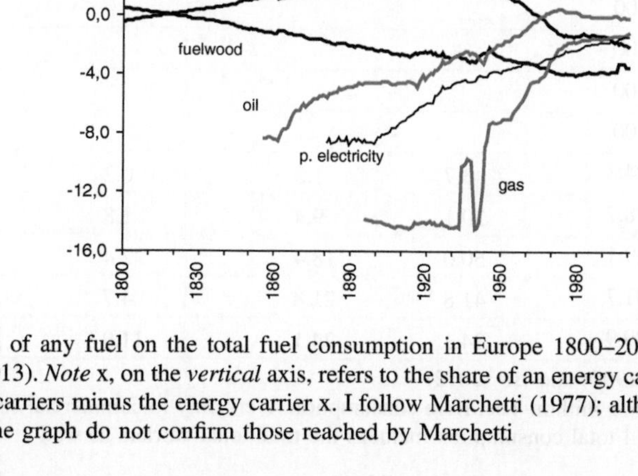

Fig. 1.5 Shares of any fuel on the total fuel consumption in Europe 1800–2000 (ln). *Sources* Kander et al. (2013). *Note* x, on the *vertical* axis, refers to the share of an energy carrier on the total of the 5 energy carriers minus the energy carrier x. I follow Marchetti (1977); although the results represented in the graph do not confirm those reached by Marchetti

energy. The same holds true for nuclear electricity, which began to develop from the late 1950s and whose primary source is the change in the atomic structure of uranium. Often, however, the expression "primary electricity" is used to single out that part of electricity not produced through fossil fuels. Today it includes solar, wind and geothermal electricity. Its share, in the form of hydroelectricity, has developed since the last decades of the 19th century. Nuclear power has been a remarkable addition since the 1970s. In 1971 it represented only 1 % of energy in Europe. In 2005 it was 13.6 %, thanks especially to the nuclearisation of the French energy system. Since the share of primary electricity in the continent was 17.2 % of primary electricity, the other sources were then negligible.

On the world scale, we find the same transition from coal to oil, to natural gas and to nuclear electricity, while photovoltaic, hydro and wind power progressed remarkably in the 1990s and the first decade of the third millennium (Table 1.5).

Table 1.5 World consumption of primary commercial energy (in Mtoe per year)

	Coal	Oil	Natural gas	Primary electricity	Total
(MToe)					
1700	3				3
1750	5				5
1800	11				11
1850	48				48
1900	506	20	7	1	534
1950	971	497	156	29	1,653
1973	1,563	2,688	989	131	5,371
1987	2,249	2,968	1,550	332	7,099
2010	3,532	4,032	2,843	1,405	11,812
(%)					
1700	100				100
1750	100				100
1800	100				100
1850	100				100
1900	94.8	3.7	1.3	0.2	100
1950	58.7	30.1	9.4	1.8	100
1973	29.1	50.0	18.4	2.4	100
1987	31.7	41.8	21.8	4.7	100
2010	29.9	34.1	24.1	11.9	100

Sources Martin (1990) and BP (2012)
Note 1 Megatoe = 1 million Toe. Here consumption refers only to commercial sources of energy, while in Table 1.4 total consumption includes the traditional carriers as well

1.4.4 The Geography of Energy Production

At the beginning of the 19th century, commercial, that is fossil, energy production was still entirely localised in Europe and especially in the north and centre. Economic growth and availability of fossil sources of energy more or less coincided. At the middle of the century, 90 % of fossil energy was still produced in Europe and 10 % in the United States (Table 1.6). Things changed during the 20th century, and especially in the second half, when oil began to play a central role in the energy systems of the developed countries. After the World War 2, Europe produced 35–40 % of world commercial energy. In particular the European production of oil has always been negligible, despite an increase of North Sea oil exploitation in the 1980s and 1990s by Great Britain and Norway. If, as a whole, the energy deficit of developed countries was only 4 % in 1950, in 1973 it had grown to about 50 %. At the end of the century, a little less than 50 % of oil production was localised, in order of importance, in Saudi Arabia, the USA, the Russian Federation, Iran and Mexico. The concentration of oil production, which is the basic source of the energy system, in specific places, resulted in a higher vulnerability of energy provisioning of developed countries. This vulnerability clearly appeared in 1973 and 1979, when the oligopoly of the main energy producers, OPEC, limited oil production and resulted in fast and remarkable price increases.

At the end of the past millennium, considerable differences existed in energy consumption per country. The geography of energy consumption is similar to the geography of growth; while the geography of energy production is not. Countries with higher per capita GDP are higher consumers (Fig. 1.6).

Among rich and poor countries the range of commercial energy consumption per head is 40 to 1. While in Niger and Mali it is 0.2 toe per capita per year, in the USA it is 8 toe. In the 1980s, on the world scale, energy consumption of market developed economies was 50 % of the total; that of centrally planned economies 20 % and that of the developing countries 30 %. At the end of the second millennium, 25 % of the world population—1.5 billion, the population, that is, of the developed economies, consumed 7,920 toe, i.e.75 % of the world consumption in one year, while 75 % of the population—4.5 billion—consumed 2,340 toe, or 25 % of the whole. With about 4.9 toe per year, an inhabitant of the most advanced

Table 1.6 Total production of commercial energy per continent (%)

	1800	1850	1900	1950	1985
Europe	99.09	90.00	61.63	35.66	38.38
America	0.91	10.00	35.71	52.38	30.73
Asia	0.00	0.00	1.72	9.99	23.11
Africa	0.00	0.00	0.12	1.24	5.83
Oceania	0.00	0.00	0.82	0.73	1.95
	100	100	100	100	100

Source Etemad and Luciani (1991)

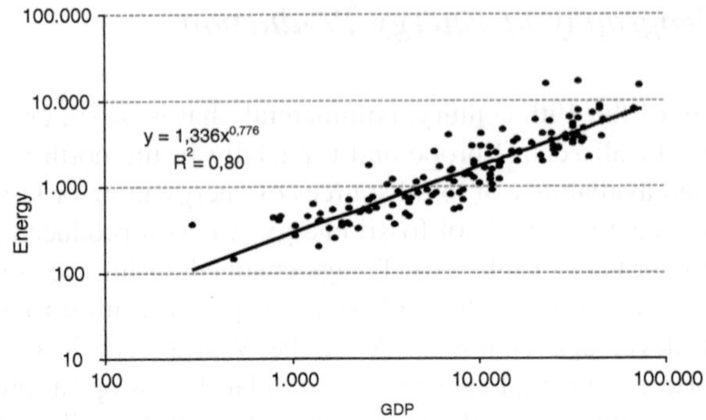

Fig. 1.6 Per capita energy consumption per country in 2009 (kg oil equivalent: koe) as a function of per capita GDP ($2005 PPP). *Source* World Bank (2011). *Note* the interpolation is drawn through a power regression, whose formula is represented in the graph. Ordinates and abscissae in log. 1 koe = 10,000 kcal

economies consumed on average 9 times more commercial energy than an inhabitant of the poorest countries—only 0.54 toe. So strong differences did not exist before modern growth. Only differences in climate and not in wealth could then imply remarkable disparities in consumption.

1.4.5 The Price of Energy

The spread of fossil fuels was fostered by their relatively low price in comparison with organic vegetable sources. During the second half of the 18th century and the first decades of the 19th, the initial progress of coal coincided with a period of rising prices of all organic vegetable sources of energy. For the same energetic content, fossil carriers were 2–3 times cheaper than the vegetable ones. If we take the curve of oil prices on the international markets, we notice that, after a couple of decades of high prices at the start of the use of oil, there was a downward curve until the 1973 crisis (Fig. 1.7).

In the 1950s and 1960s oil prices reached their lowest level. Although different sources have different prices, the trend in oil prices well represents the trend of energy prices on the whole. Data for the periods both before and after the introduction of the new fossil carriers, suggests that the fastest rate of the modern growth, occurring in the 1950s and 1960s, coincided with the lowest level of energy prices ever experienced, at least from when written information exists. On the other hand, the slower rate of growth of the world economy after 1973 depended, at least in part, on the higher price of energy and particularly oil.

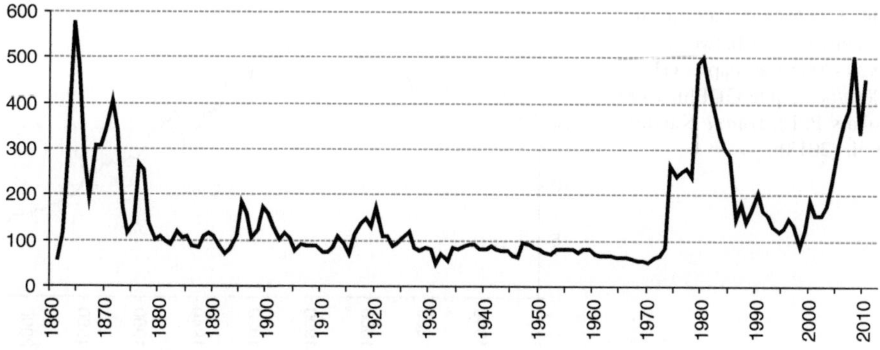

Fig. 1.7 Oil prices on the international market in 2008 euros per barrel 1861–2012. *Source* my calculations from data in https://opendata.socrata.com

1.4.6 Energy and Economy

When dealing with energy, we are interested both in the energy input into the economic system and the share of total energy actually available as mechanical work and heat. It is well known that energy cannot be created or destroyed, but only transformed (according to the first law of thermodynamics). On the other hand, it is also known that in any transformation there is a loss of *useful* energy: a large part of the energy that is consumed remains unavailable (according to the second law of thermodynamics). How great this amount is depends on the technical efficiency of the converter (as already seen in Sect. 1.3.4).

From 1800 on, not only were new fuels introduced on a wide scale, but equally important was the wider efficiency in their use. The conversion efficiency in different energy systems evolved through the following four main stages:[18]

	(%)
1. Subsistence agriculture	5
2. Advanced agriculture	15
3. Emerging industrial	25
4. Advanced industrial	35

As can be seen, modern growth implied not only a rise in the exploited energy, but also a rise in the efficiency of its exploitation. After all, machines are more efficient than animals as converters of energy.

[18] From Cook (1976, 135).

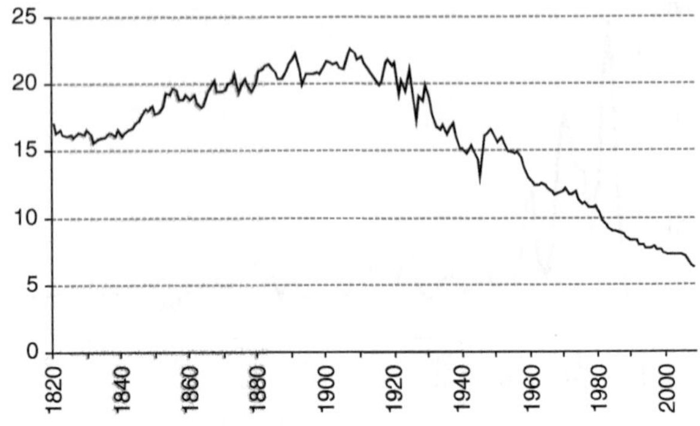

Fig. 1.8 Energy intensity in western Europe between 1820–2005 (per capita GJ (000)/per capita GDP in 1990 dollars PPP). *Source* Kander et al. (2013)

The advantages of machinery and technological change in terms of energy yield are easily visible if energy consumption (expressed in some energy measure) is divided by product (in money). The result of the ratio is the so called *energy intensity* (*E/Y*) (Fig. 1.8). The curve of energy intensity shows that, during the 19th century, some increase occurred in the energy/GDP ratio, due to the exploitation of coal by inefficient technologies especially in England, the main producer and consumer of coal. From 1900 on a remarkable decline took place. In the year 2000, the production of the same output required half the energy used some 200 years earlier.

A decline in energy intensity occurred in the second half of the 20th century in almost all world economies; although the differences were still remarkable (Table 1.7).

From 1820 until 2000, GDP per capita rose 16 times in western Europe, while energy input per head rose about 8-fold and efficiency in the use of energy doubled. A decomposition of per capita GDP proves to be useful in order to specify the relative importance of the input of energy and the efficiency of its exploitation. Per capita GDP (*Y/P*) can be represented as the result of energy consumption per capita (*E/P*) divided by the *productivity of energy* (*Y/E*):

Table 1.7 Energy intensity in different economies 1950–1990 (in Toe per $1,000 of GDP constant prices)

	OCDE	CPE	DC	OPEC
1950	0.55	1.70	0.23	0.10
1960	0.51	1.97	0.31	0.25
1970	0.52	1.66	0.38	0.22
1980	0.44	1.57	0.44	0.29
1990	0.36	1.39	0.46	0.44

Source elaboration of data from Pireddu (1990)
Note OCDE the organisation for cooperation and economic development; *CPE* centrally planned economies; *DC* developing countries; *OPEC* the organisation of the oil producer countries

$$\frac{Y}{P} = \frac{E}{P} \cdot \frac{Y}{E}$$

If we assume:

$\dot{y}$ as the rate of growth of *Y/P*;
$\dot{e}$ as the rate of growth of *E/P*; and
$\dot{\pi}$ as the rate of growth of *Y/E*;

we can specify the relative importance of *e* and π in the growth of *y*, during the period concerned; that is the years from 1820 until 2000. In fact:

$$\dot{y} = \dot{e} + \dot{\pi}$$

Our result is:

$$1.54 = 1.10 + 0.44$$

The conclusion is that, from 1820 until 2000, the annual rate of growth of per capita GDP was 1.54 %, and that *E/P* and *Y/E* grew respectively at the rates of 1.10 and 0.44 per year. The input of energy contributed more than the productivity of energy in the growth of per capita product. It was 2.5 times more important (1.10/0.44 = 2.5). Figure 1.9 shows that both per capita GDP and per capita energy consumption grew, in these last two centuries, with an almost constant and similar rate of increase. However, GDP per capita, (the higher curve) grew faster than energy (always per capita); as the interpolating exponential curves shows.

The introduction of new machines and more efficient engines was responsible for the leap in the productivity of energy. From the last decades of the 19th century, electricity contributed significantly to efficiency, together with the development of new devices which entered production plants and homes. Between 1860 and 1914, the introduction of electricity, steam and water turbines, and the internal

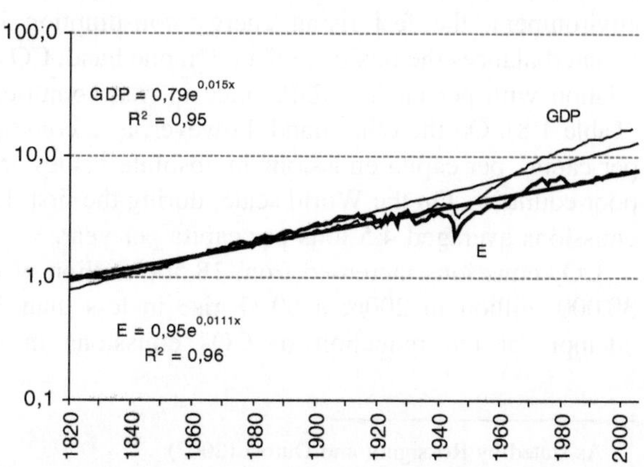

Fig. 1.9 Per capita energy consumption (the *thick curve*) and GDP per capita in western Europe 1820–2010 (1820 = 1) (exponential interpolating curves). *Source* Kander et al. (2013)

combustion engine (together with inexpensive steel, aluminium, explosives, synthetic fertilizers and electronic components), marked a technical watershed in recent economic history.

1.4.7 Energy and Environment

Among the different sources of energy, only the exploitation of wind, water and direct solar energy do not modify the environment since they do not cause a change in the molecular or atomic composition of matter. Whenever, by contrast, either the molecular or nuclear composition of matter is modified, even by the mere digestion of food, some change is introduced in the environment and some waste is produced. It is known that some environmental effects were produced by humans in past civilisations and that deforestation was not unknown in ancient economies. Lead in the atmosphere was notable in Roman antiquity due to melting metals, as the ice of Antarctica and Greenland has shown.[19] In any case, much heavier were the consequences of the environment on energy consumption by the humans than of human energy consumption on the environment. Both annual changes in temperature and rain, and long-term climatic cycles resulted in changes in the available energy, and subsequently in the level of the economic activity.

The 45–50-fold growth of energy consumption on the World scale, in the last 200 years, and the higher emissions by fossil fuels resulted in a dramatic rise in the level of gases in the atmosphere. Carbon dioxide (CO_2), water vapour, methane, nitrous oxide, and a few other gases are defined greenhouse gases. Their presence in the air has risen fast since the introduction and ever increasing use of coal and the other fossil carriers. Remarkable differences, however, exist among them: natural gas is much less polluting than oil, which is less polluting than coal. According to most paleo-climatologists the rise in temperature during the last century can be explained only as a consequence of the modern energy system and emissions of carbon dioxide into the atmosphere. Although the declining energy intensity from the 1990s results in a relatively lower impact of energy consumption on the environment, the fast rising energy consumption in absolute terms more than counterbalances the positive effect. On one hand, CO_2 emissions tend to decrease in relation with per capita GDP, since in rich countries energy intensity diminishes (Table 1.8). On the other hand, however, as a consequence of the high production per capita, per capita emissions in absolute values are much higher in rich than in poor countries. On the World scale, during the first decade of the third millennium, emissions averaged 4.5 tons per capita per year.

CO_2 emissions increased from 18,500 million of metric tons in 1980 to almost 30,000 million in 2006: a 60 % rise in less than 30 years. On the other hand, attempts at the reduction of CO_2 emissions in order to stabilize or reduce

[19] As stated by Rossignol and Durost (2007).

Table 1.8 Emissions of CO_2 in relation with GDP (kg/\$2005 PPP) and population (tons/pop.) in 2005

GDP per capita in \$2005 PPP	Level of income per capita	CO_2 kg/GDP (\$2005 PPP)	CO_2 tons/population
		2005	2005
<995	Low	0.28	0.28
996–3,945	Lower middle	0.73	2.79
3,946–12,195	Upper middle	0.48	5.26
12,196 and >	High	0.37	12.49
	World	*0.49*	*4.63*

Source World Bank (2011)

concentration of gases in the atmosphere (as in the Kyoto protocol, enforced on February 16th 2005) imply heavy consequences for the economy in the short term (and people are always more interested in the immediate negative effects on the economy rather than in the long-run positive effects on future mankind). The rise of new non polluting sources of energy is slower than it was hoped at the end of the past century.

While in a first phase of the industrialisation the CO_2 increased emissions could have played a positive role in agriculture, contributing to the fertilisation of the soils, since carbon dioxide makes crops grow faster, there is no doubt that in more recent decades the negative effects were much heavier than the positive. Although a precise quantification of the social and economic costs is hard to provide, a likely estimate for the beginning of the new millennium is in the order of \$20 per ton of carbon dioxide emitted in the atmosphere.[20] This cost would correspond to something like 0.5–5 % of GDP in advanced economic regions.

1.4.8 The Future of Energy

Forecasts of energy consumption always prove to be inaccurate.[21] Some general remarks on future developments are, however, possible. Humans have lived in organic economies since the birth of the human species, some 7–5 million years ago, until today. If humans continue to live and reproduce on the face of the Earth and enjoy the same levels of wealth we enjoy today or if they aim at increasing this wealth, our future will no longer be organic. We have seen that in about 2000–2010 per capita consumption on a global scale was around 50,000 Calories per day and that about 12 % was made up of organic vegetable sources—the old heritage of past agricultural societies—, while 80 % came from organic fossil sources—the more

[20] Nordhaus (2011).

[21] As stressed by Smil (2006).

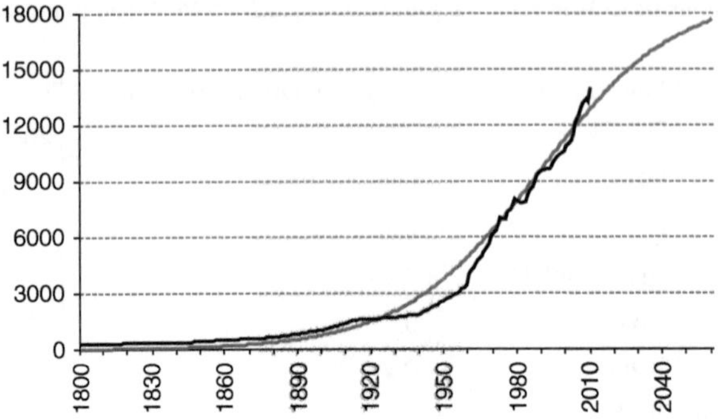

Fig. 1.10 World energy consumption (only modern sources) in Mtoe from 1800 until 2010 and prediction of future energy consumption until 2060 (*logistic curve*). *Source* see text. *Note* the coefficients of the *logistic curve* have been estimated through linear regression. The resulting equation is the following: with y = world consumption of modern sources; a = the lowest value; K the highest value; b the rate of growth; t time (starting with 1 in 1800)

recent heritage of modern growth. On a global scale, only a relatively narrow residual amount is represented by photovoltaic, wind, water and nuclear energy. These are the sources of our non-organic future. Fossil sources are diminishing and will disappear completely in one or two centuries. It is known that population rise is stabilizing, but that it will continue to grow for some decades, to reach at least 10 billion in the second half of this century. Per capita product is increasing rapidly in some countries, once poor, but now fast becoming rich. With a denser and richer world population, it will not be possible to devote land to the production of organic fuels or bio-fuels such as ethanol. Neither can we wish for the exploitation of new organic sources, with the consequences of their use inflicting such damage on the environment. A plausible trend of energy consumption in the future, drawn in Fig. 1.10 on the basis of the historical values of energy consumption in the last two centuries, suggests a relative decline of the rate of growth in the second half of the 21st century.

$$y = \frac{K}{1 + ae^{-bt}} = \frac{17.000}{1 + 3500e^{-0{,}042t}}.$$

Some environmentalists say that a decarbonisation of the economy is necessary. It will not only be necessary, but also unavoidable, in the future. More and more humans will have to learn how to deal with the non organic sources of energy, since an organic future will not be possible (or desirable). An alternative development is to return to the means of our ancestors of 3–4 centuries ago, when on a global scale, per capita consumption was one-tenth of that existing today. This means not only much less Carbon dioxide in the air, but also a smaller population and a much lower standard of living; which ultimately relies on the capacity to carry out useful work and then on energy.

Table 1.9 Energy systems, their duration, daily consumption (in kcal per capita) and yearly consumption (in GJ per capita)

Energy systems	Duration (years)	Per capita energy per day (kcal)	Per capita energy per year (GJ)
Food	7×10^6	2,000	3
Fire	5×10^5	4,000	7
Agriculture	1×10^4	5,000–6,000	8
Fossil Fuels	5×10^2	37,000	56.5

Source see text. See also, with different results, the estimates by Cook (1971)

1.5 Conclusion

The long history of energy consumption until today can be thus summarized (Table 1.9)[22]:

- the age when *food* was the only energy input extended over 5–7 million years and energy consumption per day could average about 2,000 Cal (that is 3 GJ per year);
- the age of *fire* (that is the epoch when fire was the only carrier except food) lasted some 1,000,000 years. Consumption can be established around 4,000–5,000 Cal per day or 6–8 GJ per year;
- the age of *agriculture* lasted some 10,000 years and World daily energy consumption per capita was about 5,000–6,000 Cal or 7–9 GJ per year;
- the age of *fossil fuels* lasted about 500 years and will finish this century or the next. Daily World consumption has been around 37,000 Cal or 56.5 GJ per year (according to a weighed average, based on population in the last two centuries).

The following data on energy consumption, although speculative, are not implausible.

Calculations of the number of humans since the origins are naturally uncertain and tentative depending on several assumptions; among which the epoch of the beginning of our species is of particular importance. If we accept that the hypothesis of 100 billion humans from 1 million years ago until today (an estimate proposed in the 1990s), is low and that our species was born 7–10 million years ago (such as the paleologists are nowadays inclined to believe), we must add at least 20–30 billion people. If the amount of people ever born from the origins numbers some 120–130 billion, today 5 % of our species is alive, and 18 % of the total World population has lived in the age of fossil fuels, beginning around 1600. This negligible part of the population, during the 0.01 % of time since the beginning of the species, consumed about 80 % of the energy ever consumed by humans. If the more recent period between 1700 and 2000 is observed, humans consumed 32 % of the whole wealth in fossil fuels. Consumption of coal was negligible in the 18th century

[22] Both on past and future energy consumption see the useful reconstruction by Beretta (2007).

Table 1.10 Energy consumption of fossil fuels from 1700 until 2000 and estimate of the still existing commercial energy (millions Toe)

	Toe (millions)	(%)
Consumption 1700–2000	350,000	32
Estimated reserves in 2000	750,000	68
Total	1,100,000	100

Source see text

and described an exponential curve in the following two. According to recent estimates on the future availability of fossil fuels, around the year 2000 about 70 % was remaining (although these estimates are ordinarily speculative) (Table 1.10). In any case, this considerable wealth will not last long time (100–200 years according to different estimates), seen the high and rising levels of fuel consumption.

In an important article published in 1922, Alfred Lotka established a correlation between natural selection and energy consumption. "In the struggle for existence", he stated "the advantage must go to those organism whose energy-capturing devices are most efficient in directing available energy into channels favourable to the preservation of the species".[23] Since the organisms and species with superior capacity to capture energy will increase, the energy crossing the biological system will also tend to increase. There is a tendency, in natural history, towards a flow of more and more energy through the biological sphere. In trying to survive and enjoy better living standards any living being contributes, at the same time, to the rise in the flow of energy which crosses the biological system. The human beings alive at the start of the 21st century, and still increasing, will contribute, day by day, to intensify this passage of energy through the thin biological envelope of the Earth.

References

Ayres RU, Warr B (2009) The economic growth engine. How energy and work drive material prosperity. Edward Elgar, Cheltenham

Bairoch P (1983) Energy and industrial revolution: new approaches. Revue de l'énergie 34 (356):399–408

Bartoletto S (2012) Energy and economic growth in Europe. The last two centuries. In: Chiarini B, Malanima P (eds) From Malthus' stagnation to sustained growth. Palgrave Macmillan, Basingstoke, pp 52–70

Beretta GP (2007) World energy consumption and resources: an outlook for the rest of the century. Int J Environ Technol Manage 7:99–112

BP (2012) Statistical review of world energy, June 2012

Carnot S (1824) Réflexions sur la puissance motrice du feu. Mallet-Bachelier, Paris

Cipolla CM (1962) The economic history of world population. Penguin, Harmondsworth

Cook E (1971) The flow of energy in an industrial society. Sci Am 225:134–147

Cook E (1976) Man, energy, society. Freeman, San Francisco

[23] Lotka (1922).

Cottrell WF (2009) Energy and society: the relation between energy, social change, and economic development (1st edn 1955). Greenwood Press, Westport

Etemad B, Luciani J (1991) World energy production 1800–1985. Droz, Geneve

Fernandes SD, Trautmann NM, Streets DG, Roden CA, Bond TC (2007) Global biofuel use 1850–2000. Global Biogeochem Cycles, 21

Goudsblom J (1992) Fire and civilization. Penguin, Harmondsworth

Grübler A (2004) Transitions in energy use. In: Encyclopedia of energy, vol 6. Elsevier, Amsterdam, pp 163–177

Herman IP (2007) Phisics of the human body. Springer, Berlin

Kander A (2002). Economic growth, energy consumption and CO_2 emissions in Sweden 1800–2000. Lund University, Lund

Kander A, Malanima P, Warde P (2013). Power to the people. Energy in Europe over the last five centuries. Princeton University Press, Princeton

Kostic M (2004). Work, power, and energy. In: Encyclopedia of energy, vol. 6. Elsevier, Amsterdam, pp 527–538

Kostic M (2007) Energy: global and historical background. Encyclopedia of engineering. Francis and Taylor, New York

Lotka A (1922). Contribution to the energetics of evolution. In: Proceedings of the national academy of sciences, vol VII, pp 147–151

Malanima P (1996) Energia e crescita nell'Europa pre-industriale. La Nuova Italia Scientifica, Roma

Malanima P (2006) Energy consumption in Italy in the 19th and 20th centuries. ISSM-CNR, Napoli

Malanima P (2012) The path towards the modern economy. The role of energy. In: Chiarini B, Malanima P (eds) From Malthus' stagnation to sustained growth. Palgrave Macmillan, Basingstoke, pp 71–99

Marchetti C (1977) Primary energy substitution models. Technol Forecast Soc Change 10:345–356

Martin J-M (1990) L'économie mondiale de l'énergie. La Découverte, Paris

Mokyr J (1990) The Lever of riches. Technological creativity and economic progress. Oxford University Press, New York

Nordhaus W (2011) Energy: friend or enemy? In: The New York review of books 27 Oct 2011

Perlès C (1977) Préhistoire du feu. Masson, Paris

Pireddu G (1990) L'energia nell'analisi economica. F. Angeli, Milano

Rossignol B, Durost S (2007) Volcanisme global et variations climatiques de courte durée dans l'histoire romaine (Ier s. av. J.-C. - IVème s. ap. J.-C.): leçons d'une archive glaciaire (GISP2). Jahrbuch des Römisch-Germanischen Zentralmuseums Mainz, 54(2):395–438

Sieferle RP (2001) The subterranean forest. Energy systems and the industrial revolution. The White Horse Press, Cambridge (1st German edn 1982)

Smil V (2006) Energy at the crossroads. In: Global science forum conference on scientific challenges for energy research, Paris, 17–18 May

United Nations (1956) L'utilisation de l'énergie atomique à des fins pacifiques, Conference internationale de Geneve. United Nations. Publications: 307

Warde P (2007) Energy consumption in England and Wales 1560–2000. ISSM-CNR, Napoli

World Bank (2011) World development indicators. World Bank, Washington

Wrigley EA (1988) Continuity, chance and change. The character of the industrial revolution in England. Cambridge University Press, Cambridge

Wrigley EA (1989) The limits to growth. Malthus and the classical economists. In: Teitelbaum MS, Winter JM (eds) Population and resources in western intellectual traditions. Cambridge University Press, Cambridge, pp 30–48

Wrigley EA (2010) Energy and the English industrial revolution. Cambridge University Press, Cambridge

Chapter 2
Economic History and the Environment: New Questions, Approaches and Methodologies

Enric Tello-Aragay and Gabriel Jover-Avellà

Abstract Ecological economics is enabling economic and environmental historians to enhance their understanding of economic growth, by placing it in a broader perspective of biophysical interactions between nature and society. In this chapter, several ongoing researches and historical debates are examined from this standpoint such as the missing role of energy carriers in GDP growth, the socio-metabolic profiles of past and present societies, the pre-industrial 'Smithian' responses to 'Malthusian' traps, the role of efficient land-use in breeding livestock to increase agricultural yields, the reasons why the Industrial Revolution began in a high wage and cheap energy economy, the first globalization as a socio-metabolic watershed, and the question of whether there was a general crisis of biomass energies at the coming of fossil fuels era. Research discussing long-term socio-metabolic transitions may contribute to our understanding of how economic growth actually occurred, and which ecological impacts affected the Earth's life-support systems. Equally, these projects leave room for the institutional settings or ruling actors needed to explain why growth has happened and by whom. Far from naturalising history, the use of ecology in the explanation of human history historialises ecology.

2.1 Introduction

If all the research done in the well-established scientific field of economic history had to be summed up in one word it would be 'growth'. The main subject, if not the single issue, studied by economic historians is when, where, and why economic growth has taken place. In doing so, there has been a greater tendency to rely

E. Tello-Aragay (✉)
Facultat d'Economia i Empresa, Universitat de Barcelona, Barcelona, Spain
e-mail: tello@ub.edu

G. Jover-Avellà
Departament d'Economia, Universitat de Girona, Girona, Spain
e-mail: gabriel.jover@udg.edu

M. Agnoletti and S. Neri Serneri (eds.), *The Basic Environmental History*, Environmental History 4, DOI 10.1007/978-3-319-09180-8_2

mainly or exclusively on mainstream economics as an analytical foundation. One of the earliest criticisms raised by environmentalists decades ago, and later by environmental historians is that mainstream economists, and some economic historians, have not only set aside the role played by natural resources in past and present economic growth, but have also ignored the increasingly powerful and global environmental impacts of economic growth on the planet's ecological life-support systems.[1]

Yet the misunderstanding between mainstream economics and environmental sciences goes beyond having ignored some "external" inputs and outputs that can simply be reintegrated into current macroeconomic growth analysis. As many specialists have recognized, economists have found profound and persistent problems in the explanation of long-term economic growth. These difficulties originated at the beginning of the neoclassical analytical approach. Ironically, mainstream economists intended to become the analytical physicians of the social sciences precisely as they discarded 'land' and other natural resources as relevant factors within economic theory.[2] Interestingly enough, it was also in this period when history ceased to be a basic background within economics.[3] From then onwards the standard neoclassic growth model assumed that the final value added flows of GDP are directly produced from labour and capital alone, without specifying a role for energy flows, which were only considered to be consumable intermediates.[4]

2.2 The Missing Role of Useful Work from Energy Carriers in Economic Growth

According to these neoclassical analytical assumptions, technological progress becomes exogenous and natural resource consumption is seen as a consequence, not a driver, of economic growth. Perhaps it is not so surprising that the first generation of macroeconomists who accounted for growth by means of a Cobb-Douglass production function using capital and labour as the only relevant factors, couldn't fully explain no more than a small level of growth in GDP, because the results left a large, increasing residual (Fig. 2.1). Labelling this unexplained residual "total factor productivity" (TFP), and considering it to be the contribution of technical progress to economic growth, has become common practice. However as Robert Solow stresses, by calling it "the measure of our ignorance", TFP has become an exogenous factor not taken into account by the standard growth theory.

---

1 Debier et al. (1986), Worster (1988), McNeill (2000a), Krech III et al. (2004), Hornborg et al. (2007) and Sing et al. (2013).

2 Pasinetti (1981).

3 Hodgson (2007).

4 Ayres and Warr (2005).

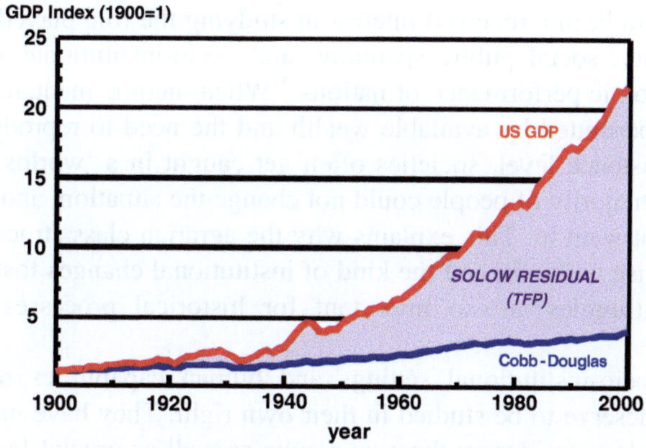

Fig. 2.1 Explained share of the actual GDP growth of the United States economy during the 20th century, and the Solow residual obtained by a conventional Cobb-Douglass production function. *Source* Ayres et al. (2009) during 100 years of economic growth, presented to the Q2 session on Energy, climate change and growth: perspectives from economic history of the 25th World Economic History Congress, Utrecht, The Netherlands. The following three-factor Cobb-Douglas production function has been used: $Y_t = A_t\,(H_t\,K_t)^{\alpha}\,(G_t\,L_t)^{\beta}\,(F_t\,R_t)^{\gamma}$, where Y_t is output at time t, a function of K_t, L_t, R_t as inputs of capital, labour and natural resources; A_t is "total factor productivity" or the "Solow residual"; H_t, G_t, F_t are the coefficients of factor contribution taken from its revenue share in the income distribution of GDP—in this case as 0.70 for L, 0.26 for K, and 0.04 for the rest. According to the constant returns to scale assumption required by this function, $\alpha + \beta + \gamma = 1$

A second wave of "endogenous growth" theories has attempted to overcome this analytical *cul-de-sac*. Nevertheless, instead of reintroducing the material and energy layers that embody and activate capital assets or enhance labour capacities, the endogenous growth theorists came from a different angle.[5] They looked towards increasingly symbolic and immaterial dimensions, such as the role played by 'human capital' endowment in long-term economic growth, and other social and cultural aspects. This approach has inspired a wide range of interesting and valuable historical research on the economic history of education, literacy and numeracy, book printing, skill premiums in the labour markets, the long-term effects of the European Marriage Pattern characterized by late weddings resulting in independent households based on a single nuclear family, and "a million of mutinies" in the everyday life of a large fraction of people which, according to Robert Lucas[6] is needed for income growth to occur in any society. These may range from nutritional standards and height increases, to the rise of contractual arrangements on weddings seen as a direct token of 'girlpower' and an indirect indicator of the habit to negotiate all sorts of business in life.[7]

[5] Ayres (2001).

[6] Lucas (2002).

[7] van Zanden (2009).

There has also been a renewed interest in studying the role played by income or wealth inequality, social public spending, and socio-institutional settings in the long-term economic performance of nations.[8] When income inequality approached the maximum permitted by available wealth and the need to reproduce the labour force at a subsistence level, societies often get caught in a 'worlds apart' lock-in state: the great majority of people could not change the situation, and the privileged minority did not want to. This explains why the agrarian class structure, the social conflicts that arise within it, and the kind of institutional changes fostered by social and political struggles are so important for historical processes of economic development.[9]

All these socio-institutional settings and human capabilities raise important questions that deserve to be studied in their own right. They have more than likely played a key role, considering them as results as well as crucial factors, that help explain *why* economic growth has taken certain directions in only some places and in only certain periods, and *by whom.* Also, if we apply here the distinction put forward by Amartya Sen and Martha Nussbaum between economic growth and human development, taking into consideration all these important questions may significantly help explain historical human development as an individual and collective increase in freedom of choice and 'empowerment'.[10] Nevertheless, it is doubtful that this can ever fill the Solow residual gap to explain *how* economic growth takes place. After several decades of endogenous growth analyses, the growth engine remains a black box.[11]

Although the historical process of human development has always included many social, institutional and symbolic dimensions, we should wonder if the empowerment of human capabilities, and the enhancement of individual and social choices, could ever be attained without relying on a greater amount of energy power able to move an increasing amounts of physical flows in a wider global scope.[12] According to both qualitative and quantitative historical evidence, physical and energy resource flows have always been a major factor in increasing the aggregated production of goods and services. A recent contribution to a never-ending debate, the Bob Allen book on *The British Industrial Revolution in Global Perspective*, has again stressed the role played by the supply of cheap coal as a driving force for the beginning of modern economic growth in England.[13]

Several economic and environmental historians have studied this link between coal and the British Industrial Revolution, or underlined the role played by the increasing access to fossil fuels for other regions of the world to industrialize and

[8] Lindert (2004), Acemoglu (2004, 2009), Acemoglu and Robinson (2006) and Aghion and Williamson (1998).

[9] Aston and Philpin (1985), Hoppenbrouwers and van Zanden (2001) and Milanovic (2005).

[10] Sen (1993, 1999).

[11] Easterly (2002) and Helpman (2004).

[12] Ayres and Warr (2005).

[13] Allen (2009).

converge with developed nations. All these studies reaffirm what Nicholas Georgescu-Roegen wrote many years ago in *Energy and Economic Myths*: "Now Economic history confirms a rather elementary fact—the fact that great strides in technological progress have generally been touched off by a discovery of how to use a new kind of accessible energy. On the other hand, a great stride in technological progress cannot materialize unless the corresponding innovation is followed by a great mineralogical expansion. […] This sort of expansion is what has happened during the last one hundred years".[14]

The failure to explain how the growth engine actually works is relevant from an environmental standpoint because the energy flows or material throughputs moved by the economy all over the planet are put aside. The principal ways through which the economy affects the ecosystems are exactly these same energy and material flows. Moreover some recent developments made by Robert Ayres and Benjamin Warr seem to open a promising new way to address the unsolved problem of the long-term growth accounts, without encountering the Solow residual.[15] This approach considers economic growth as an open multi-sector processing system of materials, energy and information, which moves forward in a perpetual disequilibrium, beginning with the extraction of natural resources and ending with the consumption and disposal of wastes. Since the Industrial Revolution, radical innovations in energy conversion technology have been among the most potent drivers of growth and structural change, which have put in motion much positive feedback by means of reducing energy costs. The substitution of increasingly cheap mechanical, thermal and chemical useful work (or 'exergy') for increasingly expensive human labour and capital has played a key role as a driver of economic growth (Fig. 2.2).

Considering that this evidence strongly suggests that 'exergy' (or the useful work actually performed by all energy converters which empower human labour and capital goods at its disposal) should be taken as a factor of production, Ayres et al. have been able to almost fit the empirical GDP historical series of the United States, Japan and other countries during the 20th century by including the useful work performed by all energy converters after discounting energy loses, together with the standard labour and capital factors, either in a conventional Cobb-Douglass or in a linear-exponential (LINEX) production function where all factors become mutually dependent, and where empirical elasticities do not equal cost share (Fig. 2.3).

14 Georgescu-Roegen (1976).

15 Ayres and Warr (2005).

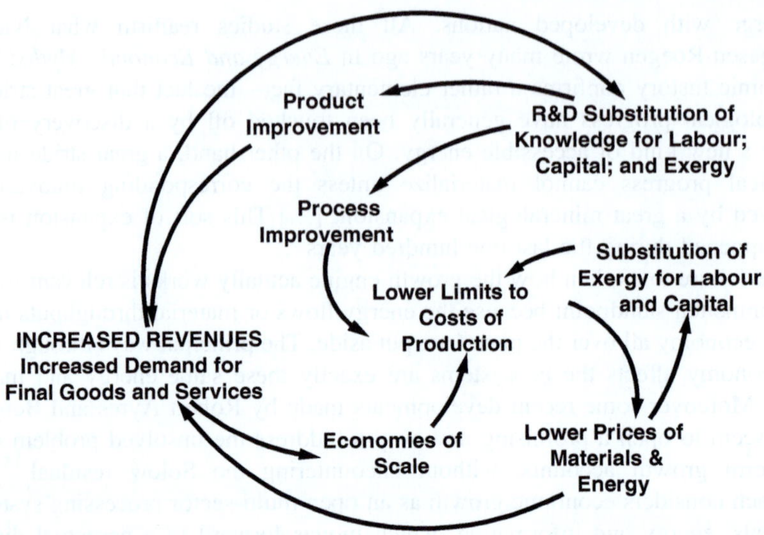

Fig. 2.2 The substitution of exergy for labour and capital seen as the key factor of lowering costs and increasing revenues in the virtuous cycle driving historical economic growth. *Source* Ayres and Warr (2005)

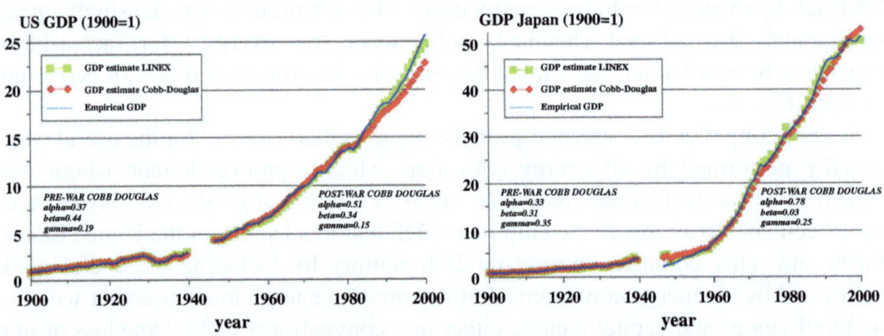

Fig. 2.3 The explanatory capacity of a LINEX or Cobb-Douglass production function which includes useful work together with labour and capital, confronted with the historical GDP series of the United States and Japan (1900–2000). *Source* Ayres (2008). Besides the standard Cobb-Douglass function, the following LINEX production function has also been used: $Y_t = U\exp\left\{a\left(2 - \frac{L+U}{K}\right) + ab\left(\frac{L}{U} - 1\right)\right\}$ which includes capital (K), labour (L) and useful work (U). Considering that there is an apparent inconsistency between very small factor payments directly attributable to physical resources—especially fossil fuels—and the obvious importance of final useful energy (or exergy) as a factor of production, this approach abandons the neoclassical assumption that the productivity of a factor of production must be proportional to the share of that factor in the national income. Alternatively, it considers that available useful work, either mechanical, chemical or thermal, multiplies the joint productivity of any combination of capital and labour throughout all value-added stages of the whole set of production chains (Ayres 2001, pp. 817–838; Ayres and Warr 2005, pp. 181–209 and Ayres et al. 2009). Therefore, the neoclassical identification of marginal productivities with factor shares is here replaced by a statistical assessment of the equation parameters

For the United States and Japan, a = 0.12. For the United States b = 3.4, and for Japan b = 2.7. It corresponds to $Y = K_{0.36} L_{0.08} U_{0.56}$ (i.e., useful work performed by energy sources could explain as much as 56 % of actual GDP growth experienced during the 20th century, while growth of capital stock would account for 36 and 8 % would come from the increase in labour capabilities).

It is too early to tell if this new way to account for the long-term economic growth, just now being opened from an ecological economics standpoint, will consolidate and gain acceptance among the majority of mainstream economists and economic historians. For the moment, even admitting the increasing relevance of environmental global concerns, mainstream developing economists and economic historians continue to consider primary energy as only another input or intermediate good that can always be substituted in the market. All of this explains why there is a growing suspicion among ecological economists and environmental historians that ignoring the environmental impacts of economic growth comes from the same analytical foundation that has forgotten the role played by natural resources in human economy and ecology, and both seem to be tightly related to the persistent inability by mainstream economists to fully explain how economic growth actually works.

2.3 From Economic History to Social Metabolism and Beyond

The rise of ecological economics is enabling economic and environmental historians alike to share and, at the same time, enhance their respective long-term understanding of economic growth by placing it in a broader perspective of biophysical interactions between human economies and natural systems in the biosphere. This socio-metabolic approach has been summarized by the Institute of Social Ecology in Vienna as follows: "The central theme underlying this research is the notion that most, if not all, global sustainability issues have to do with the fact that about two-thirds, if not three-quarters of the world population are currently in the midst of a rapid transition from agrarian society to the industrial regime. This transition is fundamentally changing societal organization, economic structures, patterns of resource use and so on, thereby probing the limitations of the planet Earth in many ways, among others by using up exhaustible resources, altering global biogeochemical cycles, depleting diversity and degrading Earth's ecosystems".[16] A new set of questions, methods and accounts arise from this, focus on the main socio-ecological transitions experienced in the interplay between nature and societies:

1. Was there a 'characteristic metabolic profile' of agrarian societies? Was such a metabolic profile connected to, and dependent on, certain land-use patterns?

[16] Fischer-Kowalski et al. (2007).

2. What happened when these socio-ecological agrarian regimes started to change? Which were the major drivers of change? Which pressures upon the environment gained momentum with industrialization and urbanization based on burning cheap fossil fuels, and which pressures receded? Which changes within natural systems could be observed during the socio-ecological transitions?
3. How much did the course of the socio-ecological transitions depend on the historical context, either local, regional or worldwide? Do common patterns exist?
4. How does the interplay between different spatial scales and levels of society work and interact with nature? Does globalization matter?

This approach has led to detailed quantitative studies of the energy and biophysical flows that link human economic activities with their ecological foundations, opening new ways of accounting: Material and Energy Flow Analysis (MEFA), the reconstruction of energy balances of economic systems and sectors, the estimation of energy returns on energy inputs (EROI), the study of nutrient and water cycles, the extent of the human appropriation of the ecological net primary production (HANPP) or the historical evolution of ecological footprints. These have established themselves as leading lines in current research.[17] There have also been attempts within the European Union to standardize these methods of ecological-economic accounting in established systems of National Accounts (EUROSTAT 2001).

As Fridolin Krausmann, Heinz Shandl and Rolf Peter Sieferle have written, "In this way, industrialization appears as a process of continuous increases in labour productivity and energy efficiency as well as growing industrial output resulting in continuous economic growth. Besides impelling social change and creating material wealth it has fundamentally changed the human domination of the Earth's ecosystems and brought along a plethora of environmental problems. A major claim of ecological economics is to broaden our understanding of economic processes and how they are embedded in nature by taking a biophysical perspective which conceptualizes economic processes also as natural processes in the sense that they can be seen as biological, physical and chemical processes. […] In this context, a historical understanding of the long-term development of society-nature interactions is of vital importance. […] We understand the industrialization process as a qualitative transition which transforms the agrarian socio-ecological regime into an industrial regime thereby establishing a distinct and fundamentally new pattern of society-nature interaction and material and energy use".[18]

The following scheme (Fig. 2.4) summarizes the key features of the two last main socio-ecological transitions from a solar land-based socio-metabolic regime (a) towards the coal stage of industrialization, combined with a set of 'advanced organic agricultures' which optimised traditional low-input agrarian systems (b); and then to a new stage of the oil and electricity driven technologies of the second Industrial

[17] Martinez-Alier (2011) and Krausmann et al. (2012).

[18] Krausmann et al. (2008), pp. 187–188.

Revolution that fuelled mass production and consumption, together with a reversal in the traditional relationship between the agricultural and non-agricultural sectors (c), by means of massive fossil energy subsidies for all economic activities and transport which fostered a boom in worldwide trade.

Perhaps the most interesting feature of this socio-metabolic approach is that it establishes a clear and accountable link between local and regional environmental problems with regards to the input side of nature-society interaction, based on resource-use together with related land-use changes; and from the output side, with local and global environmental problems derived from polluting emissions along the economic throughput chains. As Krausmann, Schandl and Sieferle have put it, "Taking a biophysical view it becomes evident that it will not be possible to accomplish global industrialization without an alternative pathway for the metabolic transition. Scarcity of oil and gas will increasingly become an issue and declining energy prices, a major precondition for the industrialization of the industrial core, are unlikely to prevail for latecomers. Before energy scarcity and rising energy prices become a major problem, the world is faced with rising greenhouse gases in the atmosphere contributing to global warming and destabilization of the world climatic system to a large and unknown extent. […] In the light of the historical process, the need for a new, sustainable, industrial socio-ecological regime with lower per capita material and energy turnover and a lower share of non-renewable energy and materials becomes a vital need for the global system".[19]

One of the aims of this broader ecological-economic perspective is to explore the connections, on all levels, between value-added flows in the market sphere, and the biophysical and energy flows or climatic suitability that sustain them from their ecological base. Measuring the energy and material dimensions of what GDP growth actually means for natural systems can provide us with new answers to previous questions regarding what triggers economic growth, what growth in fact involves, and what consequences it has for both social and natural environments. This standpoint connects the understanding of economic growth with the new studies on Global Warming and Climate Change which, during the last thirty years, have enhanced the focus on climate history. The IPCC concern about Global Warming has led to a development of new indicators and methodologies that have had a dramatic impact on all areas of knowledge, especially in a long-term historical perspective. Many recent studies have broadened the methodological possibilities open to climate historians, aimed at understanding the evolution of climate and its impacts on past and present times.[20]

Moreover, the analysis of biophysical flows linking economic performance with the carrying capacity of ecosystems necessarily leads to the study of changes in terrestrial land covers by human land-uses. Together with pollution and bio-invasions, this changing face of the Earth by human landscapes is precisely the main origin of the crisis of biodiversity at present. Putting together biophysical flows moved by

---

[19] Krausmann et al. (2008), p. 199.

[20] Brázdil et al. (2005) and Costanza et al. (2007a).

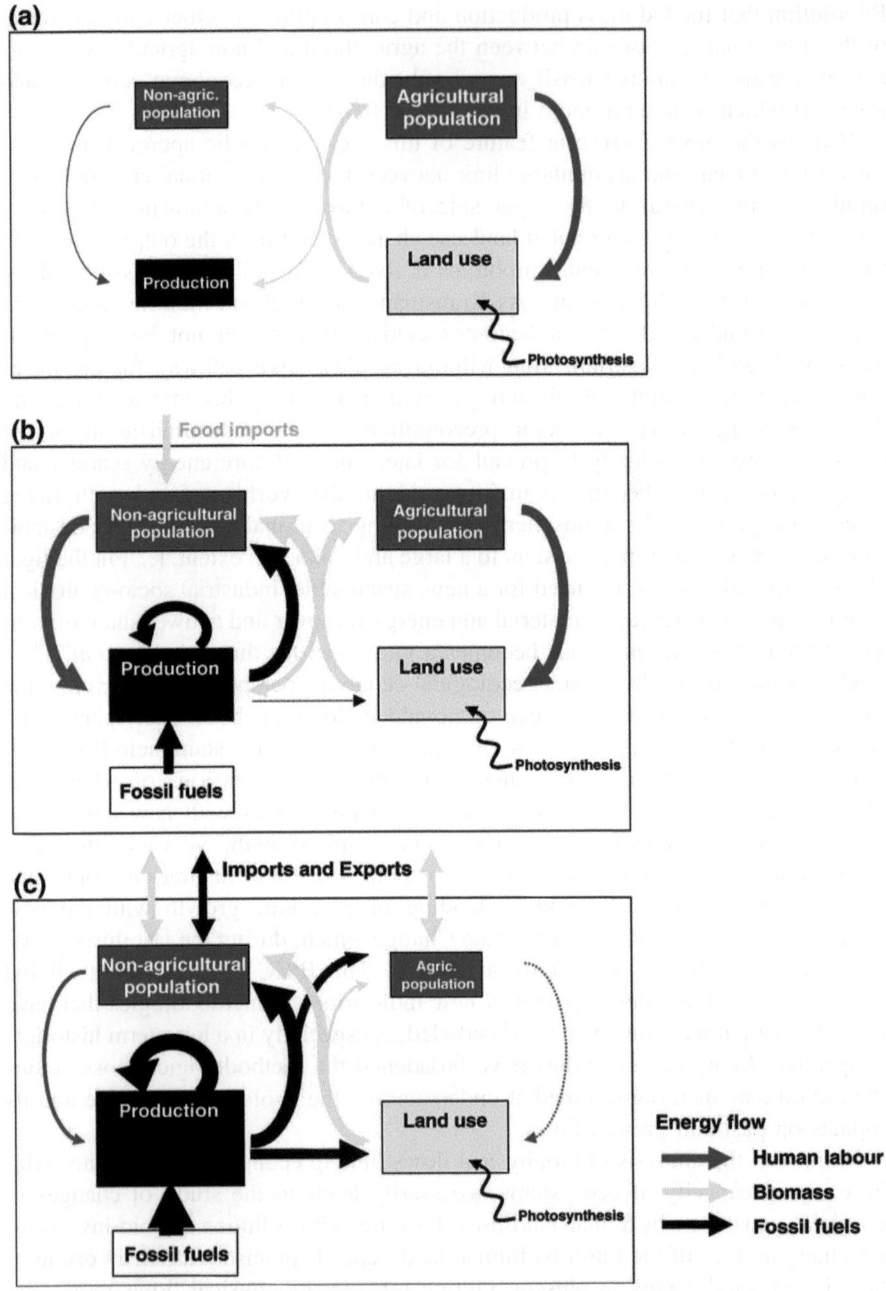

Fig. 2.4 The changing relation of energy, land and labour during the stages of the socio-ecological transitions. *Source* Krausmann et al. (2008)

human societies with the land-use changes made by them leads to the study of Global Change, a crucial meeting point for all scientific disciplines interested in the sustainability of human-nature interaction.[21]

This emerging socio-metabolic perspective does not entail prior assumptions concerning the causal direction in the ecological-economic interaction or phenomena.[22] The common attitude among practitioners of the emerging 'sustainability science' is a cautious, multidimensional and transdisciplinary approach which, from a co-evolutionary background, can admit that sometimes the driving forces originate from within the economic sphere and leave their ecological footprint on the surrounding environment; whereas in other cases researchers highlight the role played by the availability of energy, water, raw materials or climatic conditions and variability, either as a limiting factor or as a source for economic growth. Neither does such an approach entail the making of any deterministic presumptions; rather it is dependent on the type of enquiry being undertaken and on its historical or geographical scope.[23]

When environmental historians seek to discover the ecological impact of economic growth, usually from a short or medium-term perspective, they typically adopt market or state economic forces as the main driving force. But by adopting a long-term, comparative historical perspective, they also raise questions about the role played by the availability of energy, land, water and raw materials in accounting for historical economic growth processes or catching-up paths. On occasion, both approaches can be adopted simultaneously within the same research strategy, as Astrid Kander demonstrates in her study of the long-term relationship between energy, economic growth and greenhouse emissions in Sweden since the beginning of the 19th century—a research strategy that has been adopted within a broader comparative analysis between different countries and regions of the world undertaken by the members of the Energy-Growth and Pollution Network and the Institute of Social Ecology.[24]

Whereas it is true that all these new perspectives and methods provided by ecological economics and environmental history are greatly expanding the toolbox of economic historians, it is no less true than among scholars devoted to the study of past organic economies there was already a long tradition of taking bio-geographic, agro-ecological, energetic and landscape factors into account. However, their explanatory relevance has tended to decrease with the shortening of the time perspective from which economic historians seek to understand the present. The practitioners of prehistory and ancient, medieval or early modern history have never

[21] Cronon (1983, 1991), Crosby (1986), Cuff and Goudie (2009) and Hornborg and Crumley (2007).

[22] Costanzaet al. (2007b), pp. 522–527.

[23] Kates et al. (2001) and Haberl et al. (2006).

[24] Kander (2002).

failed to analyze changing environmental conditions as a key dimension to understanding the evolution of any human society.[25] Nevertheless, until recently, their presence has tended to vanish between the historians devoted to modern and contemporary times.

This growing lack of interest cannot be attributed to the loss of relevance of such environmental factors, as it is from mid-twentieth century onwards when the impact of human societies on the face of the Earth has become more intense, global and dangerous. The reason is ideological, and derives from a way of seeing reality that has characterized the two major socio-economic visions of the 20th century.[26] Within the mainstream approach to these two great visions, natural environments were considered as a set of restrictions and limitations that development would overcome. By seeing economic growth as a "liberation" of environmental constraints, the relevance of their study was considered inversely proportional to the degree of technological progress. Hence the explanatory weight of environmental factors was seen to decrease with the time-distance to the present covered by the analysis, in open contradiction with the degree of human degradation of Earth's ecosystems. Thus, this long-lasting Faustian vision of the modern Unbound Prometheus has paid a learned ignorance to the environmental dimension, until the obvious signs of a global ecological crisis have forced many to rethink.[27]

Since it is impossible to summarize within this text all lines of research which are currently changing the old visions of economic growth that formerly remained disconnected from environmental constraints and effects, we will take only a few relevant issues and ongoing debates as examples to illustrate the new emerging trends.

2.4 The Socio-metabolic Profiles of Past Organic and Present Industrial Economies

Until the mid 1980s a tradition of historical studies of pre-industrial agrarian economies was highly skewed by a pessimistic reading of the classical economists, especially Malthus and Ricardo. The work of B.H. Slicher van Bath, Michael Postan, Wilhelm Abel or David Grigg emphasized the difficulties experienced by traditional societies to increase agricultural output per capita because of the limits imposed by technological backwardness, and the inevitable arrival of diminishing returns spurred by population growth.[28] This tradition has been revisited and

[25] Bloch (1955–1956), Slicher van Bath (1963), Campbell and Overton (1991), Overton (1996) and Allen (2008).

[26] Thompson (1991) and Scott (1998).

[27] Landes (1969) and Landes (1998).

[28] Postan (1973), Abel (1980) and Grigg (1982).

revised by E. Anthony Wrigley by means of a fruitful dialogue with the classical texts which, along with limits, also noted the advances in productivity that could be achieved within the pre-industrial economies—particularly through trade specialization and urbanization.[29] Taking into account the approach of Adam Smith, and the range of changes or adaptations in demographic patterns studied by the Cambridge Group for the History of Population and Social Structure, Wrigley's work has contributed to better identify the actual limits and feasible possibilities to remove them ahead.

While Anthony Wrigley has never been directly interested in the interaction between society and nature as such, his most important contribution has opened a very important bridge between economic and environmental history by placing the emphasis on the characteristics of energy supply based on capturing solar energy through photosynthesis. In order to stress its relevance, he has coined the term 'organic economy' to highlight the fact that when any economic activity had to rely on the tiny fraction of solar energy that is being stored in the form of biomass through photosynthesis: "[…] neither the process of modernization nor the presence of a capitalist economic system was capable of guaranteeing sustained growth […]"; though he adds, "[…] both could help to ensure that the possibilities for growth offered by such economies were exploited effectively".[30]

The results found by reconstructing the long-term historical series of energy intensities of Sweden, Italy, United Kingdom, the Netherlands and Spain,[31] seem to confirm the view forwarded by Anthony Wrigley. From a long-term perspective, and when animal work and human labour are included together with fossil fuels and other modern energy carriers, there is no apparent single inverted-U Kuznets curve in the historical trend of the energy consumption per unit of GDP. What appears is rather something that resembles a downward staircase, or a winding path following an N-shape form (Fig. 2.5). This outcome clearly shows that in earlier pre-industrial times the energy cost per GDP unit was higher than that in the subsequent industrial period. The reason for this seems very simple: before the arrival of large amounts of fossil fuels, considerable amounts of primary energy were needed to obtain a single unit of value added into the market by means of the energy conversion of biomass. This is exactly what Wrigley hypothesised.

Long-term historical series of energy intensity per unit of GDP also show an increasing convergence between countries. While in a biomass-based energy system climatic conditions and natural resource endowment entailed big regional differences in the amount of primary energy consumed, the common adoption of a new set of coal or oil-based converters and technologies led to greater parity. Yet convergence in energy intensities might not be complete, because latecomers do not always use a factor endowment as appropriate as leading countries to the adoption of new technologies deployed. While the initial delay allowed them to adopt more

[29] Wrigley (2010).

[30] Wrigley (2004). See also Kander et al. (2013).

[31] Kander (2008), Warde (2007) and Gales et al. (2007).

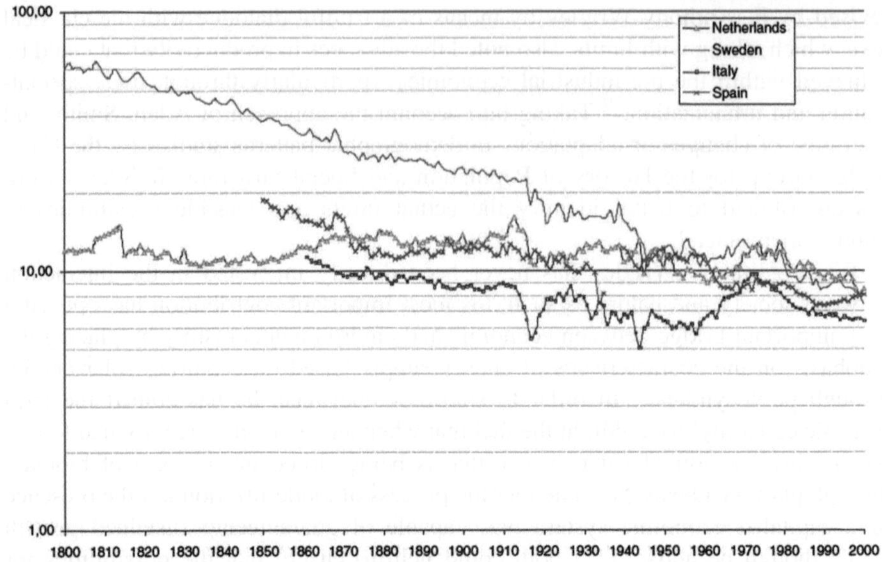

Fig. 2.5 The long-term fall of energy intensity in Sweden, the Netherlands, Italy and Spain from 1800 to 2000, in Mj per dollar (constant 1990 $ at ppp). *Source* Gales et al. (2007, p. 234)

advanced designs of these technologies, this often encouraged a further innovative adaptation that may increase efficiency (see below the comparison between the UK and Austria in Fig. 2.10).

These historical series stress again the role played by mechanical, thermal or chemical useful work provided by increasingly cheaper fossil energy sources in the contemporary increase of labour productivity and capital deepening. As long as the initial high energy costs of production, based on biomass converters, could not be reduced, the capacity of the economy to grow would have been strictly limited. Increasing competition between all lines of production for the same resource base, derived from the solar radiation converted and stored in the soil by vegetal land covers, would lead to a necessary curtailment of growth in a steady state. Wrigley was to draw this asymptotic assumption by studying demographic tendencies in pre-industrial societies, and also through a careful reading of Adam Smith and Thomas Robert Malthus. From an environmental history standpoint, whose foundations were laid by Nicholas Georgescu-Roegen, Rolf Peter Sieferle has adopted a similar perspective of what he calls the 'socio-metabolic regime' based on an indirect agrarian control of solar energy.[32] Wrigley's and Sieferle's approaches do not deny that trade specialization could have greatly helped to optimize the use of available organic resources. Nonetheless, only through the substitution of the renewable solar flow captured in the soil through photosynthesis with the

[32] Sieferle (2001).

subterranean stock of fossil fuels, were the energy limits towards modern economic growth finally removed. A structural change in the whole resource base of a pre-industrial organic economy was required.

As Joan Martínez Alier and Marina Fisher-Kowalski have reminded us, the history of these basic ideas on the socio-metabolic foundations of long-term economic growth is quite long,[33] extending back from Georgescu-Roegen through Frederick Soddy, Wilhelm Ostwald, Otto Neurath, Stanley Jevons, Leopold Pfaundler, Edward Sacher and Sergei Podolinski, all the way to Karl Marx who was the first social scientist to coin the term 'social metabolism'.[34] However, until very recently these ecological-economic insights have been ignored by a mainstream approach to either Liberal or Marxian economics which placed all the limits or stimulus for modern economic growth almost exclusively in institutional settings. Backwardness became the standard answer to the question as to why some regions lagged behind the world economic growth race.

The socio-metabolic approach can be very helpful to overcome this backwardness paradigm, and also in understanding why economic historians must deal with three different types of economic growth, labelled by Jan De Vries as 'Malthusian', 'Smithian' and 'Schumpeterian' (in the terminology previously suggested by William Parker).[35] Although Adam Smith shared the same pessimistic outlook as Malthus and Ricardo for long-term economic growth based on an organic resource base, the term 'Smithian' can be used to describe the type of growth that exploited all existing possibilities, stimulating growth through a better allocation of available organic resources, thus temporarily escaping the 'Malthusian fate' and giving rise to different 'advanced organic economies'. What has been called a 'consumer revolution' or an 'industrious revolution' in some European or Asian countries during the 17th and 18th centuries may be understood from this point of view.[36] The term 'advanced organic economy' coined by Wrigley could also be used to characterize the different paths taken by agrarian development in a wider range of European regions and countries during the 18 and 19th centuries, before the full industrialization of agriculture under the so-called 'green revolution' paradigm became widespread from the 1950s onwards.[37]

This approach also seeks to relate 'Schumpeterian' growth with the increasing burning of fossil fuels during the onset of the Industrial Revolution—Sieferle's 'hidden forest'.[38] This entailed the introduction of a completely new socio-metabolic regime, with a different energy and material flow exchange with ecosystems based on other types of land usage. But during this first stage of industrialization, the agricultural sector remained basically organic, at least until what Jan Luiten van

---

[33] Martínez Alier (1990) and Fisher-Kowalski (1998).

[34] Sacristán (1992) and Foster (2000).

[35] De Vries (2001).

[36] De Vries (2008a) and Sugihara (2003).

[37] Kjaergaard (1994) and Krausmann et al. (2008).

[38] Sieferle (2001).

Zanden labelled 'the first green revolution'—which was initially fostered by innovative responses to the European agrarian crisis experienced between 1870 and 1914,[39] when the cheap wheat and corn exported from the United States, Canada, Australia, Argentina or Russia flooded European markets.[40] From then onwards organic manure started to be supplemented by increasing amounts of external inputs of mineral and fossil origin. However, up until the 1950s they remained very small, as long as the diffusion of industrial fertilizers, the adoption of tractors or newly selected seeds and animal varieties served as a complement rather than as a full substitute for organic sources, crop rotations and animal work. Therefore, the radical turnaround in the agrarian sector did not occur until the second half of the 20th century (see the differences between schemes b and c in Fig. 2.4). Before the massive use of fossil fuels within the agrarian system, the main means of increasing agrarian outputs in the various European bioregions still lay in the development of several types of 'advanced organic agricultures'.[41]

2.5 Why the Industrial Revolution Began in a High Wage and Cheap Energy Economy

How did a Schumpeterian-type of modern economic growth began? A long-lasting historiographical tradition has been collecting data on agricultural land usages and yields from medieval times onwards, and has shown two main features: (1) up to 1800, yields and labour productivity remained low but stable in the long run in most European and Asian regions which had experienced comparatively old agrarian colonization and attained high population densities; and (2) only a few regions seem to have been able to overcome the Malthusian-Ricardian constraints, taking advantage of all existing possibilities to optimise traditional low-input organic systems in order to achieve higher agrarian yields per unit area without diminishing agricultural labour productivity at the same time. As far as the ongoing debate on the Great Divergence between Western Europe and Eastern Asia allows to tell for the moment, prior to 1800 these upper outlying cases were the Dutch and English economies, the only ones where wages and standards of living seem to have increased above the rest of the World at the time.[42]

Why did these changes develop in some regions whilst not in others? Why did it take so long to be adopted or emulated by other regions of the World? These questions raise two sets of issues: (1) which kind of land entitlements and institutional settings created an incentive structure that encouraged a long run increase

[39] Van Zanden (1991).

[40] Koning (1994).

[41] Leach (1976) and Naredo (2004).

[42] Pomeranz. (2000) and Allen et al. (2005).

in labour productivity; and (2) how could yields per unit of land be increased, thus overcoming the Ricardian-Malthusian fate of long-term growth in an already organic economy. The first question asks for the agency of change, by looking at those who made them and what they made them for. The second set of questions looks at how they did it, taking into account the available choices offered by their bio-physical and technological context. In order to attain a complete historical answer, both agency and structure must be combined in a single interpretation encompassing natural as well as social environments.

Following the interpretive lines proposed by Bob Allen in *The British Industrial Revolution in Global Perspective* (2009), the outset of the Industrial Revolution in England at the end of the 18th century could be summarized as follows. Within the framework of new institutions that emerged from the defeat of Royalists in the English Civil War (1641–1651) and the Glorious Revolution (1688), together with the agrarian changes towards a highly productive 'advanced agriculture' mainly introduced by the yeomanry at the time, the rise of British colonial hegemony and overseas uncontested power after the defeat of Holland navy in the three wars from 1652 to 1674 enabled the United Kingdom to develop a particularly successful industrious revolution, which turned the country into a textile export economy—during the first half of the 18th century 85 % of the value of English exports were already manufactured goods.[43]

This commercial expansion and industrious revolution spurred urbanization and converted London in the single biggest city at the top of the hierarchy in the urban centre of gravity around the North Sea.[44] Up to a point the English agricultural, commercial, industrious and urban improvements helped to achieve the above mentioned increase in wages and pre-industrial standards of living, well beyond the ones existing at the time in the rest of Europe and Eastern Asia, thus sustaining a distinctive although not exclusive consumer revolution.[45]

At the same time the amount of energy needed to heat the homes of Londoners and other English urbanites in the more and more deforested isle of Great Britain encouraged the replacement of increasingly expensive firewood or charcoal by cheaper coal.[46] As Paul Warde notes, considering that energy embodied in labour, capital and transport services required that coal supply remained mainly 'organic', up to a point its primary difference in price with firewood was probably determined by rents because wood competed for space with other uses: "The changing point at which coal-use became more economic than wood use was probably thus determined by the general level of rents, and these in turn were determined by the necessity of producing by far the least efficient output in energetic terms, food. It is likely that it was not the scarcity of wood but the relative scarcity of food that made

[43] Allen (1992, 2009).

[44] De Vries (1984) and Wrigley (1987).

[45] McKendrick et al. (1983) and De Vries (2008).

[46] Wrigley (1987).

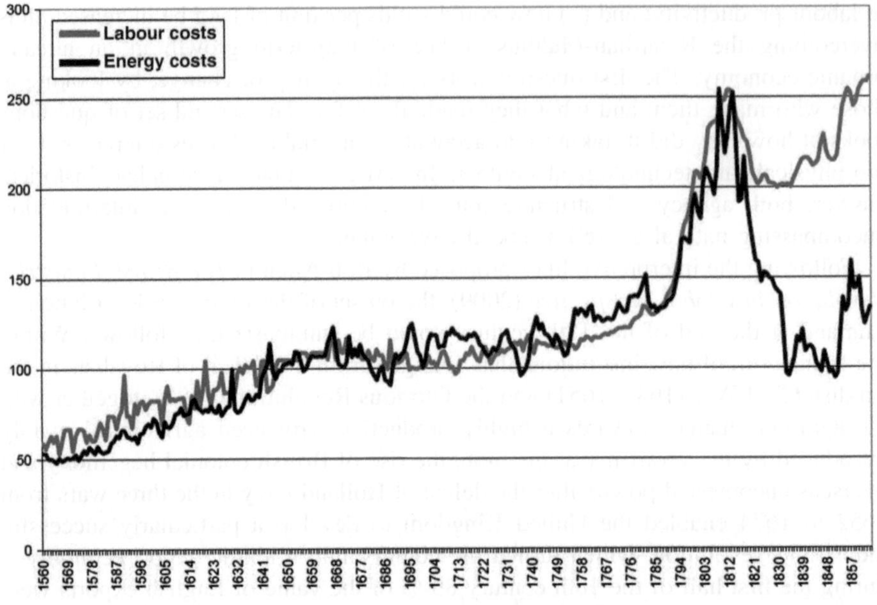

Fig. 2.6 Long-term trends in labour and energy costs at nominal prices in southern England (1560–1860, 1700 = 100). *Source*: Warde (2007, p. 87, 2009). The series has been calculated from the changing mix of energy carriers shown in graph 7, taking into account the labour, capital and land required to get a unit of energy (considering that a single woodcutter could prepare about 1.1 tonnes of firewood a day or 3.3 million Kcal, compared with a coal miner who could extract as much as 2.5 tonnes of coal or 17.5 million Kcal). Until the 1820s, overland travel of firewood as well as coal mainly depended on 'organic' muscle power. This meant that in practice obtaining coal at some distances from the coalfield was largely determined on the prices of human food and animal feed. Therefore, fossil fuel did not enjoy a great advantage over firewood until the use of steam engines, and new transport facilities such as canals or railways become widespread (Warde 2007, pp. 83–86)

coal more attractive fuel".[47] Finally, the combination of higher wages and exceptional availability of cheap fossil fuels created exactly the economic context where relative factor prices led entrepreneurs and financers to invest in the new type of capital goods able to perform thermal, mechanical or chemical useful work that opened the road to a new Schumpeterian-type of economic growth (Fig. 2.6).

After 1800 an unprecedented acceleration of technological change powered by cheap coal ensued, and became a formidable weapon in the hands of new industrial bosses which enabled them to earn and reinvest considerable profits while keeping wages well below the contemporary increase in labour productivity. The new factories powered by steam engines or waterwheels were aimed at centralizing and mechanizing the production processes, in order to replace comparatively expensive human labour with capital goods as well as to control and master the fierce

[47] Warde (2007).

traditional independence of the English labouring people. This may help to explain why the previously high British pre-industrial real wages became stagnant, or even temporarily decreased, during the first phase of the Industrial Revolution. Bob Allen has labelled 'Engels' Pause' (1780–1830/40) this long-lasting gap between wages and productivity growth, which corresponds to the pessimistic outlook about the standards of living during the first phase of British industrialization.[48] The pessimistic view has also been reassessed using new biological and social evidence,[49] like the fall in the heights of military conscripts' and the average life expectancy, or the increase in infant mortality rates, child labour and income inequality.[50]

It is interesting to notice that macroeconomic accounts of economic growth during the British Industrial Revolution have tended to reduce its revolutionary character, by dismissing that growth rates experienced any sudden acceleration compared with previous pre-industrial ones.[51] However, and at the same time, the beginning of the new Schumpeterian-type of industrial economic growth meant a revolutionary turnaround in socio-metabolic terms, as it has been reassessed by the study of the first energy transition from a biomass solar-based energy system towards another, based on burning the underground stock of fossil fuels. Moreover, the historical series on primary energy consumed and the energy intensity per unit of GDP produced in England and Wales (Fig. 2.7), reconstructed from 1560 onwards by Paul Warde, have again shown the outlying character of the English economy which had started substituting coal for biomass energy carriers well before the Industrial Revolution began.

Thus, as Anthony Wrigley pointed out (1988), the continuity of a gradual increase in economic growth rates combined with the chance of having large accessible coal deposits in England and Wales and gave rise to the big energy change of the British Industrial Revolution. Historical comparative analysis has also shown that GDP convergence kept pace with the energy transition to fossil fuels, and this evidence opens up the question whether this relationship was a consequence of economic growth over energy consumption, or rather that the convergence in economic terms would have to wait until each nation or region could find their own path towards the new mineral-based energy system.[52]

While the abovementioned interpretive lines seem to offer an explanation about as to why the first Industrial Revolution was British, they do not offer a full and satisfactory outline as to why it did not also take place in other places throughout Europe and Asia, where convergence with the industrial growth of the United Kingdom encountered numerous obstacles. Bob Allen's interpretive outline stresses

---

48 Allen (2009).

49 Hobsbawm (1964) and Thompson (1968).

50 Williamson (1997), Horrell and Humphries (1995), Crafts. (1997), Feinstein (1998) and Komlos (1998).

51 Crafts and Harley (1992).

52 Krausmann et al. (2008).

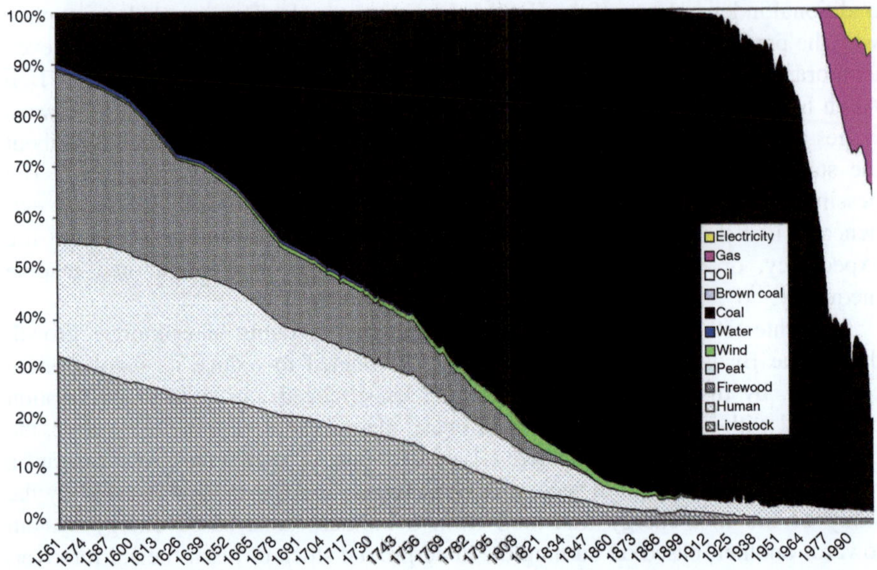

Fig. 2.7 Long-term energy transition in England and Wales, as percentages of each carrier (1560–2000). *Source* Warde (2007, p. 74)

from the beginning the role played by mercantilist policies—under colonialism and imperialism—in fostering British commercial development. Without this commercial empire that followed and undermined the previous Dutch trade expansion, the population of London could not have grown from approximately 50,000 to 200,000 between 1500 and 1600, then doubled in the next century, and reached nearly one million by 1800. In the meantime the fraction of the English population living in settlements of more than 10,000 people increased from 7 to 29 %, whilst the share of the workforce in agriculture dropped from about 75 to 35 %.[53] This, in turn, raises the question to what extent was the new type of Schumpeterian industrial growth started by the British Industrial Revolution an actual possibility for any other nation at the time. Adopting an ecological-economic approach may help to properly address this question by taking into account the environmental load displacement or ecological footprint that the Industrial Revolution entailed.[54]

Following the idea of a 'ghost acreage' won by Europe through the colonization of America, already put forward by Eric Jones and then Kenneth Pomeranz stressed again, we may wonder about the role of environmental endowments and resource availability in the Great Divergence between Europe and Asia before 1800.[55]

---

[53] Omrod (2003).

[54] Hornborg et al. (2007).

[55] Goody (2004) and Emmer et al. (2006).

Table 2.1 A first rough estimates to some components of ecological footprint entailed by the British Industrial Revolution (1801–1831)

		1801	1811–1815	1827–1831
Coal	Millions of tonnes burnt	13.9	n.a.	22.6
	Equivalent woodland area as total surface of England and Wales[a]	92.7 %	n.a.	150.7 %
Sugar	Footprint hectares (thousand)	436.3	464.6	604.8
	As percentage of contemporary cropland in England and Wales[b]	9.5 %	9.6 %	13.4 %
Cotton	Footprint hectares (million)	n.a.	9	23
	As percentage of contemporary cropland in England and Wales[c]	n.a.	154.4 %	322.8 %
Timber	Footprint hectares (thousand)	666.5	n.a.	n.a.
	As percentage of contemporary cropland in England and Wales[d]	18.5 %	n.a.	n.a.

Source our own, from Sieferle (2001, pp. 14–15), Pomeranz (2000, pp. 313–315) and Grigg (1982, p. 38)

[a] Estimated translating the energy content of coal burnt into cubic meters of firewood, and then assessing the woodland area needed to annually grow this amount of firewood in England and Wales

[b] Estimated assuming that the average caloric intake with sugar imported from the colonies had to be replaced by cereal cultivated in England and Wales

[c] Estimated assuming that average cotton imports had to be replaced from wool produced by sheep bred in the pastureland of England and Wales

[d] Estimated assuming that average timber imports from America and the Baltic had to be replaced by woodlands of England and Wales

Table 2.1 shows the approximate ghost acreage of English consumption of coal, sugar, cotton and timber during the first half of the 19th century:

As can be seen, in spite of the initial tiny amounts of GDP which they might have represented before 1800,[56] the access to these four key natural resources would have immediately outstripped the biological carrying capacity of any 'advanced organic economy'—particularly cotton and coal, which together with iron formed the basic triad of the English Industrial Revolution. The only way out of these land-related constraints was the unique combination enjoyed by the United Kingdom of coal mined from the underground with the 'ghost acreage' provided through an exceptional network of worldwide trade flows nucleated into a single European region. No other nation or region of the world could have had such access to analogous natural resources, until railroads and steam vessels diminished travel costs and opened the way to the first globalization from 1870 onwards.[57] It is worth remembering, as William McNeill pointed out,[58] that this was a very exceptional

[56] Van Zanden (2009).

[57] Williamson (2006).

[58] Mc Neill (1982).

combination of commercial networks and resource availability focused within the British Isles, which allowed for a high wage and cheap energy economy. And this combination was assembled also thanks to a persistent and successful pursuit of military power.

The military aspect of the question leads directly from economic growth towards the contemporary experience of underdevelopment. The 'ghost acreage' of industrialization entailed many different and persistent impacts upon peoples, communities, nations and landscapes of many non-European countries, usually not taken into account within the history of growth of the developed regions of the World. These external impacts of modern economic growth in Western countries deeply affected the subsequent historical path followed for the rest of Humankind. Not only were their lands fissured by mines or plantations that spoiled natural resources and left the burden of heavy pollution everywhere, but their social lives were distorted as their institutions were submitted to foreign colonial or neo-colonial rules.[59] As Ramachandra Guha and Joan Martínez Alier have pointed out, the popular resistance to the exploitation of their natural resources, destruction of their modes of life, and externalization of ecological impacts by the rich, has been the true origin of contemporary environmentalism, often known as the environmentalism of the poor.[60]

2.6 Land-Use and Livestock Breeding as a Crucial Metabolic Hinge for Yield Increase

Precisely because the natural resources required by industrialization in the United Kingdom soon surpassed the bio-capacities of the British Isles, they exercised an increasing series of pressures upon the rest of the World. While in many underdeveloped regions these mainly entailed a set of difficulties and distortions, in other parts of the World a combination of pressures and incentives became apparent which might produce different outcomes, depending on natural and social endowments, institutional settings and public policies, together with the role assigned to the region in the global division of labour established by the British economy.[61] The convergence path followed by the small but growing group of developing nations, and the under-developing divergence of the rest, both ensued the beginning of a Schumpeterian-type of economic growth under British leadership. What were the reasons behind the enduring fortunes of some, and misfortunes of many others? Once again, this brings the research back to the Malthusian-Ricardian constraints that were tightly related to the prevailing land-labour ratios.

59 Gadgil (2000) and Gadgil and Guha (1993).

60 Guha and Martínez Alier (1997).

61 Warde (2009).

Examining one part of the answer within the evolving nature-society interface may help to understand how yield increases can be achieved in some still basically agricultural economies and societies, while not in others. This requires the black-case of the agrarian system to be open in order to look carefully into its agro-ecological engine to understand how matter, water and energy were processed into fertile soil allowing plants to grow. When applied to organic agrarian economies, the quantitative reconstruction of energy and biophysical flows requires that they be located carefully within the territory. It must be noted that almost all energy or biophysical flows mobilized by past agrarian societies were highly dependent on their land-use pattern. Land usage became pivotal, in those organic economies, for any socio-metabolic exchange with nature.

The importance of an integrated management of the three main components of any agrarian system, that is, cropland, woodland, pastureland and the key role played by livestock in linking the three, are readily apparent to all agrarian historians studying pre-industrial societies. However, until recently, very few attempts had been made to connect the energy and material flow analysis with land-use systems. Today the most important research programme seeking to relate socio-metabolic flow analysis with land usages is that being undertaken by the Institute of Social Ecology at the University of Vienna (http://www.iff.ac.at/socec/). Many studies and publications on the changing face of human colonization of terrestrial ecosystems come from the extensive international research programme Land-Use Land-Cover Change (http://www.geo.ucl.be/LUCC/lucc.html). The LUCC examines the transformations undergone by Earth's vegetal cover over the centuries, in order to identify the main driving forces behind global socio-environmental change, and also to assess its socio-ecological impact. Perhaps environmental and economic historians, working together to understand economic growth in past agrarian societies and present industrial ones, can help merge these two approaches, i.e. the accounting of biophysical flows combined with a closer and more analytical examination of the land-use systems in which they take place.

Thanks to the energy balances calculated by many scholars from the mid 1970s onwards, we now know that energy returns on energy inputs were higher in earlier organic agricultures than those attained following the widespread adoption of the 'green revolution' after the Second World War—as wee see in schemes b and c in Fig. 2.4. However, there is an important element to this seemingly paradoxical discovery which should focus our attention on the land-use system: How were these pre-industrial societies able to attain such a high energy performance, while being so heavily dependent on livestock bioconversion which is so inefficient? Why has industrialized agriculture become so inefficient in processing energy and material flows, and thus so pollutant, having at its disposal a wider range of more efficient converters? Some recent results obtained analysing the link between energy-use performance and land-use management in some Mediterranean local case studies suggest that a great deal of the answer lies in the loss of landscape efficiency.[62]

[62] Tello et al. (2006) and Marull et al. (2010).

Many past agrarian societies were able to build up and maintain sound land-use management which enabled them to attain high energy returns on energy inputs (EROI), in spite of the then inevitable losses in available bio-converters.[63] However, this was not for the sake of any positive environmental externality that we might discover now and praise them for. They did so by necessity that is, because of their own energy poverty in available carriers and sources. Thus, there is no contradiction in the lack of primary energy for economic growth in an 'organic economy' and the typically high energy performance attained by pre-industrial agrarian systems, as identified by Anthony Wrigley, Rolf Peter Sieferle, Paolo Malanima or Paul Warde.

We can understand this, following Vaclav Smil, taking into account that under pre-industrial conditions any final output had to be obtained through a set of energy production chains that relied on firewood burnt in highly inefficient fireplaces, animal feed bio-converted inefficiently into power traction and manure, vegetal and small animal human food consumption, and a small amount of raw materials also inefficiently transformed into costly industrial goods, and so on.[64] With only a few alternatives available in order to increase end-use efficiency, i.e., through the adoption of better converters (such as stoves instead of fireplaces), these societies had to rely on the highest primary energy output they could attain if they wanted to achieve even a small increase in the amount or the variety of final consumable goods. The main way of achieving this was to develop and maintain an efficient land-use pattern. Only through the increasing integration of crop production and livestock breeding could past organic agricultures hope to achieve even a modest rise in their agricultural and forestry output, enabling them to diversify and enrich both household and market economies. Herein lies possibly the most important reason explaining why, through sound landscape management, past advanced organic agricultures were more energy efficient than the majority of current agricultural systems (Table 2.2).

The improvement of past organic agricultures became a key condition for any advance in urbanization, as can be seen when looking at cropland needed to provide staple food for any growing city. For example, when only 20 % of agricultural output could be sold and carried to urban centres, a city with 250,000 inhabitants such as 1820 Vienna, required an agricultural hinterland of around 22,000 km^2 (or nearly 88 km^2 of cropland per thousand inhabitants). Thanks to an agrarian surplus that grew by almost 60 %, in 1910 Vienna could be supplied by 24,000 km^2 of cropland when the city already housed two million people.[65] Feeding Paris required a footprint of around 60,000 km^2 with a population of 660,000 in 1784, and a similar area was needed when the population exceeded two million between 1876 and 1881, and even in 1921 when it reached three million.[66] This meant a decrease

---

[63] Agnoletti (2006).

[64] Smil (1999, 2001).

[65] Krausmann (2013).

[66] Billen et al. (2009).

Table 2.2 Summary of the energy balances of the agrarian system in a Catalan area near Barcelona (Spain) towards 1860 and in 1999–2004, compared with the contemporary agricultural landscapes

		Towards 1860	In 1999–2004
Primary solar energy fixed in the useful agrarian cover (UAC, 1,000GJ, GJ/ha/yr)	Cropland	146.3	187.3
	Pastureland	34.4	–
	Woodland	87.2	211.0
Final output by sector (FO, 1,000GJ, GJ/ha/yr)	Crops	38.6	135.9
	Livestock	2.9	144.5
	Forestry	129.5	69.1
Livestock weight units (LU of 500 kg, Kg/ha)		983 (41)	23.833 (1,021)
Cattle feed (1,000GJ, GJ/ha/yr)		68.7 (5.7)	1,095.7 (103)
Manure or fertilizers applied (1,000GJ, GJ/ha/yr)		23.9 (2)	55.5 (4.7)
Total inputs consumed (TIC, 1,000GJ, GJ/ha/yr)		102.4 (8.5)	1,625.8 (139.1)
External inputs consumed (EIC, 1,000GJ, GJ/ha/yr)		6.6 (0.2)	1,574.4 (134.9)
Total final output (TFO, 1,000GJ, GJ/ha/yr)		171.0 (14.3)	349.5 (29.9)
Energy return on total inputs (EROTI = TFO/TIC, 1,000GJ, GJ/ha/yr)		1.67	0.21
Energy return on external inputs (EROEI = TFO/EIC, 1,000GJ, GJ/ha/yr)		66.6	0.22
TFO/primary solar energy fixed in the AUC (%)		64	88
EIC/primary solar energy fixed in the AUC (%)		1	395

(continued)

Table 2.2 (continued)

Land mosaics towards 1860	Caption	Land covers in 1999–2004
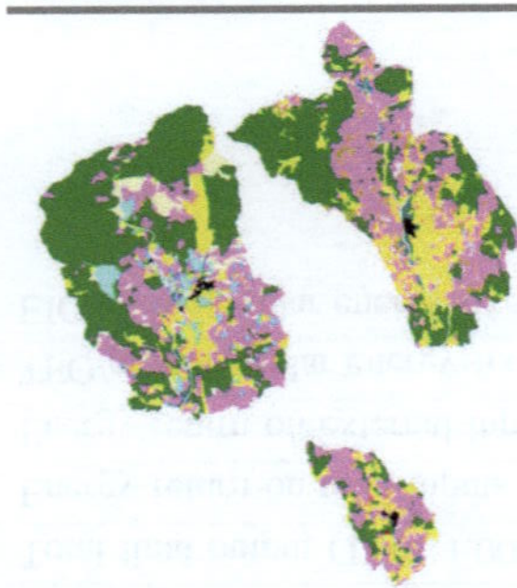	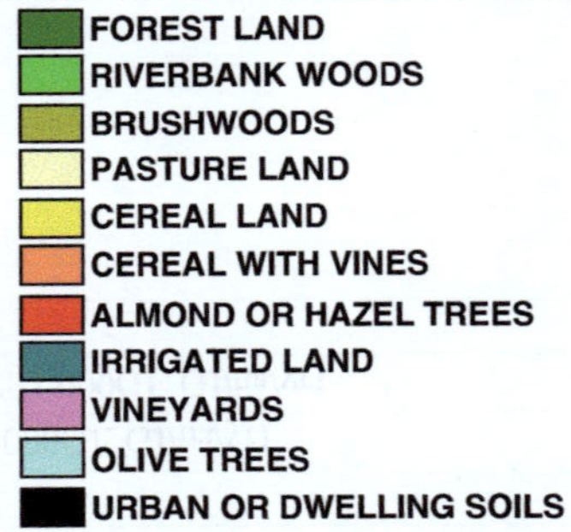	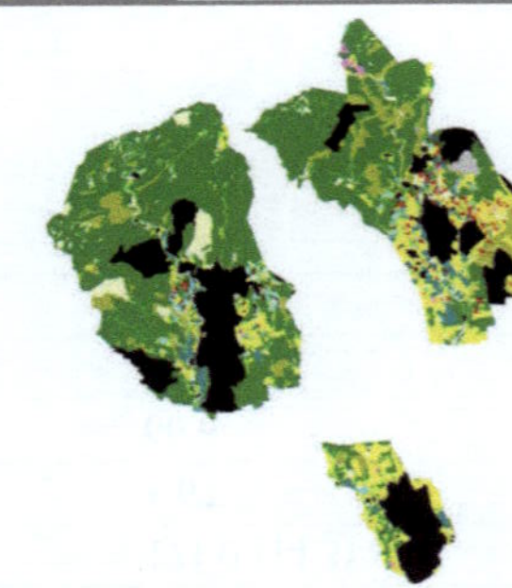

Source our own, from Cussó et al. (2006b, pp. 471–500)

from some 92 m^2 of cropland, pasture and woodland to only 20 for every thousand inhabitants, thus illustrating the role played by the advancement of organic agricultures for urban growth.

2.7 Nature-Society Interaction Between a Malthusian Trap and a Smithian Response

To what extent could the development of global market networks have increased opportunities to improve prevailing land-use systems? This question involves, of course, the defining feature in what William Parker and Jan De Vries have labelled a Smithian-type of growth.[67] David Grigg has described the socio-metabolic way of achieving this improvement, characterizing market specialization as a way of taking advantage of the 'ecological optimum' of different soils or regions according to their agronomic aptitudes and limiting factors.[68] Market diversification and specialization meant that these regional 'ecological optimums' could be exploited, while simultaneously other local agro-ecological constraints or 'Liebig minimums' could be overcome through imports. Economic history provides many examples of the link between resource endowment and trade patterns pointed out by the Heckscher-Ohlin model, such as vineyard or olive oil specialization in the Mediterranean regions.[69]

However, if we adopt an ecological-economic perspective we can easily see that market networks might also become a double-edged sword. For example, in the abovementioned local west-Mediterranean case study we found a high energy return on energy inputs of 1.67 attained in mid-19th century Catalonia (Spain); nonetheless, energy efficiency could only have been maintained if agricultural landscapes were kept poly-cultural and combined with some amount of woodland or brushwood. The final energy balance depended on certain key factors, for example a great deal of branches pruned from the vineyards or olive trees were used as fuel, thus serving as a substitute when there was a reduction of firewood supplies, following the loss of woodland due to the growing plantations of wood crops. This was probably the case while the agrarian system remained poly-cultural and vine-growing was a partial specialization that coexisted with many other land usages in a diverse agrarian mosaic. But if increase in demand for wine triggered a complete regional specialization in a single export cash-crop, as was the case during the infestation of the French vines by the *Phylloxera* plague between 1867 and 1890, it would have led to a shortage in livestock and manure. In such circumstances, the pruning of vineyards and other wood crops would have been used as a poor

67 Parker and Jones (1975) and De Vries (2001).

68 Grigg (1982).

69 Pamuk and Williamson (2000), Pinilla and Ayuda (2007) and Badia and Tello (2014).

substitute for manure, by burning them in the so-called Mediterranean *hormigueros* —small charcoal kilns made in fields covered with topsoil.

These Spanish *hormigueros* were traditionally used as a complementary fertilizing method which produced Potassium, and incorporated charcoal into the soil in order to improve its bacterial populations.[70] This was part of the ancient European Mediterranean culture of fire,[71] that took advantage of the summer water stress which meant that large quantities of dead biomass could not fully decompose and tended to accumulate in forests or scrubland until ignited by lightning. This accounts for the fact that Mediterranean woodlands have always coexisted with natural fires.[72] Thus, the removal of this dead biomass from scrub and forests in order to burn it on cropland was a sound human adaptation to natural conditions. In order to provide soil nutrients, however, the thermal process was a less efficient way of decomposing biomass compared to the humid method by means of compost or manure. Furthermore, it was a very labour-intensive fertilizing system. These features would explain why the *hormigueros* were only practised in the most populated and cultivated regions of the Iberian Peninsula. In any case, a complete replacement of the traditional poly-cultural mosaics by a vineyard monoculture would have entailed a serious bottleneck in nutrient availability, together with a rapid reduction in the energy return on energy invested (EROI) up to an index of around one or below. Of course, these local tendencies could have been offset by fertilizer imports, or by substituting tractors for animal traction but, once again, this would have meant a higher amount of external inputs and a further decrease of EROI.[73]

As Martínez Alier reminded us, the theoretical foundations of this double-edged market sword were forwarded as early as 1902 by Leopold Pfaundler, in his attempts to assess Earth's maximum capacity for sustaining human needs. He argued that any estimate would depend on whether we were to aggregate the maximum local capacities of each small territory, where limiting factors vary; or whether we were to consider Earth as one territory, assuming that any local resource would be available globally from any place without transport restrictions. Pfaundler suggested that a reasonable answer would lie somewhere between the two extremes, noting that transport always consumes energy and produces environmental impacts.[74] Looking at the abovementioned Catalan case study, the low energy return on energy inputs of 0.21 that we have found presently, appears to be in keeping with Pfaundler's argument.[75] The most noticeable feature at present is the fact that current energy flows are not in proportion to the land area in which the agricultural systems are placed. The metabolic chains operate in a monoculture pattern or in linear livestock breeding systems that have become virtually unconnected with the surrounding

70 Tello et al. (2012).

71 Pyne (1997).

72 Grove and Rackham (2001).

73 Cussó et al. (2006a).

74 Martínez Alier (1990).

75 Cussó et al. (2006b).

agro-ecosystem.[76] The majority of external inputs–fertilizers, oil, fodder, and the like—merely pass through a territory that operates as an inert platform. Ironically, the massive fodder imports and consumption of cheap fossil fuels have turned most of the woodland area into a derelict space which is increasingly prone to devastating wildfires. Here again, we see a very close link between low energy performance and inefficient land-use, both of which give rise to increasing levels of pollution and landscape degradation—the ecological imprint of what we now know as globalisation.[77]

Thus, any discussion on the relationship between markets and agro-ecological efficiency or environmental impacts cannot be viewed in black and white terms. Depending on the type and extent of the markets under consideration, trade might promote more efficient land usage and biophysical flows, or the contrary. A number of current approaches to the relationship between human development and markets in developing countries seem to draw similar conclusions, considering both the social as well as the natural environments.[78] While a network of local and regional markets was, and still remains, an important tool for sustainable human development, a direct connection to globalized markets is often little more than a trap.[79] To gain a better understanding of this relevant question, more studies on past agrarian economies need to be undertaken from this standpoint.

However, it is already apparent that globalization matters, when considering the relationship between the sort or scope of trade flows and their environmental effects. Therefore, transport must be taken into account as a key component in the ecological side that lies beneath any example of 'Smithian' growth fostered by significant increases in urban population, their consumer baskets, and market development. Marina Fischer-Kowalski, Fridolin Krausmann and Barbara Smetschka tackle this important issue from a socio-metabolic approach, and conclude that "the volume of transport necessarily rises faster than both the size of the society (in terms of population of urban centres and their hinterland) and its material wealth, and this not only constrains but limits the possible size of urban populations. The core mechanism behind these limits is the agrarian energy metabolism: in order to overcome distances, agrarian societies need more land to feed the human and animal labour power required for transportation. So they have to enlarge their territory, thereby again increasing the distances that have to be overcome. Fossil fuels provide a two-edged benefit: they allow to span larger distances, and to manage reproduction within a smaller area. So under industrial conditions, size-constraints for urban centres and for freight transport disappear: transport volumes 'explode'".[80]

---

[76] Goodman and Redclift (1991).

[77] Fischer-Kowalski and Amann (2001).

[78] Shiva and Gitanjali (2002).

[79] Aoki and Hayami (2001) and Chang (2002).

[80] Fischer-Kowalski et al. (2004).

In an interesting study of the relationships between ecology, economy and state formation in early modern Germany, Paul Warde has characterized the type of socio-environmental changes brought about by the dual development of wider market networks and the political strength of state rule. According to Warde, a previous 'territorial ecology' sustained by local agrarian communities began to be undermined by a 'transformational ecology' triggered by a new set of merchants, tax-collectors and state-rulers that operated at a wider scale. As Paul Warde says, "The 'territorial ecology' implies a repeatable set of actions happening at a particular place. It is a process that reinforces the 'integrity' of a particular way of doing things. The 'transformational ecology', put bluntly, does not. Eventually it must result in the disturbance of local processes: it is a problem generator".[81] His argument implies that "tracing the 'integrity' of, and 'disturbance' to, systems of resource flows, is one of the most useful tasks historians can undertake. It is precisely because the results of ecological interaction can only be determined empirically that ecology should be historical".[82] Perhaps we could generalize this approach by saying that, when studying the various paths of economic growth taken by early pre-industrial societies, we should adopt the working hypothesis put forward by John McNeill: "In any case, human history since the dawn of agriculture is replete with unsustainable societies, some of which changed not to sustainability but to some new and different kind of unsustainability.[83]"

2.8 From One Unsustainable Path to Another: The First Globalization as a Watershed

Rolf Peter Sieferle, Robert Shiel and other scholars have expressed their suspicions concerning two major forces that perhaps led past organic agrarian societies away from sustainability. The first was the lack of manure to sustain crop yields in a highly intensive organic agriculture, and the second deforestation. According to Shiel, estimates made for the European Atlantic bioregions, with no water stress, practising a three-course crop rotation, and where the livestock grazed on pastures kept separate from cropland, show that the highest level of nitrogen availability would have been achieved when less than 15 % of useful agrarian area was sown with grains.[84] Higher cropland proportions would lead either to diminishing returns–after some decades during which the nitrogen reserves stored in the soil would be exhausted—or to new 'Boserupian' innovative responses aimed at improving seeds or varieties and achieving a closer integration between cropland

[81] Warde (2006a), p. 284.

[82] Warde (2006a), p. 19.

[83] McNeill (2000).

[84] Shiel (1991).

and livestock breeding, as took place in England with the well-known Norfolk four-course rotation.[85]

Here it would be interesting to apply the synthesis proposed by Ronald Lee, of Malthus' and Boserup's approaches.[86] Lee suggests that past technological advances usually occurred within a 'Boserupian space' comprising a limited set of combinations of population densities and technological capacities, as a response to 'Malthusian' tendencies towards diminishing returns. Only from time to time did some historical discontinuities occur that pushed forward technological capacities from one 'Boserupian space' to another.[87] However, this immediately raises the question: What factors induced these large but unusual technological shifts? As Bruce Campbell and Mark Overton have stressed, an important side of that issue lies in knowing how differently pre-industrial societies solved the fertilizing trap, within several historical contexts and natural endowments.[88] Much more research is needed in the historical reconstruction of nutrient cycles, based on the methodological tools that agronomists can offer to historians. This long-term agro-ecological research must be undertaken without forgetting the diverse natural and climatic conditions in which very different kinds of 'advanced organic agricultures' developed. For example, owing to the summer water stress in Mediterranean regions the strongest limiting factor was not the nutrients but the soil water content.[89]

In order to fill the nutrient gap opened up by population growth, or a higher amount in trade and taxes, past 'advanced organic agricultures' had to rely on a more intense land-use which, in turn, required a higher labour intensity and a longer time-span of investment.[90] The Chinese way of solving this problem clearly shows the dilemma usually faced between keeping high yields per unit of land without diminishing labour productivity at the same time.[91] Here again the role played by European colonization of America in abolishing this land constraint can be seen. Any 'ghost acreage' assessment would be very limited without extending the time perspective until the beginning of the 20th century, in order to take into account the American grain exports from the Great Plains where settlers stopped using fertilizer for nearly 60 years up to the 1930s.[92]

The overseas exports of cheap cereal coming from the United States and other 'new Europes', leading to the well-known European agricultural crisis of the late 19th century, were based on large scale soil mining of nutrients stored in the

85 Shiel (2006).
86 Boserup (1965).
87 Lee (1986).
88 Campbell and Overton (1991).
89 Bevilacqua (2001) and Garrabou (2005).
90 Allen (2008).
91 Elvin (2009).
92 Cunfer (2005) and Cunfer and Krausmann (2009).

Fig. 2.8 Decline in nitrogen by percentage of soil and years of cropping in several case studies of the Great Plains (United States, 1870–1930). *Source* Cunfer (2004, p. 546)

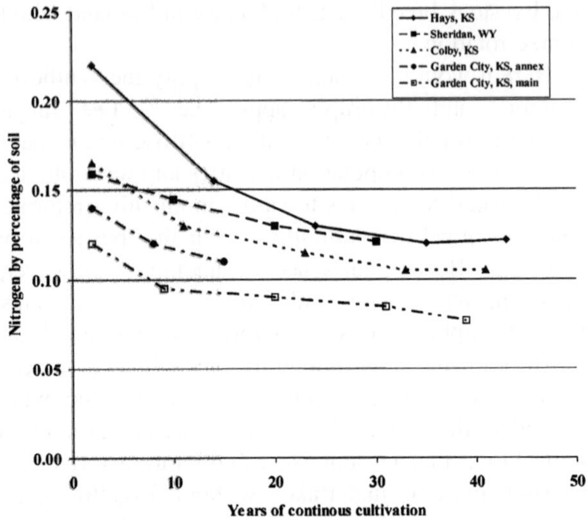

previously unbroken sod of the Great Plains (Fig. 2.8). In order to compete with these cereal imports embodied with the unpaid 'virtual soil' mined, European farmers were forced to further increase yields per unit of land. Thus, the first globalization linked an unsustainable extensive cropping system on one side of the Atlantic, with an increasingly unsustainable intensive farming on the other.

Eventually, this double-sided process led to an agro-ecological and an economic final crisis of the two sorts of 'advanced organic agricultures'. As Geoff Cunfer has explained, referring to the case of Rooks County in Kansas: "By the late nineteenth century, farmers had pushed into the upper Midwest, the Great Plains, and the Pacific Northwest. In the early twentieth century, only California remained to be tapped for agriculture. The slow wave of westward plowing left behind a secondary wave of abandoned farms. Farmers adopted the old Indian system of swidden agriculture to solve the fertility dilemma. Traditional American farming relied on the existence of an ever-new frontier. Played-out fields eventually grew back to forest or became low-intensity pasturage. Thus, it is no surprise that when the latest wave of American farmers rolled into western Kansas in the 1870s, they implemented a farm system that mined soil nutrients. They applied manure as it was available and occasionally rotated legume crops when convenient, but they had no strategy to sustain cropping for the long term. By the 1930s, Rooks County's fields had been planted, cultivated, and harvested 60 times without rest. Soil nitrogen was about half what it had been at sod breaking, and crop yields were declining steadily. Moreover, the western frontier had disappeared. All of the arable land in Rooks

County—and in the nation for that matter—had been identified and plowed. Soil nitrogen and organic carbon drifted steadily downward, and with it yields and profits. Faced with this problem, farmers implemented a dramatic innovation in soil-nutrient management. Rather than revisiting ancient strategies, farmers (and the industrial nation behind them) appropriated cheap fossil-fuel energy to import enormous amounts of synthetically manufactured nitrogen into their fields".[93]

Being a non-renewable resource, dependence on these chemical fertilizers synthesized from fossil fuels already entailed problems of sustainability. Besides this, they became aggravated later on by the impacts of pollution resulting from excessive and inefficient use of new industrialized cropping systems, that became territorially disjointed from livestock breeding and forestry.[94] Yet before the 1950s the difficulties found in the acceleration of the nutrient throughput remained a major issue for European and American farmers, who kept applying manure as a basic resource and used mineral or chemical fertilizers as a complement.[95] During the spectacular economic growth in the second half of the 20th century, on the contrary, the application to the soil of higher doses of synthetic fertilizers was not only aimed at maintaining acceptable levels of yields, but also to continually increase them up to what was proved feasible. The complete substitution of manure by synthetic fertilizers put an end to the old integrated management of cropland, livestock breeding and forestry, thus entirely upsetting the agrarian social metabolism, and turning farms into an extended network of diffuse pollution and landscape degradation.[96]

Once again we found that economic growth and ecological degradation became the two faces of the same coin. As John McNeill has written, "Environmental change of the scale, intensity and variety witnessed in the twentieth century required multiple, mutually reinforcing causes. The most important immediate cause was the enormous surge of economic activity. Behind that lay the long booms in energy use and population. The reasons economic growth had the environmental implications that it had lay in the technological, ideological and political histories of the twentieth century".[97] The spread of the so-called green revolution added to the diffusion of electrification and forest industrial mass-production and transport, thus allowing a further range of Southern and Central European countries to quickly converge with the United Kingdom and the United States. As the example of Austria shows, catching up in terms of GDP per capita went hand in hand with the convergence in energy consumption and global pollution (Fig. 2.9):

---

[93] Cunfer (2004).

[94] Galloway et al. (2004).

[95] Tisdale and Nelson (1956).

[96] Marull et al. (2008).

[97] McNeill (2000).

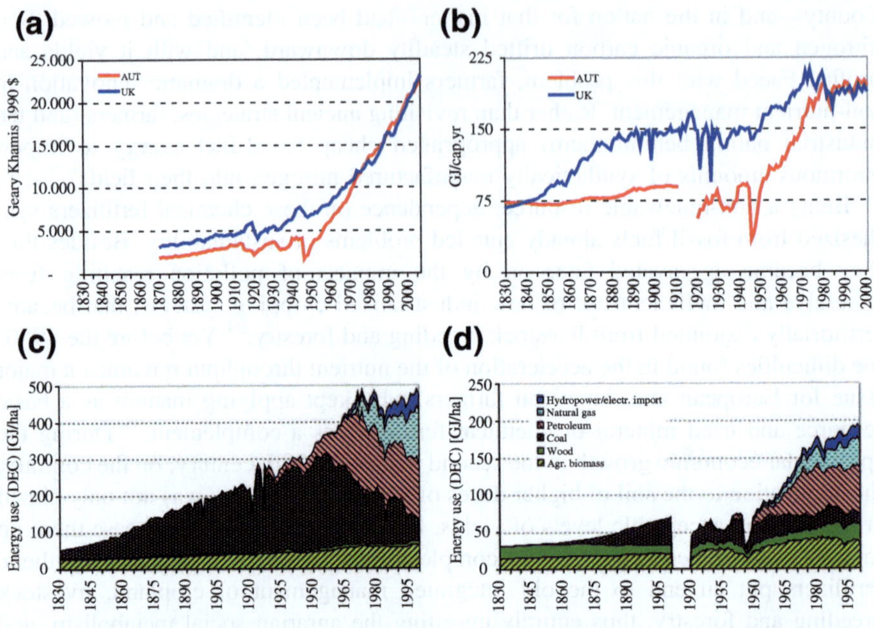

Fig. 2.9 Historical divergence and convergence paths followed by the United Kingdom and Austria in per capita GDP and energy consumption (1830–2000) **a** GDP per capita in real terms **b** Primary energy consumed per capita **c** Energy transition in the UK (per hectare domestic energy consumed by sources) **d** Energy transition in Austria (per hectare domestic energy consumed by sources). *Source* Krausmann et al. (2008, p. 191)

2.9 Was There a General Crisis of Biomass Energy Carriers in Europe?

Looking at deforestation, the second factor that, we can presume, led pre-industrial societies towards greater unsustainability, Rolf Peter Sieferle is in little doubt that Europe did suffer an increasing shortage of wood. He claims that "the historically decisive escape from the wood crisis of the 18th century was the substitution of wood by coal. In the end this process resulted in such an enormous breakthrough in energy supply that any other attempts to substitute and conserve appear marginal by comparison. But for contemporaries it was only one way out among others—they were unaware of its epoch-making importance".[98] This is a controversial issue that deserves to be studied further to clarify whether there was a true widespread wood crisis in Europe or not before the large-scale resort to fossil fuels; and if so, what the main factors were that brought this about.

Landscape photographs taken between the 1870s and 1920s, and the first aerial photos made shortly afterwards, show an apparent deforestation and rejuvenation of

[98] Sieferle (2001).

the surviving forests throughout Europe. Subsequently, and in parallel with the massive consumption of fossil fuels, the forests started to grow again during the second half of the 20th century and today cover a surface area that is perhaps greater than at any other period in the last millennium. The current research on LUCC, applying GIS to historical cadastral maps and aerial photographs, might help historians to confirm the general trend towards a new forest transition.[99] The study on the human appropriation of aboveground net primary production of biomass (HANPP) has demonstrated that in Austria "HANPP decreased continuously from 60 % in 1830 to 48 % in 1970 and then started to increase again slightly, up to 51 % in 1995. This means that today about 23 % more biomass (i.e., 129 PJ/yr or 7 Mt of biomass) remains in terrestrial ecosystems than in 1830".[100] A more recent study shows there was, in the United Kingdom, a HANPP decline from 74 % in 1800 down to a level of around 65 % in the late 19th and early 20th centuries, followed by an increase up to the late 1950s and a new decline to a value of 67 % in the year 2000.[101] Although much more research is needed on this issue, it seems clear that up to a point past 'advanced organic economies' could have exerted a greater direct pressure on European forests than in more recent times, when fossil fuels consumption has globalized ecological footprints and displaced environmental load onto the rest of the world or the atmosphere, subject to global warming emissions.[102]

Here again the question of which driving forces underlay this trend arises. Clearly, increasing population densities must have represented a challenge for any 'organic-based' economy. Nevertheless, deforestation must also be linked with market networks and urbanisation, in spite of the fact that many scholars have paid no direct attention to this. To obtain a final energy unit of charcoal, five times more firewood had to be burnt in a charcoal-burner with an energy loss of nearly 60 %. Taking into account that fuel wood extracted from forests could have been consumed either as firewood or charcoal, any switching in consumption from the latter to the former would have had considerable impact on the primary energy needed. What then would have been the use of transforming firewood into charcoal? The main reason was to allow available terrestrial means of transporting heavy goods to travel greater distances, without consuming more energy carrying the fuel than the energy actually carried.[103] The rural population could easily obtain enough firewood from neighbouring forests, coppices, brushwood or wood crops and orchards. Despite the heavy water content of wood, they could easily carry it home for short distances. But cities needed to be provided with much greater quantities of fuel wood coming from quite distant locations. Even the slightest increase in urbanization would have meant

[99] Kovář (1999) and Agnoletti (2006).
[100] Krausmann (2001).
[101] Musel (2008).
[102] Haberl et al. (2001).
[103] Sieferle (2001).

a rapid growing of fuel weights, and of the distances needed to be covered, thus fostering the shift from firewood to charcoal provision.

We have already seen the close connection that existed between the growth of London and other British industrial cities, and coal extraction in England and Wales.[104] But how did other European cities cope, since they could not expect to be supplied with coal in the same way at the time? We know, for example, that in order to provide Madrid with charcoal during the 18th century the annual output of almost all woods in an area of 70,000 km^2 was required, which represents nearly 15 % of the total area of Spain.[105] More than 17,000 tons were transported every year within a range of 100 km to supply a population of 164,000 inhabitants in Madrid in 1787, with an average of 154 kg of charcoal per inhabitant a year (0.4 kg/inhabitant/day). Assuming the usual efficiency in a charcoal burner, this meant 2 kg of primary firewood per inhabitant a day mainly for domestic purposes. Adding another half a kilo burnt for different industrial activities, we would reach 2.5 kg/inhabitant/day in the pre-industrial city of Madrid. This figure would have been 66 % higher than the average consumption of fuel wood estimated by Paolo Malanima in the pre-industrial Mediterranean Europe, and more than double the minimum supply recorded in Sicily.[106]

We must bear in mind, however, that up to a point charcoal, as well as firewood, could have been kept exploited as a sustainable renewable source, as long as they were made out of small logs shredded, pollarded or cut from coppice-woods in the North-Atlantic regions, and lopped from *dehesa*-types of open forests turned into wood pastures in the Mediterranean South.[107] Even the pruning of vines and olive or almond tress could have been used that way.[108] According to Rolf Peter Sieferle, "at first sight there was no shortage of fuel. It was always possible, and with little effort, to produce firewood […] by establishing coppices. In general, it can be said that the fuel aspect was only part of the wood crisis, and that part most easily open to a traditional solution". He concludes that "the wood crisis of the eighteenth century was in the first place a timber crisis. The enormous consumption of firewood in combination with agricultural uses of woodlands made it increasingly difficult to find old tree stands that were suitable for construction".[109]

The cautious scepticism of A.T. Grove and Oliver Rackham goes even further when they oppose the 'Ruined Landscape' myth with the hypothesis that, instead of a true deforestation, human impacts over Southern Europe mainly altered different types of the ever dynamic forest and shrub covers that characterize the Mediterranean environment (open-tall *dehesa*-type of savannah instead of a thick-short

[104] Allen (2009).
[105] Bravo (1993).
[106] Malanima (2001).
[107] Clément (2008).
[108] Grove and Rackham (2001).
[109] Sieferle (2001).

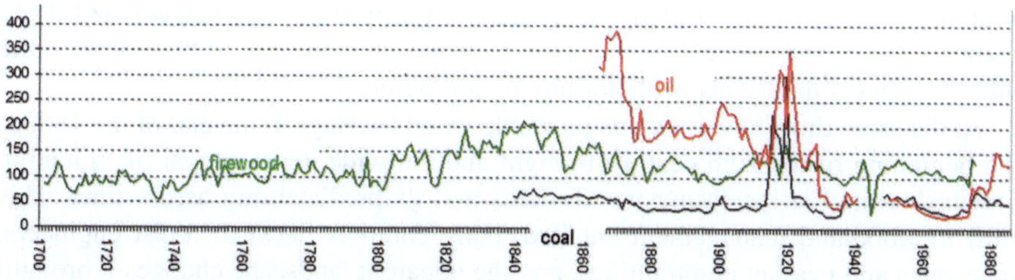

Fig. 2.10 Long-run trends in real prices of firewood, coal and oil in Italy (1700–1985 in 1911 ITL per TOE). *Source* Malanima (2006, pp. 70–71)

forest, *maquis*, etc.).[110] In his overview on world-wide deforestation Michael Williams suggests that, in Europe, before the coming of the fossil fuel era, timber and firewood shortages were more of a local or regional feature than a general one, and he discards charcoal consumption in industrial uses as a relevant factor.[111] On the opposite side, Paolo Malanima bears no doubts when he states that "from the mid-eighteenth century onward, while Europe's population was growing faster, energy availability was decreasing. The result was a sharp per capita decline in energy consumption".[112] This included both food intake and fuel wood availability, which seems to fit well with the anthropometric height decrease of Europeans born between 1770 and 1820. "The decline of forest is borne out—according to Malanima—by the quick rise of the price of firewood, which was usually faster than the overall growth of agricultural prices. In Western European cities, between 1700 and 1800, firewood prices increased by more than three times".[113] The long-term evolution in Italian prices of firewood, compared with prices of coal and oil, clearly fits with the 'exergetic' economic growth theory proposed by Bob Ayres and Benjamin Warr (Fig. 2.10; see also Fig. 2.2):

In a nuanced and detailed overview of the wood shortage debate in pre-industrial Europe, Paul Warde assesses that "if the European population in 1500 was around half that in 1800, and if there were general scarcities in 1500, survival could only have been possible in 1800 as a consequence either of a radical alteration in the domestic fuel economy, or a greatly increased woodland area or productivity. As there is very little evidence for any of these things we must be suspicious of any claims for a general scarcity at any time before the late eighteenth century. Western Europe had a population of around 122 millions by 1820, and if annual domestic demand is set at about three cubic meters per hectare, a coppiced area of 407,700 km^2 would have been required for a sustainable supply. This approximates to the area of modern Germany and Switzerland combined, something under a fifth of western Europe (excluding

[110] Grove and Rackham (2001).

[111] Williams (2003).

[112] Malanima (2006), pp. 101–121.

[113] Malanima (2006), pp. 116–118.

Scandinavia). As it is doubtful that many areas of Europe were this well wooded at any point in the period, the case for a general wood shortage by 1820 appears quite plausible, but is hardly plausible for any period before 1750".[114]

At the same time, however, this plausible wood shortage at the end of the 18th or the beginning of the 19th century brought about a wide development of 'scientific forestry', aimed at increasing wood yields and its predictability across time. This shift in woodland management entailed many conflicts between forest engineers, state-rules and peasant communities, and the apparent landscape changes it brought about raised deep social unrest all over Europe as well as in colonial regions such as India under British rule.[115] Yet, as Paul Warde concludes, "in dealing with general scarcity, forestry was fairly successful. [...] The nineteenth century augmentation of wood yields demonstrated that there was plenty of scope for productivity increase within the economy after the Napoleonic age, but equally, that the ability to raise consumption per head and indeed income levels was limited. [...] When Jevons in 1865 turned to the question of the exhaustion of coal reserves [...], most of Europe still looked to wood as its primary source of thermal energy. That this could still be the case after a period of enormous population growth is a tribute to the capacities of the preindustrial ancient regime, and an indicator that Europe, for all its late eighteenth-century problems, remained distant from any ecological frontier". [116] It must be added immediately, though, that it was precisely during the late 19th century when written sources and the first landscape photographs provided direct and indirect evidence of a peak in deforestation all over Europe, just before the start of a fast reforestation wave fostered by rural abandonment.

Was there or was there not a general biomass energy crisis at the beginning of the fossil fuel era? This remains a significant, open historical question that deserves to be extensively looked into in the future by reconstructing land-use statistics or surveys, and making GIS analysis of land-cover changes from aerial photographs and cadastral maps. The aim should be to extend the land accounts that the EEA have started to assemble for the last decade of the 20th century as far back in time as possible,[117] similar to the historical series of main land uses in the United Kingdom and Austria reconstructed from 1830 onwards by Fridolin Krausmann, Heinz Schandl and Rolf Peter Sieferle (Fig. 2.11):

In the meantime, the available evidence suggests that timber, firewood or charcoal scarcity might have been more of a regional situation than a general continental phenomenon. As Paul Warde has suggested for the English case, it is likely that scarcities became a true economic problem when, together with fuel, they increased food or feed prices and land rents as well.[118] It would have been enough, however, that these regional scarcities affected general trends of energy prices to become a

[114] Warde (2006b), pp. 38–39.

[115] Guha (1991).

[116] Warde (2006b), p. 52.

[117] European Environment Agency (2006).

[118] Warde (2007).

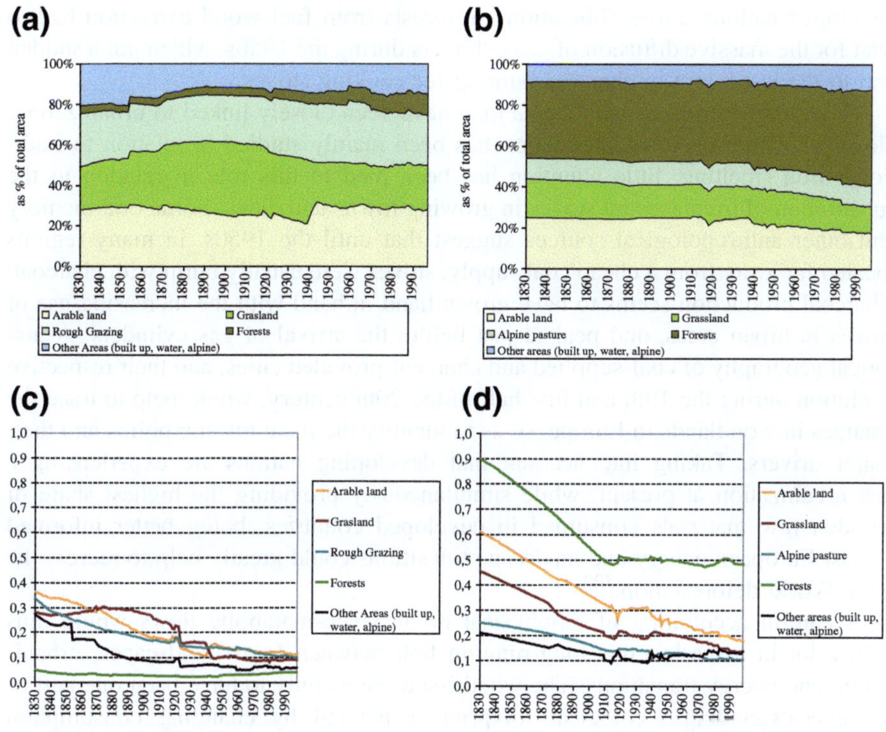

Fig. 2.11 Land use changes in The United Kingdom and Austria (1830–2000 as % of total area) **a** As % of total area in the United Kingdom **b** As % of total area in Austria **c** Per capita land-uses in the United Kingdom **d** Per capita land-uses in Austria. *Source* Krausmann et al. (2008, pp. 187–201)

driving force that urged the energy transition from land-based biomass energy carriers to mineral-based fossil fuels (see Figs. 2.6 and 2.2). The same way the Stone Age did not end because of a shortage of stones, the way out of a biomass-based energy system had no reason to wait for a devastating deforestation.

Almost all known energy transitions have required a long of time to attain an overall turnaround of the preceding mix of energy sources (see graphs 7, 11.3 and 11.4 as examples). Over a period of time all energy sources, old and new, can continue to grow at different and sometimes similar rates. This was the case of many woodland products whose consumption increased during the 19th and the first third of the 20th century, despite the contemporary growth in coal burning. This happened not only in Nordic countries such as Finland, but also in Mediterranean ones such as Spain, due to the maintenance of many traditional uses, together with the appearance of new applications related to industrialization and urbanization—railway sleepers, mine roof supports, posts for telegraph and power lines, furniture, packages of fruit and wood pulp to make paper.[119] In several

119 Myllyntaus and Mattila (2002) and Iriarte-Goñi and Ayuda (2008).

developed nations a true "liberation" of forests from fuel wood extraction had to wait for the massive diffusion of gas cylinders during the 1950s, which put a sudden end to the last boom in charcoal burning for cooking stoves.

The fate of European woodlands may have been closely linked to urbanization. However, while charcoal production has been mainly studied in relation to traditional iron smelting, little attention has been paid to this role in relation to the substitution of fireplaces by stoves in growing towns and cities. Some oral memory and other anthropological sources suggest that until the 1950s, in many regions located far away from a cheap coal supply, stoves were usually burnt with charcoal. Charcoal production seems to have grown hand in hand with the increasing use of stoves in urban areas, and peaked just before the arrival of gas cylinders. A historical geography of coal-supplied and charcoal-provided cities, and their respective evolution during the 19th and first half of the 20th century, would help to trace the changes in woodlands in Europe, so as to identify the main turning points and their major drivers. Taking into account that developing nations are experiencing a fast urbanization at present, while simultaneously providing the highest share of wooden raw materials consumed in developed countries, being better informed of past European energy and landscape transitions could greatly help to redress the huge World deforestation.[120]

A careful accounting of biophysical urban socio-metabolic flows would thus help a lot in assessing this long-running link between forest landscapes, urbanization and energy transitions. The wood issue seems only to have been one part of the ever-expanding 'ecological footprint' generated by changing consumption patterns as industrialization and urbanisation processes got underway. In addition to firewood, charcoal and timber, the historical approach to urban social metabolism includes the consumption of cereals, meat, milk and other food products, together with the excretion of wastes, which constitutes another interesting field of research for human ecology, ecological economics and environmental history.[121]

2.10 Concluding Remarks

Despite being neither complete nor exhaustive, the examples presented in ongoing debates and current research show a great interest of a further dialogue and interdisciplinary collaboration between environmental and economic history. These examples also show, however, that the strongest barrier lies in the mainstream approach to economic growth and macroeconomic theories which neglect the role of energy and other natural resources. Yet, adopting a common bio-physical and socio-metabolic approach, linked to land-use and global environmental changes,

[120] Williams (2003).

[121] Stanhill (1984), Schmid Neset (2005) and Billen et al. (2009).

would help achieve a greater understanding of how economic growth actually occurs, and the role ecological impacts entail on life-support systems of the Earth.

Neither traditional mistrust nor the suspicion of biased ecological or economic determinism should prevent the necessary rapprochement between economic and environmental history, which is being opened up by the development of ecological economics. In order to put aside any suspicion about causal primacy, it is worth distinguishing from the beginning those explanations that tell us what growth is or how it happens from those that explain why growth occurs.[122] The socio-metabolic research on long-term transitions in the interaction between nature and society may significantly contribute to enhance the understanding of what economic growth is, and how it takes place. Nevertheless, this would leave space for other dimensions, such as institutional settings or ruling actors needed to explain why growth happens, by whom, and for what purpose. Thus, the development of this new biophysical standpoint does not mean that we can afford to neglect institutional, social, cultural and political factors.[123] On the contrary, as Joan Martinez Alier has written, "far from naturalising history, the introduction of ecology into the explanation of human history historialises ecology. This is because human ecology, that is, the relationship between human societies and nature, cannot be comprehended without an understanding of the history of human beings and their conflicts".[124]

Acknowledgments This essay has been written in the framework of the linked research projects on *Sustainable farm systems: long-term socioecological metabolism in western agriculture* funded by The Social Sciences and Humanities Research Council of Canada, and the Spanish one HAR2012-38920-C02-02 directed by Enric Tello at the University of Barcelona. We thank Leah Temper for her careful revision of the English version.

Glossary

EROI	Energy returns on energy inputs
MEFA	Material and Energy Flow Analysis
HANPP	Human Appropriation of the ecological Net Primary Production
EUROSTAT	The Statistical Office of the European Communities located in Luxembourg. Its main responsibilities are to provide statistical information to the institutions of the European Union (EU) and to promote the harmonisation of statistical methods across its member states

122 North (1999).

123 Teich et al. (1997) and Sörlin and Warde (2009).

124 Martínez Alier (1998).

Production function	It is a function that relates the output to the inputs or factors of production used in a production process. The Cobb-Douglass is the most standard in which the output (Y) is produced with two factors, labour (L) and capital (K), and the remaining growth share not explained by the variation of both is explained with the Total Factor Productivity (A). That is, $Y = AL^{\beta}K^{\propto}$ where $\beta + \propto = 1$ and account for the output elasticities of capital and labour, respectively
TFP	The Total Factor Productivity measures the fraction of economic growth that cannot be explained by the contribution assigned to the increases in capital, labour and land. As it is commonly considered that it grasps the efficiency gains obtained through the combination of factors that participate in a production process taken together, and it is taken as a measure of an economy's long-term technological change
Exergy	The useful work actually performed by all energy converters which empower human labour and capital goods at its disposal
IPCC	The World Meteorological Organization (WMO) and the United Nations Organization (UNO) created the Intergovernmental Panel on Climate Change (IPCC) in 1988. The IPCC summarizes the technical, biophysical, socio-economical information to understanding and measure the risk of climate change
LUCC	Land-Use Land-Cover Change programme (http://www.geo.ucl.be/LUCC/lucc.html) examines the transformations undergone by Earth's vegetal cover over the centuries, in order to identify the main driving forces behind global socio-environmental change, and also to assess its socio-ecological impact

References

Abel W (1980) Agricultural fluctuations in Europe: from the thirteenth to the twentieth centuries. Methuen, London

Acemoglu D (ed) (2004) Recent developments in growth theory, vol 2. Edward Elgar, Cheltenham

Acemoglu D (2009) Introduction to modern economic growth. Princenton University Press, Princeton

Acemoglu D, Robinson JA (2006) Economic origins of dictatorship and democracy. Cambridge University Press, Cambridge

Aghion Ph, Williamson JG (1998) Growth, inequality, and globalization: theory, history, and policy. Cambridge University Press, Cambridge

Agnoletti M (ed) (2006) The conservation of cultural landscapes. CABI Publishing, Wallingford/Cambridge

Allen RC (1992) Enclosure and the Yeoman. Clarendon Press, Oxford

Allen RC (2008) The nitrogen hypothesis and the English agricultural revolution: a biological analysis. J Econ Hist 68(1):182–210
Allen RC (2009) Engels' pause: technical change, capital accumulation, and inequality in the British industrial revolution. Explor Econ Hist 46(4):418–435
Allen RC, Bengtsson T, Dribe M (eds) (2005) Living standards in the past: new perspectives on well-being in Asia and Europe. Oxford University Press, Oxford
Aoki M, Hayami Y (eds) (2001) Communities and markets in economic development. Oxford University Press, Oxford
Aston TH, Philpin CHE (eds) (1985) The Brenner debate: agrarian class structure and economic development in pre-industrial Europe. Cambridge University Press, Cambridge
Ayres RU (2001) The minimum complexity of endogenous growth models: the role of physical resource flows. Energy 26:817–838
Ayres RU (2008) Sustainability economics: where do we stand? Ecol Econ 67:303
Ayres RU, Warr B (2005) The economic growth engine: how energy and work drive material prosperity. Edward Elgar, Cheltenham
Ayres R, Eisenmenger N, Krausmann F, Schandl H, Warr B (2009) Energy use and economic development: a comparative analysis of useful work supply in Austria, Japan, the United Kingdom and USA
Badia-Miró M, and Tello E (2014) Vine-growing in Catalonia: the main agricultural change underlying the earliest industrialization in Mediterranean Europe (1720–1939). Eur Rev Econ Hist 18:203–226
Bevilacqua P (2001) Demetra e Clio. Uomini e ambiente nella storia. Roma: Donzelli Editori. González de Molina, M. 2002. Environmental constraints on agricultural growth in 19th century Granada (Southern Spain). Ecol Econ 41:257–270
Billen G, Barles S, Garnier J, Rouillard J, Benoit P (2009) The food-print of Paris: long-term reconstruction of the nitrogen flows imported into the city from its rural hinterland. Reg Environ Change 9:13–24
Bloch M (1955–1956) Les Caractères originaux de l'histoire rurale française. Armand Colin, Paris
Boserup E (1965) The conditions of agricultural growth. Aldine/Earthscan, Chicago
Bravo J (1993) Montes para Madrid. El abastecimiento de carbón vegetal a la villa y corte entre los siglos XVII y XVIII. Caja Madrid, Madrid
Brázdil R, Pfister Ch, Wanner H, Von Storch H, Luterbacher J (2005) Historical climatology in Europe—the state of the art. Clim Change 70(3):363–430
Campbell BMS, Overton M (eds) (1991) Land, labour and livestock: historical studies in European agricultural productivity. Manchester University Press, Manchester
Chang H-J (2002) Kicking away the ladder. Development strategy in historical perspective. Anthem Press, London
Clément V (2008) Spanish wood pasture: origin and durability of an historical wooded landscape in Mediterranean Europe. Environ Hist 14(1):67–87
Costanza R, Graumlich LJ, Steffen W (eds) (2007a) Sustainability or collapse? an integrated history and future of people on Earth. The MIT Press, Cambridge
Costanza R, Graumlich L, Steffen W, Crumley C, Dearing J, Hibbard K, Leemans R, Redman Ch, Schimel D (2007b) Sustainability or collapse: what can we learn from integrating the history of humans and the rest of nature? Ambio 36(7):522–527
Crafts NFR (1997) Some dimensions of the quality of life during the British industrial revolution. Econ Hist Rev 50(4):617–639
Crafts NFR, Harley CK (1992) Output growth and the British industrial revolution: a restatement of the Crafts-Harley view. Econ Hist Rev 45(4):703–730
Cronon W (1983) Changes in the land: Indians, colonists, and the ecology of New England. Hill and Wang cop, New York
Cronon W (1991) Nature's metropolis: Chicago and the Great West. Norton, New York
Crosby AW (1986) Ecological imperialism: the biological expansion of Europe. Cambridge University Press, Cambridge

Cuff DJ, Goudie AS (2009) The Oxford companion to global change. Oxford University Press, Oxford
Cunfer G (2004) Manure matters on the Great Plains Frontier. J Interdisc Hist 34:539–567
Cunfer G (2005) On the great plains. Agriculture and environment. Texas A&M University Press, Texas
Cunfer G, Krausmann F (2009) Sustaining soil fertility: agricultural practice in the old and new Worlds. Global Environ J Hist Nat Soc Sci 4:9–43
Cussó X, Garrabou R, Tello E (2006a) Social metabolism in an agrarian region of Catalonia (Spain) in 1860–1870: flows, energy balance and land use. Ecol Econ 58:49–65
Cussó X, Garrabou R, Olarieta JR, Tello E (2006b) Balances energéticos y usos del suelo en la agricultura catalana: una comparación entre mediados del siglo XIX y finales del siglo XX. Historia Agraria 40:471–500
De Vries J (1984) European urbanization, 1500–1800. Methuen, London
De Vries J (2001) Economic growth before and after the industrial revolution. A modest proposal. In: Prak M (ed) Early modern capitalism. Economic and social change in Europe, 1400–1800, Routledge, London, pp 177–194
De Vries J (2008) The industrious revolution: consumer behavior and the household economy, 1650 to the present. Cambridge University Press, Cambridge
Debier J-C, Deléage JP, Hémer D (1986) Les Servitudes de la puissance: une histoire de l'énergie. Flammarion, Paris
Easterly W (2002) The Elusive quest fro growth: economists' adventures and misadventures in the tropics. The MIT Press, Cambridge
Elvin M (2009) Why intensify? The outline of a theory of the institutional causes driving long-term changes in chinese farming and the consequent modifications to the environment. In: Sörlin S, Warde P (eds) Nature's end. History and the environment. Palgrave Macmillan, New York, pp 273–303
Emmer PC, Pétré-Greouilleau O, Roitman JV (eds) (2006) A Deus ex Machina Revisited. Atlantic colonial trade and European economic development. Brill, Boston
European Environment Agency (2006) Land accounts for Europe 1990–2000. Towards integrated land and ecosystem accounting, EEA Report no 11
Feinstein CH (1998) Pessimism perpetuated: real wages and the standard of living in Britain during and alter the industrial revolution. J Econ Hist 58(3):625–658
Fischer-Kowalski M, Amann Ch (2001) Beyond IPAT and Kuznets curves: globalization as a vital factor in analysing the environmental impact of socio-economic metabolism. Popul Environ 23 (1):7–47
Fischer-Kowalski M, Haberl H (eds) (2007) Socioecological transitions and global change. Trajectories of social metabolism and land use. Edward Elgar, London, p 24
Fischer-Kowalski M, Krausmann F, Smetschka B (2004) Modelling scenarios of transport across history from a socio-metabolic perspective. Rev J Fernand Braudel Cent Study of Econ Hist Syst Civiliz XXVII 4:307–342
Fisher-Kowalski M (1998) Society's metabolism. The intellectual history of materials flow analysis. Part I, 1860–1970. J Ind Ecol 2(1):61–78
Foster JB (2000) Marx's ecology. Materialism and nature. Monthly Review Press, New York
Gadgil M (2000) The use and abuse of nature: incorporating this fissured land, an ecological history of India and ecology and equity. Oxford University Press, Oxford
Gadgil M, Guha R (1993) This fissured land: an ecological history of India. Oxford University Press, Oxford
Gales B, Kander A, Malanima P, Rubio M (2007) North versus South: energy transition and energy intensity in Europe over 200 years. Eur Rev Econ Hist 11:219–253
Galloway JN, Denetener FJ, Capone DG, Boyer EW, Howarth RW, Seitzinger SP, Asner GP, Cleveland CC, Green PA, Holland EA, Karl DM, Michaels AF, Porter JH, Townsend AR, Vörösmarty CJ (2004) Nitrogen cycles: past, present and future. Biogeochemistry 70:153–226

Garrabou R (2005) Conflict and environmental tension in the adoption of technological innovation in the agrarian sector. In: Sarasúa C, Scholliers P, Van Molle L (eds) Land, shops and kitchens. Technology in the food chain in Twenty-century Europe, Brepols, Turnhout, pp 30–41
Georgescu-Roegen N (1976) Energy and economic myths: institutional and analytical economic essays. Pergamon Press, New York
Goodman D, Redclift M (1991) Refashioning nature: food, ecology and culture. Routledge, London
Goody J (2004) Capitalism and modernity: the great debate. Polity Press, Cambridge
Grigg D (1982) The dynamics of agricultural change. The historical experience. Hutchinson, London
Grove AT, Rackham O (2001) The nature of Mediterranean Europe. An ecological history. New Haven and London, Yale U. P
Guha R (1991) The unquiet woods: ecological change and peasant resistance in the Himalaya, Expanded edn. University of California Press, Berkeley
Guha R, Martínez Alier J (1997) Varieties of environmentalism: essays North and South. Earthscan, London
Haberl H, Erb KH, Krausmann F, Loibl W, Schulz N, Weisz H (2001) Changes in ecosystem processes induced by land use: human appropriation of net primary production and its influence on standing crop in Austria. Global Biogeochem Cycles 15(4):929–942
Haberl H, Winiwarter V, Andersson K, Ayres RU, Boone Ch, Castillo A, Cunfer G, Fischer-Kowalski M, Freudenburg WR, Furman E, Kaufmann R, Krausmann F, Langthaler E, Lotze-Campen H, Mirtl M, Redman ChL, Reenberg A, Wardell A, Warr B, Zechmeister H (2006) From LTER to LTSER: conceptualizing the socioeconomic dimension of long-term socioecological research. Ecol Soci 11:2–13
Helpman E (2004) The mystery of economic growth. Belknap Press of Harvard University Press, Cambridge
Hobsbawm EJ (1964) Labouring men: studies in the history of labour. Weidenfeld and Nicolson, London
Hodgson GM (ed) (2007) The evolution of economic institutions: a critical reader. Edward Elgar, Chletenham
Hoppenbrouwers P, van Zanden JL (eds) (2001) Peasants into farmers? The transformation of rural economy and society in the low countries (middle ages-19th century) in light of the Brenner debate. Brepols, Turnhout
Hornborg A, Crumley C (eds) (2007) The World system and the Earth system: global socioenvironmental change and sustainability since the Neolithic. Left Coast Press, Walnut Creek
Hornborg A, McNeill JR, Martínez Alier J (eds) (2007) Rethinking environmental history. World-system history and global environmental change. Altamira Press, New York
Horrell S, Humphries J (1995) The exploitation of little children: child labor and the family economy in the industrial revolution. Explor Econ Hist 32(4):485–516
Iriarte-Goñi I, Ayuda MI (2008) Wood and industrialization: evidence and hypotheses from the case of Spain, 1860–1935. Ecol Econ 65(1):177–186
Kander A (2002) Economic growth, energy consumption and CO_2 emissions in Sweden 1800–2000. Almqvist and Wiksell International, Lund
Kander A (2008) Is it simply getting worse? Agriculture and Swedish greenhouse gas emissions over 200 years. Econ Hist Rev 61(4):773–797
Kander A, Malaima P, Warde P (2013) Power to the people. Energy in Europe over the last five centuries. Princeton University Press, Princeton
Kates RW, Clark WB, Corell R, Hall JM, Jaeger CC, Lowe I, McCarthy JJ, Schellnhuber HJ, Bolin B, Dickinson NM, Faucheux S, Gallopin GG, Grüber A, Huntley B, Jäger J, Jodha NS, Kasperson RE, Mabogunje A, Matson P, Money H, Moore B III, O'Riordan T, Svedin U (2001) Sustainabilty science. Science 292:641–642
Kjaergaard Th (1994) The Danish revolution, 1500–1800. An ecohistorical interpretation. Cambridge University Press, Cambridge

Komlos J (1998) Shrinking in a growing economy? The mystery of physical stature during the industrial revolution. J Econ Hist 58(3):779–802
Koning N (1994) The failure of agrarian capitalism: agrarian politics in the UK, Germany, the Netherlands and the USA, 1846–1919. Routledge, London
Kovář P (ed) (1999) Nature and culture in landscape ecology. The Karolinum Press, Charles University, Prague
Krausmann F (2001) Land use and industrial modernization: an empirical analysis of human influence on the functioning of ecosystems in Austria 1830–1995. Land Use Policy 18:21
Krausmann F (2013) A city and its Hinterland: Vienna's energy metabolism 1800–2006. In: Sing SJ, Haberl H, Chertow M, Mirtl MY Schmid M (eds) Long term socio-ecological research: studies in society-nature interactions across spatial and temporal scales, Springer, Berlin, pp 247–268
Krausmann F, Schandl H, Sieferle RP (2008) Socio-ecological regime transitions in Austria and the United Kingdom. Ecol Econ 65:187–201
Krausmann F, Gingrich S, Haberl H, Erb K-H, Musel A, Kastner Th, Norbert Kohlheb N, Niedertscheider M, Schwarzlmüller E (2012) Long-term trajectories of the human appropriation of net primary production: lessons from six national case studies. Ecol Econ 77:129–138
Krech III S, McNeill JR, Merchant C. (2004). Encyclopaedia of world environmental history, vol 3. Routledge, New York
Landes DS (1969) The Unbound Prometheus: technological change and industrial development in western Europe from 1750 to the present. Cambridge University Press, Cambridge
Landes DS (1998) The wealth and poverty of nations: why some are so rich and some so poor. W. W. Norton, New York
Leach G (1976) Energy and food production. IPC Science and Technology Press, Guildford
Lee RD (1986) Malthus and Boserup: a dynamic synthesis. In: Coleman D, Schofield RS (eds) The state of population theory. Forward from Malthus, Basil Blackwell, Oxford, pp 96–110
Lindert PH (2004) Growing public: social spending and economic growth since the eighteenth century. Cambridge University Press, Cambridge
Lucas R (2002) Lectures on economic growth. Harvard University Press, Cambridge
Malanima P (2001) The energy basis for early modern growth, 1650–1820. In: Prak M (ed) Early modern capitalism. Economic and social change in Europe, vol 55, Routledge, London, pp 1400–180
Malanima P (2006) Energy crisis and growth 1650–1850: the European deviation in a comparative perspective. J Global Hist 1:101–121
Martínez Alier J (1990) Ecological economics: energy, environment and society. Basil Blackwell, Oxford
Martínez Alier J (1998) Ecological economics as human ecology. Fundación César Manrique:122, Madrid
Martinez-Alier J (2011) The EROI of agriculture and its use by the Via Campesina. J Peasant Stud 38(1):145–160
Marull J, Pino J, Tello E (2008) The loss of landscape efficiency: an ecological analysis of land-use changes in Western Mediterranean agriculture (Vallès county, Catalonia, 1853–2004). Global Environ J Hist Nat Soci Sci 3:112–150
Marull J, Pino J, Tello E, Cordobilla MJ (2010) Social metabolism, landscape change and land use planning. The metropolitan region of Barcelona as a referent. Land Use Policy 27(2):497–510
Mc Neill WH (1982) The pursuit of power: technology, armed force, and society since A. D. 1000. The University of Chicago Press, Chicago
McKendrick N, Brewer J, Plumb JH (1983) The Birth of a consumer society: the commercialization of eighteenth-century England. Hutchinson, London
McNeill J (2000) Something new under the sun. An environmental history of the twentieth century. Penguin, London
Milanovic B (2005) Worlds apart. Measuring international and global inequality. Princeton University Press, Princenton

Musel A (2008) Human appropriation of net primary production (HANPP) in the United Kingdom, 1800–2000: a socio-ecological analysis. Social Ecology Working Paper 101, Viena
Myllyntaus T, Mattila T (2002) Decline or increase? The standing timber stock in Finland, 1800–1997. Ecol Econ 41(2):271–288
Naredo JM (2004) La evolución de la agricultura en España (1940–2000). Publicaciones de la Universidad de Granada, Granada
North DC (1999) Understanding the process of economic change. Institute of Economic Affairs, London
Overton M (1996) Agricultural revolution in England: the transformation of the agrarian economy 1500–1850. Cambridge University Press, Cambridge
Omrod D (2003) The rise of commercial empires. England and the Netherlands in the age of Mercantilism, 1650–1770. Cambridge University Press, Cambridge
Pamuk S, Williamson JG (eds) (2000) The Mediterranean response to globalization before 1950. Routledge, London
Parker and Jones (eds) 1975 (De Vries J 2001). Economic growth before and after the Industrial Revolution. A modest proposal. In: Prak M (ed) Early modern capitalism. Economic and social change in Europe, 1400–1800, Routledge, London, pp 177–194
Pasinetti L (1981) Structural change and economic growth: a theoretical essay on the dynamics of the wealth of nations. Cambridge University Press, Cambridge
Pinilla V, Ayuda MI (2007) The internacional wine market, 1850–1938. An opportunity for export growth in Southern Europe?. In: Campbell G, Guibert N (eds) Wine, society and globalization. Multidisciplinary perspectives on the wine industry, Palgrave Macmillan, New York, pp 179–199
Pomeranz K (2000) The great divergence. China, Europe and the making of the modern world economy. Princeton University Press, Princenton
Postan MM (1973) Essays on medieval agriculture and general problems of the medieval economy. Cambridge University Press, Cambridge
Pyne SJ (1997) Vestal fire. An environmental history, told through the fire, of Europe and Europe's encounter with the World. University of Washington Press, Seattle, pp 81–146
Sacristán M (1992) Political ecological considerations in Marx. Capitalism Nat Socialism 3 (1):37–48
Schmid Neset T-N (2005) Environmental imprint of human food consumption (Linköping, Sweden, 1870–2000). Linköping University, Linköping
Scott JC (1998) Seeing like a state: how certain schemes to improve the human condition have failed. Yale University Press, New Haven
Sen AK (1999) Development as freedom. Oxford University Press, Oxford
Sen AK, Nussbaum MC (eds) (1993) The quality of life. Clarendon Press, Oxford
Shiel RS (1991) Improving soil productivity in the pre-fertiliser era. In: Campbell BMS, Overton M (eds) Land, labour an livestock: historical studies in European agricultural productivity, Manchester University Press, Manchester, pp 71–73
Shiel RS (2006) Nutrient flows in pre-modern agriculture in Europe. In: McNeill J, Winiwarter V (eds) Soils and societies. Perspectives from environmental history, pp 217–242. White Horse Press, London
Shiva V, Gitanjali B (eds) (2002) Sustainable agriculture and food security: the impact of globalisation. Sage Publishers, New Delhi
Sieferle RP (2001) The subterranean forest. Energy systems and the industrial revolution. The White Horse Press, Cambridge
Sing SJ, Haberl H, Chertow M, Mirtl M, Schmid M (eds) (2013) Long term socio-ecological research: studies in society-nature interactions across spatial and temporal scales. Springer, Berlin
Slicher van Bath BH (1963) The Agrarian history of Western Europe: A.D. 500–1850. Arnold, London
Smil V (1999) Energies: an illustrated guide to biosphere and civilization. MIT Press, Harvard

Smil V (2001) Enriching the Earth: Fritz Haber Carl Bosch and the transformation of world food production. MIT Press, Cambridge
Sörlin S, Warde P (eds) (2009) Nature's end. History and the environment. Palgrave Macmillan, New York
Stanhill G (1984) Energy and agriculture. Springer, Berlin
Sugihara K (2003) The East Asian path of economic development. A long-term perspective. In: Arrighi G, Hamashita T, Selden M (eds) The resurgence of East Asia. 500, 150 and 50 Year perspectives, Routledge, London, pp 78–123
Teich M, Porter R, Gustafsson B (1997) Nature and society in historical context. Cambridge University Press, Cambridge
Tello E, Garrabou R, Cussó X (2006) Energy balance and land use. The making of and agrarian landscape from the vantage point of social metabolism (the Catalan Vallès county in 1860/70). In: Agnoletti M (ed) The conservation of cultural landscapes. CABI International Publishing, New York, pp 42–56
Tello E, Garrabou R, Cussó X, Olarieta JR (2012) Fertilizing methods and nutrient balance at the end of traditional organic agriculture in the Mediterranean Bioregion: Catalonia (Spain) in the 1860s. Human Ecol 40:369–383
Thompson EP (1968) The making of the English working class. Peguin, Harmondsworth
Thompson EP (1991) Customs in common. The Merlin Press, London
Tisdale SL, Nelson WL (1956) Soil fertility and fertilizers. Macmillan, New York
Van Zanden JL (1991) The first green revolution: the growth of production and productivity in European agriculture, 1870–1914. Econ Hist Rev XLIV 2:215–239
Van Zanden JL (2009) The long road to the industrial revolution. The European economy in a global perspective, 1000–1800. Brill, Leiden-Boston
Warde P (2006a) Ecology, economy and state formation in early modern germany. Cambridge University Press, Cambridge
Warde P (2006b) Fear of wood shortage and the reality of the Woodland in Europe, c.1450–1850. Hist Workshop J 62:38–39
Warde P (2007) Energy consumption in England and Wales 1560–2004. ISSM-CNR, Naples
Warde P (2009) Energy and natural resource dependency in Europe 1600–1900. Working paper of the Brooks World Poverty Institute, University of Manchester
Williams M (2003) Deforesting the earth. From prehistory to global crisis. The University of Chicago Press, Chigaco
Williamson JG (1997) Industrialization, inequality, and economic growth. Edward Elgar, Cheltenham
Williamson JG (2006) Globalization and the poor periphery before 1950. The MIT Press, Cambridge
Worster W (ed) (1988) The ends of the earth. Perspectives on modern environmental history. Cambridge University Press, Cambridge
Wrigley EA (1987) People, cities and wealth. The transformation of traditional society. Basil Blackwell, Oxford
Wrigley EA (2004) Poverty, progress, and population. Cambridge University Press, Cambridge, p 29
Wrigley EA (2010) Energy and the industrial revolution. Cambridge University Press, Cambridge

Chapter 3
Environmental History of Soils

Verena Winiwarter

Abstract Soils are complex ecosystems. They are the basis of human sustenance and have been changed by humans for millennia. Their environmental history needs to incorporate pedological, historical and archeological data. A primer on important concepts of soil science introduces the complexity of soils and their interactions with humans. Many societies developed soil classification systems, testing methods for soil quality and a multitude of measures for soil fertility maintenance. They also developed landscaping techniques such as terracing to enhance the utility of their soils. In a comparative approach, these three fields of soil knowledge and their development during pre-industrial times are discussed for the history of China, Mexico, Mesoamerica and Amazonia as well as for India. Ghana and the Nile valley serve as two examples from Africa, and finally the situation for the Mediterranean and Europe north of the Alps is presented. Human influence on soils has been both beneficial and detrimental. Anthrosols, soils that have been significantly changed by humans, are part of the ecological inheritance of societies; they can be much more fertile than the unchanged land. Salinization through irrigation and human-induced enhanced erosion are the two most widely known negative influences of humans on soils, making it much less fertile. Under conditions of industrial societies, the nitrogen cycle has expanded to encompass the air. Unsustainable soil use leading to compaction and pollution poses a threat to soils. All soil histories are local, because soils are so varied. Unlike other fields of environmental history, the environmental history of soils is still in its infancy. Providing long-term data on sustainable and unsustainable use of soils in the past is a daunting task for environmental historians for the next years and decades.

V. Winiwarter (✉)
Institut Für Soziale Ökologie, Alpen-Adria-Universität Klagenfurt, Klagenfurt, Austria
e-mail: verena.winiwarter@uni-klu.ac.at

M. Agnoletti and S. Neri Serneri (eds.), *The Basic Environmental History*,
Environmental History 4, DOI 10.1007/978-3-319-09180-8_3

3.1 Introduction

A comprehensive environmental history of world soils has yet to be written. It would have to combine pedological, historical and archeological perspectives and encompass a multitude of case studies. In such a story, a set of actors new to history would play important roles: soil biota are among the main players. Earthworms do not write history, but they are extremely important in making it, a fact recognized by Charles Darwin, in a book he considered as one of his most important:

> The plough is one of the most ancient and most valuable of man's inventions; but long before he existed the land was in fact regularly ploughed, and still continues to be thus ploughed by earth-worms. It may be doubted whether there are many other animals which have played so important a part in the history of the world, as have these lowly organized creatures.[1]

Looking at environmental history from a soil perspective reveals several striking cases of unsustainable soil use, but also by a steady stream of human knowledge acquisition and technical ingenuity to deal more sustainably with this prime resource. The biblical proverb that we all come from the soil and shall return to it holds true in the very literal sense of the word: Our deceased bodies are decomposed by specialized soil organisms, releasing nutrients for the growth of vegetation and hence, all life. Only some human cultures hold soils sacred,[2] only some cultures have learned to produce fertile soils from barren ones,[3] but all cultures have developed some sense of the importance of soils.

Soils are central to the biogeochemical cycles of the world; they interact with the hydrosphere as well as with the atmosphere, and are themselves part of the biosphere.[4] The soil sphere is called the pedosphere, recognizing its unique characteristics. Dirt, although a recent popular book on soils wishes to suggest otherwise, is different from soils[5]: Dirt is under fingernails; soil is the living matrix of life on which we walk.

While concern about soils on the part of scientists has a long history, with contributions such as Bennet's and Chapline's plea to combat erosion of 1928 standing out,[6] a self-aware environmental history of soils is a relatively young phenomenon.[7] But readers will find discussions pertaining to the environmental history of soils in the context of soil science, agricultural history, anthropology and archaeology. In the following paragraphs, a soil science primer offers the necessary

[1] Darwin (1883).

[2] Winiwarter and Blum (2009).

[3] Lehmann et al. (2003).

[4] De Deyn and Van Ruijven (2005).

[5] Montgomery (2007).

[6] Bennett and Chapline (1928).

[7] McNeill and Winiwarter (2004, 2006).

basics. The subsequent chapters trace soils as a material entity and soils in the mind through historically and geographically distinct cases. The final paragraphs discuss implications for sustainability.

3.2 A Soil Science Primer

Soils are varied and manifold. According to the most widely accepted attempts at classifying them, 12 soil orders and a multitude of sub-orders and classes can be discerned.[8] While the air can be viewed as largely homogeneous, and its origin is of little concern for historians, one must understand the formation of soils, because their resulting qualities are so different and because cultivation is a major factor interacting with soil development.

The natural history of soil is called pedogenesis, an evolutionary development of soils over time which was first described fully by the Russian soil scientist Dokuchaev in the late 19th century.[9] The human history of soils is the history of their cultivation. Taken together, natural and human histories create the history of human interaction with soils, their environmental history.

Hans Jenny first detailed the factors of soil formation: climate, parent rock material, topography and organisms interact to form soils.[10] Soils form the surface layer of the earth in a range from several meters thickness to a few centimeters.[11]

There is no single definition of soil that all soil scientists accept, but most would agree that soils are three-dimensional entities composed of mineral and organic matter, with their own architecture comprising micro- and macropores through which water and air circulate, and particles of different sizes and surface textures, which form a multitude of quite different habitats for microbial and macroscopic soil organisms. Particle size is an important soil characteristic, with sand, silt and clay being the three categories most often discerned in order of decreasing particle size. A typical soil (if such a thing exists) consists of roughly 25 % each of air and water, 45 % mineral particles, and 5 % soil organic matter (SOM), most of which is comprised of large organic compounds called humus. The rest of SOM is roots and soil organisms. Processes in soils can be physical (such as aggregate formation), chemical (such as nutrient dissolution and leaching) or biological (such as earthworm digestive action). Taken together they control a major part of global biogeochemical cycles, in particular the cycling of reactive nitrogen, and of carbon and its compounds.

Soil processes (in all three senses) depend very much on surfaces, and many involve exchanges at active surfaces such as clay minerals offer. The origin of life

[8] Blaser (2004).

[9] Evtuhov (2006) for a longer history of pedology see Feller et al. (2006).

[10] Jenny (1941) Compare the overview about soils in: Muoghalu (2003).

[11] For example, Pidwirny and Heimsath (2008).

itself has been associated with the active surfaces of clay minerals.[12] Besides surfaces, much in soils depends on the organic constituents. The rhizosphere, the soil region in direct contact with plant roots, is a zone not only of increased microbiological activity, but its own chemical characteristics. These influence nutrient uptake and thus, the perceived fertility of the soil. SOM content is decisive for water uptake and storage ability, influences pore structure and microbial activity and hence is crucial for the role of soils as sinks or sources of greenhouse gases. Cultivation lowers SOM content. Agricultural techniques such as manuring or plowing in stubble are geared at restoring SOM in cultivated soils.

3.2.1 Soils and Their Fertility

Agriculture intervenes into the biodiversity of ecosystems. It transforms them in a planned way by management of the agro-ecosystem, e.g. by crop selection. It also influences associated biodiversity, made up from organisms which colonize the agro-ecosystem after it has been set up by the farmer. The combination of both is responsible for ecosystem functions in an agro-ecosystem.[13] Much of this associated biodiversity is that of the soil, which only came to be recognized with the development of soil microbiology in the second half of the 19th century.[14] One cubic centimeter of soil can contain more than 1,000,000 bacteria. A hectare of pasture land in a humid mid-latitude climate can contain more than a million earthworms and several million insects.[15] Biological and chemical activity is concentrated in the uppermost 10–15 cm of soil,[16] but there is more to soils than the uppermost layer. Pedogenesis does not create uniform mixtures of particles. Most soils are multi-layered, 'soil profiles' over depth serve as the main discriminator between soil types. Most existing overviews for a general readership give details about soil types by profile.[17]

Questions of soil fertility are more important for the historian, as it is the productive relation with the soil that is decisive in human history. Patzel et al. have shown that the concept of soil fertility itself is not historically stable.[18] Nowadays productive soils are conceptualized as systems governed by both natural and anthropogenic factors. In Fig. 3.1 factors influencing yield in an agro-ecosystem are shown. The natural fertility depends on the factors identified by Jenny, of which all but time are depicted, minerals and morphology being combined into one factor.

---

[12] Orgel (1998).

[13] Altieri (1999).

[14] Eldor (2007), Berthelin et al. (2006).

[15] Pidwirny and Heimsath (2008).

[16] Coleman et al. (2004).

[17] For example, Beach et al. (2006), Pidwirny and Heimsath (2008) and Nortcliff (2009).

[18] Patzel et al. (2000).

Long-lasting interventions by humans change a soil profoundly, so that the resulting fertility of the cultivated soil can be much greater than the natural one (e.g. in the case of plaggen soils). While this cultivated fertility can be considered an acquired long- or at least mid-term characteristic of soils, the yearly yield will depend on short term influences of both natural and anthropogenic origin. If human interventions lower fertility, one speaks of anthropogenic soil degradation.

Soil ecosystems are complex in many ways this primer cannot adequately address. As but one example, Fig. 3.2 shows factors influencing the availability of nutrients. Not all nutrient pools in the soil, are available to plants, and the soluble fraction can be quite small, but on the other hand fully mobile ions run the highest risk of being leached, nutrient management thus tries to create large amounts of easily exchangeable nutrients which are bound to surfaces.

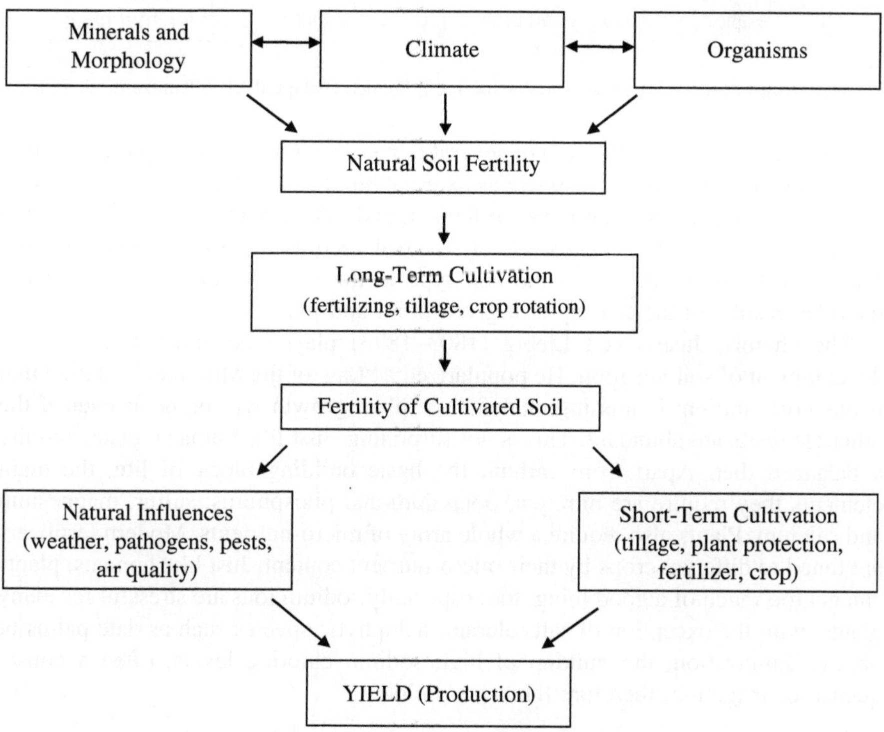

Fig. 3.1 Factors influencing yield in an agrosystem (After Gisi et al. 1997)

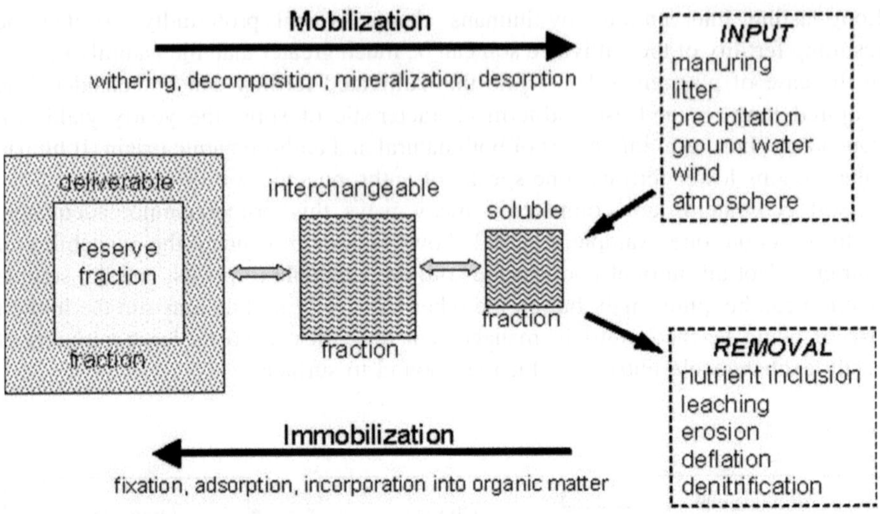

Fig. 3.2 Nutrient behavior in soils under human influence (Adapted from Kuntze et al. 1994)

Nutrient balances are the sum of several dynamic equilibria, with (1) fertilization and deposition from the atmosphere as external input parameters, (2) mineralization as most important factor particular to the soil and (3) nutrient export by harvesting products and through various kinds of removal such as leaching or wind erosion. The other factor particular to the soil type is the rate of immobilization. Their interplay results in the fertility at a given point and place.

The chemist Justus von Liebig (1803–1873) played an epochal role in the development of soil nutrition. He popularized a "Law of the Minimum", stating that if one crop nutrient is missing or deficient, plant growth will be poor, even if the other elements are abundant. This is not surprising. Just like humans, plants require a balanced diet. Apart from carbon, the basic building block of life, the main elements they require are nitrogen, potassium and phosphorus, sulfur, magnesium and calcium. Plants also require a whole array of micro-nutrients. Modern fertilizers are tuned to different crops by their micro-nutrient content. Just like humans, plants can get too much of a good thing, too: especially sodium ions are stressful for many plants, with the exception of salt-tolerant, halophytic species such as date palms or barley. Salinization, the buildup of high sodium chloride levels, often a consequence of irrigation, therefore threatens yields.

3.2.2 Soil Functions and Threats to Soils

Soils perform several key functions, apart from their role in biomass production. They are filters and buffers and perform transformations between the atmosphere, the ground water and the plant cover, strongly influencing the water cycle at the earth's surface as well as the gas exchange between terrestrial and atmospheric systems. Soils are also a biological habitat and gene reserve, supporting a large variety of organisms. Soils contain more species in number and quantity than all aboveground ecosystems. Therefore, soils are a main basis of biodiversity. Soils are also the physical basis for technical, industrial and socio-economic structures and their development. Independent of all aforementioned functions, soils are a source of raw materials, e.g. of clay, sand, gravel and minerals in general, as well as a source of geogenic energy and water. Furthermore, soils are a cultural heritage, protecting valuable paleontological and archaeological remnants.[19] The roles soils are expected to play for humans often exclude one another, leading to conflicts about land-use such as those encountered between nature protection and infrastructure development or quarrying.

Next to human-induced erosion and salinization, nutrient depletion is the most prevalent damage to soil ecosystems inflicted by humans.[20] Commonly, all these processes are subsumed as 'soil degradation'.[21] In the UNCCD definition, degradation is defined as "reduction or loss of the biological or economic productivity and complexity of rain fed cropland, irrigated cropland, or range, pasture, forest and woodlands resulting from land uses or from a process or combination of processes, including processes arising from human activities and habitation patterns, such as: (i) soil erosion caused by wind and/or water, (ii) deterioration of the physical, chemical and biological or economic properties of soil; and (iii) long-term loss of natural vegetation".[22]

Land-use resulting in the covering of soils with concrete and other materials to use them for infrastructural purposes is a further major threat to soils, particularly close to urban agglomerations. It is important to keep in mind that soils like all ecosystems, are dynamic entities. Soils can be changed through management-induced or through natural processes.

3.2.2.1 Erosion

Apart from impairments of soil quality, the mobility of soil as such is an important issue. A term often used for soil mobility is 'erosion'. Erosion is a natural process which shapes the earth in an interplay with other processes such as volcanism and tectonics. Through the action of water and wind, mountains are reduced to sand, and within geological times, sediments undergo metamorphosis and uplift and end as sandstone mountains, beginning the cycle anew. Erosion can benefit agriculture, as its result, alluvial deposits or aeolian sediments such as loess are prime land for

[19] Blum (2008).

[20] Compare particulary Sect. 5.2 in Koehler (2005).

[21] Lysenko (2004).

[22] Section 1.2 in: Juergens (2006).

cultivation. About 100–200 tons/Km^2 of new soils are currently formed annually by weathering processes.[23]

Erosion processes can often reach dangerous velocity and extent due to human intervention. Enhanced erosion is a worldwide problem, but particularly pronounced in tropical and subtropical climates. Between 1958 and 2001, a terrace in the central loess plateau of China lost 3,400 $m^3\ km^{-2}\ a^{-1}$ of soil. A fluvial catchment on clayey substratum in the Transkei region of the Eastern Cape Province in South Africa displayed erosion of 5,400 T $km^{-2}\ a^{-1}$ between 1949 and 1975. In the loess region of the Palouse, Washington and Idaho, USA, about 7,600 T $km^{-2}\ a^{-1}$ were eroded between 1980 and 1998 and on deeply weathered crystalline rocks in Brazil 17,000 m^3 or 23,000 T $km^{-2}\ a^{-1}$ were displaced between 1850 and 1979.[24] These measured Brazilian soil destruction rates are more than 100 times higher than the average rate of regeneration of soil material by weathering[25]. Erosion processes like this are potentially able to remove the entire soil cover in a few centuries and would then prevent agricultural use on the long term. But the upscaling of such results is not easy. Continent-wide estimations seem to be rather doubtful as they are not based on representative data, much remains incompletely understood.[26]

Humans have been aware of soil movement for a long time. In some places (such as the Andes and central Mexico), soil erosion was stimulated by humans so that soil could be collected and concentrated to create agricultural surfaces. In other locations (e.g. Central and West Africa, northern Mexico), soil management systems were designed to minimize or prevent soil erosion associated with tillage and vegetation, or to contain soil movement within a field by using vegetative boundaries.[27] Where large scale crop production developed to supply distant markets, soil erosion was often ignored, went unchecked by human intervention, and led to large scale soil loss.[28,29]

3.3 Human Interaction with Soils

3.3.1 Overview

Human interaction with soils predates agriculture. Non-agricultural peoples used specific soils for medicinal purposes and to make pigments, while clay materials were formed into containers. Red ochre served in burial rituals since Paleolithic

---

23 Arnold et al. (1990).

24 All data from Bork (2006).

25 Arnold et al. (1990).

26 Pimentel et al. (1995), Boardman (2006), Cogo and Levien (2006) and Flanagan (2006).

27 Reij et al. (1996).

28 Showers (2006).

29 The preceding chapter 3.2.2.1 Erosion is based on Winiwarter et al. (2012).

times.[30] Paleolithic hunters, using fire for hunting, unintentionally influenced soil ecosystems via vegetation. Human-induced amplification of erosion has accompanied human life ever since the advent of a sedentary life-style.

Important origins of agriculture lay in the semi-arid south-west corner of Asia, the Levant and Fertile Crescent east of the Mediterranean where water scarcity was a limiting factor. Hence, irrigation techniques, next to the hoe or ard, were the main early interventions into soils. Short-term and long-term interventions can be discerned. Dams, irrigation networks and terraces are among the longer-lasting. Mechanical (such as hoeing, plowing and harrowing), chemical (fertilization and soil amendments) and biological (weeding, mulching, and irrigating) cultivation techniques influence soils on a shorter term. Salinization due to irrigation under arid conditions was a problem for the early civilizations in the Indus and Euphrates/Tigris valleys: soils there are still impacted. Archaeological excavations have produced evidence for manuring for Bronze Age European settlements in areas as far apart as Estonia, the Netherlands and Switzerland. The practice seems to have been widespread.[31]

From the early civilizations onwards, aside from their main use in agriculture, soils were the basis of infrastructure. Roads and buildings sealed soils; cisterns, cellars, sewers and waste pits all needed and changed soils.[32]

Figure 3.3 shows major types of interaction between humans and soils and their impact on soil processes. Again, parent material, climate, vegetation and relief are the natural factors determining which kind of soil is available. Human influences differ with soil use: grazing animals on rangelands have influences different from those on either cropland or forest. Infrastructural use leads to a fourth, apparently different impact. Grazing can lead to changes in micro-relief and to compaction; it also leads to nutrient input through animal feces. Humans using fire on pastures to prevent perennial growth exert an influence on plant species, and hence, on soils. Grazing animals likewise have an effect on vegetation, leaving poisonous or thorny plants to grow. The impact on soils is widespread, as grazing areas are usually large. Its details depend on the animal species and grazing patterns and density, but the overall impact is relatively small. Human influence on the soils of croplands is much more pronounced. Cultivation techniques such as hoeing, plowing and harrowing change the physical characteristics of soils. Fallowing, leaving soils to rest from cultivation, was a time of intensive plowing at least in Roman antiquity. Wild plant species growing on the land during fallow periods also influence the soil. Soils were and are changed also by crop selection (the most striking example are legumes, with their nitrogen-fixating abilities), by residue management (e.g. by burning or plowing stubble) and, most importantly, by nutrient inputs through fertilizing agents. Monoculture of any kind does change soils. Water management plays a decisive role in irrigation agriculture. Some slow-growing crops leave land

[30] Einwögerer (2005).

[31] Bakels (1997), Reintam and Lang (1999).

[32] This summary is based on Winiwarter (2006a).

bare for elongated periods, and thus can enhance erosion. From a soil standpoint, harvest is a removal of nutrients, and hence, plays an important role in balances. Soil ecosystems are nowadays influenced by agrochemicals, most of which can and do enter the soil.

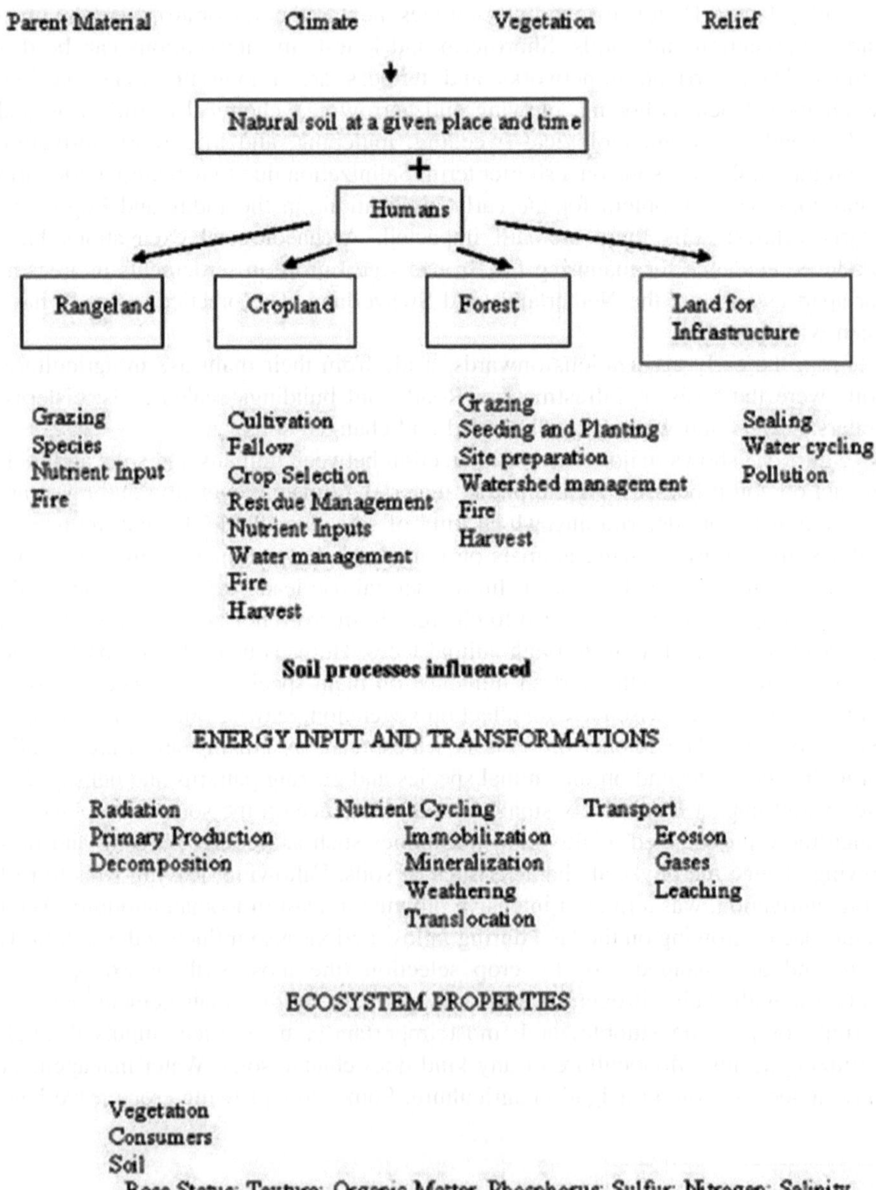

Fig. 3.3 Major types of interaction between humans and soils and their impact on soil processes. (Adapted from Coleman et al. 2004)

Forests have been used in a multifunctional form prior to the 18th and 19th century in Europe. Grazing animals in forests, particularly pigs, led to an overturning of the uppermost soil layer as well as to nutrient inputs. Tree species are inadvertently selected by grazing animals. Those saplings animals like eat are disadvantaged. Later on, in Europe starting with the late Middle Ages, forests would be seeded and planted with desirable species; in particular a change from deciduous trees to conifers took place, which changed soil characteristics profoundly.

Sealing of soils is the most profound change, a complete sealing means the end of subterranean ecosystems, with solely the mineral matrix being left, but changes to the water regime by partial sealing also change soil ecosystems. Infrastructural use of soils can lead to soil pollution, and so often does mining. Soil remediation techniques using plants which concentrate pollutants and a variety of chemical treatments are used to restore such soils.

The lower part of Fig. 3.3 lists the processes in soils which are changed by the various impacts of humans. Among the most important are influences on nutrient status and nutrient cycling, weathering, texture and organic matter content.

Over the millennia, humans did change soils profoundly, on purpose and inadvertently. We know of many examples where interventions did not have the desired effect, or had massive side effects. Procuring food from the soil was always a challenge, and societies responded to it by acquisition of knowledge and by development of sophisticated cultivation techniques.

3.3.2 Soils in Agricultural Societies

Procuring a surplus from agricultural operations was never an easy task. Being a successful farmer depended on good decisions as to crops, timing of operations, diligent use of different fields according to their soil type, success in breeding, as well as treatment of human and animal ailments, and yet was always at the mercy of weather and pests such as locust swarms or animal maladies. The body of knowledge necessary to be a good farmer is vast. Therefore, all major agricultural societies collected and systematized knowledge about the management of agricultural operations, and hence about soils, which will be discussed in the following overview. These collections are one of the main sources for the reconstruction of agricultural systems of the past, in addition to archeology or, in some cases, ethnological studies ('ethnopedology').

All humans living in agricultural societies possess knowledge about how to produce harvests. Knowledge about soils falls into three broad categories: Without a stable vocabulary about the phenomenon one wishes to transmit knowledge about, transmission is not possible. Hence, systematic nomenclature and classification, that is, observing characteristics of soils and their differences with regard to agriculture and naming them, is a major part of soil knowledge. The second category is knowledge about the testing of soils, basically about ways and means to distinguish fertile from infertile soils. This was of particular interest if virgin land was to be put under the plow. The third category comprises all techniques to sustain or amend soils, be it fertilization, lithic mulching or crop rotation, to name but a few.

Readers have to be reminded that translation of technical texts requires not only philological, but also technical expertise. While editors and commentators of ancient works try to render the contents as well as possible, ambiguities do remain. A translation is a rendering of an ancient system, which as a whole might be very different from the current system, into the terminology of this current system. This never works perfectly. Also, extant texts are almost never originals, historians work from often distorted and incomplete later copies or collections. Inconsistencies within texts therefore are rather the norm than the exception. It is difficult to infer practices from these written texts, but this is still the luckiest case. Despite its complexity, much agricultural knowledge was orally transmitted, and never documented in texts. We can only infer such knowledge from its surviving physical effects. It is with these limitations in mind that readers should consult the following sections.

3.3.2.1 China

China is one of the oldest agricultural civilization from which written testimony survives. Chinese agricultural systems comprised few animals in comparison to Europe, millet, rice and various vegetables such as cabbage being main staples. The Chinese developed intensive uses of small-scale plots, such as combinations of fish ponds with mulberry trees and sericulture, and tended to invest into intensive, horticultural ways of making optimal use of the land, one of them being the 'pit cultivation' system unique to China.

Joseph Needham, the great historian of Chinese science, has devoted part of his book on botany to the discussion of early soil science, presenting the oldest extant testimony of human concerns with soils in agricultural societies. The *Kuan Tzu Ti Yuan Phien* which was probably written between the 5th and 2nd century BCE, is probably the oldest extant text which classifies soils systematically. Three productivity classes are discerned, and within each, six sub-classes are distinguished, each with its own name. The author names the tree and plant species which grow best on each soil, and informs about the yields to be expected. Farmland, literally 'irrigable land', is divided in five soil types and then classified by means of the depth of the water-table—a remarkably systematic approach, and one which is based on invisible characteristics, hence testifies to a kind of experimental approach. The descriptions are very detailed. A soil called *hsi thu* is described as fertile loess of silty texture, ca. 10 m above the water table, well suited for cereal crops. After a list of plants growing on such soil, the author denotes people living on such land as robust. When such land is dry, it gives out a ringing sound corresponding to the musical note *chio* when knocked on. Other soils described within the category of farmland are given with a list of characteristics: *chhi lu* is said to be reddish, crumbly, hard and fertile, particularly suited for hemp cultivation, *huang huang* is yellowish, brittle, salty and alkaline soil, occurring on land liable to flooding, useful only for the cultivation of millet. *Chhih chih* soils are argillaceous and rather saline, with a much higher groundwater table of ca. 4 m, but apparently, the author is convinced that wheat and soybean can be grown there. The *hei chih* soils, dark, sticky, saline clay soils with groundwater at 2 m, can support the growth of rice and wheat. The *Kuan Tzu* is not a stand-alone contribution. Even earlier than this work, in the 5th century BCE, the *Yü Kung*, a geographical treatise, detailed the soils of the ancient Nine Provinces of China. The Chou Li, a collection of texts about the administration of the country, probably from the 3rd century BCE, discussed taxation in relation to soil type and hence, quality.[33]

Apart from naming soils, one needs also to test them for quality, as looks can be deceptive. There is only scant evidence on soil testing in that part of the Chinese literature which is available for the non-specialist. Needham speculates that the terminology itself, *hsi thu* soils as opposed to *hao thu* soils bears witness of Chinese knowledge about a test of soil quality which has also been described by Roman agricultural writers (see below). The quality of the soil can be determined by digging a pit and refilling the pit with the excavated earth. *Hsi* means ‘much’ or ‘full’, *hao* means ‘little’ or ‘lacking’. This way, they describe soil quality according to the pit-test. If the material proved to be too much for the pit (as *hsi thu* denotes), the soil would have been fertile, whereas in the opposite case, soil not filling the pit would be infertile.[34]

Over five-hundred agricultural works are known for China before the end of imperial rule (1912), with an outstanding contribution to agricultural science dating as early as the 6th century CE. The *Chhi Min Yao Shu* (Essential Techniques for the Peasantry) starts with a section on clearing and tilling the land, documenting the importance of soil matters in agriculture. Like some other books of the genre, it was obviously written by a practitioner who combined his own experience with the wisdom of the already existing works. It is a main source for details about agriculture in the north of China. Together with several extant overviews and concentrating on particular topics, the *Chhi Min Yao Shu* allows us to reconstruct soil management practices in detail. This 6th century manual is the oldest proof of green manuring, of the use of the nitrogen fixating power of legumes to boost the growth of subsequently planted other crops; the manual recommends particularly melons, mallows and other vegetables. Manures comprised animal droppings collected with the bedding straw of their stables; nitrogen-rich silkworm excrement was valued highly, as was human excrement (nightsoil). Diligently prepared hemp waste, oilcake, or the cake left over from bean-curd making were added to the Chinese fertilizer repertoire during the Sung and Ming dynasties (from the 10th century onwards). The fertilizer repertoire was regionally differentiated, including animal

[33] Needham et al. (1984).

[34] Needham et al. (1986).

bones and hoofs, chicken feathers and lime in various places, a total of 80 different substances has been counted. Seeds were prepared for sowing by mixing them with manure or compost (for which very detailed and complex recipes exist) and non-organic soil amendments such as river mud were recommended, the latter for wheat cultivation. As late as the 1930s, marine fish too small to market as food were dried and sold as fertilizer, a practice which was certainly unsustainable with regard to the fisheries, but benefitted the nutrient balance of soils.[35] Chinese farmers and writers knew that the application of the wrong type or amount of fertilizer could do more harm than good and that some fertilizers could 'burn' crops.

Rice paddies needed and had to be treated with special care. Application of fertilizing agents was time-sensitive. One manual gives a detailed prescription involving four different types of fertilizers. "You should not put down manure too early or its strength will not last [...] Only at sowing time must river mud be applied as a base, and although its strength is lasting and dissipates slowly, by midsummer you should apply a little potash or oil-cake, which also dissipates slowly and is long-lasting. Only at the end of the summer or the beginning of autumn should you apply nightsoil, by which time it will have the double effect, so that the rice panicles will grow very long".[36]

The creation of rice paddies is a long-lasting intervention into soils. Paddy fields cover large areas of the Chinese south, bearing witness to a technique of human creation of fertile soil. A hardpan soil layer on the bottom of the paddy is consciously created to prevent water seepage, with a system of ditches, ridges and dykes to control water flow. Yields are higher in paddies which are several years old. Other special types of fields known in Chinese history are poldered fields (*yü thien*), diligently ridged spaces to reclaim swampy areas such as in the Yangtze delta, where the lakes had been turned into fields by the 12th century CE. On marshy lake shores, floating fields were created from wooden frames, filled with mud and water-weed, enabling the planting of crops on artificial soils.

Terraces are a feature of many agricultural systems. In China, they were considered beneficial to prevent erosion, conserve soil moisture and nutrients, and improve yields. Terracing allows the cultivation of slopes otherwise unfit for irrigated agriculture and hence, enlarges the cultivable area. The Chinese agricultural writers distinguish between stone-walled terraces and fields formed from piled-up soil. Terraces are labor intensive to build and maintain, but the overall improvement was obviously considered to outweigh the effort.

Chinese agriculture was based on cereals and vegetables, with pig, poultry and fish as the main sources of animal protein. Interventions into soils were undertaken consciously, and the extant manuals are proof of intricate and detailed technical knowledge, based on systematical observations of key features of soils, both their relation to the water-table and their situation with regard to the relief of the landscape. A system of soil classification was the basis for decisions on crops, rotations and the timing, amount and quality of fertilizer to be applied, but also for the siting

---

35 Muscolino (2008).

36 Bray and Needham (1984).

of engineering works for terraces and similarly treated fields. Nutrient recycling techniques for cereal as well as for specialized types of cultivation such as sericulture (by using silkworm excrement as fertilizer) are highly developed in the manuals. Although we know much less about the practice of the Ancient Chinese, we do know that in their world-view soils held a special place. The Chinese are the only agricultural people who worshipped the soil as such, not in the form of fertility deities, but the material as it was. The soil cult did not prevent erosion, which occurred time and again over the long course of Chinese history, often as a side-effect of agricultural expansion. The altar of the Earth and Harvests in Sun Yat-sen Park in Bejing, however, which was used from the 14th century CE onwards is an architectural testimony to concern about soils.

There might still be reverence paid to soils on village altars today, but the picture Lindert[37] and others paint of current trends in erosion, pollution and other types of soil degradation is not positive. As scholars like James Reardon Anderson have pointed out, many of the trends which were enhanced in communist, industrial China and are now continuing, have been started in the late 18th and 19th centuries, when China expanded agriculture to lands that had formerly been pasture or forest.[38] Due to large-scale interventions China has lost much of its forest cover, leading to increased runoff and erosion. Industrialization has taken its toll in terms of heavily polluted soils unfit for cultivation. With changing dietary habits, increasing population and extremely fast industrialization, the pressure on China's soils is likely to increase in the coming decades.

3.3.2.2 Mesoamerica

It is pretty clear from archeological evidence that humid tropical lowlands were the major settings for the origin and development of agricultural systems in Mesoamerica. Maize and manioc cultivation were the drivers for the diffusion of swidden agriculture from 5,000 to 7,000 years ago, giving the region a long history of cultivation, and thus, of interaction with soils. Beach, Luzzader-Beach and Dunning have detailed the history of soils in the area in an overview article.[39]

Our knowledge in terms of classification, testing and amending is much more scarce for the Americas than for China. This is due in part to destruction of many written sources by the conquering Europeans. The few extant sources are not manuals about agriculture but basically survey texts and maps with glyphs denoting soil quality. Nevertheless, Barbara Williams has been able to reconstruct Aztec soil knowledge in some detail, which is the best documented of the entire continent.[40] The classification in the *Codice de Santa Maria de Asunción* shows 104 variants for

---

37 Lindert (2000a).

38 Reardon-Anderson (2005).

39 Beach et al. (2006).

40 Williams (2006).

soils over only 200 ha. Peasant farmers today apply four classes, and modern taxonomy ascribes five soil phases to the soils of this area. The 104 variants probably represent 18 taxa and three separate class levels. Classification of soils was done in a multi-dimensional system. One important denominator was grain size, stony, sandy, clayey, silty and the volcanic *tepetate* soil particular to the area were distinguished, each of them is described in some detail. The distinction between material that could be crushed in the hand and that which could not, a main difference in the classification system, is an experimentally derived denominator for Aztec soil taxa.

Tepetate (best rendered as 'soft rock') is a hardened volcanic tuff, which can be found on the surface only after erosion has taken place. It is considered as marginal land, but cultivable, if pulverized. A depiction of a man beating up *tepetate* soil is preserved in a 16th century description made by the Franciscan friar Bernardino de Sahagún, one of the main sources for Aztec agriculture. Loess soils, the result of Pleistocene wind erosion, were denoted by a special glyph, which was described as "that which swirls up, that which sweeps up", possibly showing knowledge about the Aeolian origin of these soils, but definitely about their inclination to wind erosion. Soil classes between sand and clay, commonly called loams in English, are not distinguished by special names, but rather by combining the names of the two constituents, at least such conjunctures are used by Nahuatl speakers today. Each of the main soil classes has several sub-classes, such as sandy clay or gravelly clay, bother denoted by the combination of glyphs. Color does not play an important role in the taxonomy of the Aztecs, but topography does, the soil on slopes has its own name. Woodland and reedy soils are both characterized by their content of organic matter, showing an understanding of its centrality for cultivation. The soil taxonomy comprises a specific word for humus, *tlazollali*. It was understood to come from the compost pile and to turn into fertile soil. Another term which bears witness to Aztec understanding of soils as dynamic entities is a specific word for alluvial soils, those that are the result of fluvial sediment deposition (*atoctli*), and again, several subclasses for this soil type are denoted. Moisture retention by soils, a particularly important feature in arid or semi-arid lands such as those of the Aztecs, was included in the nomenclature system, with two names for such soils. Even a notion for anthrosols, soils which are co-created by humans, has come down to us in Nahuatl: *callalli* denotes a soil where a house has been, it is considered to be fertile.

All classification systems found in agricultural societies also include a concept of infertile soils. The Nahuatl word for them is *tequixquitlalli*, described as 'salty, bitter, corrosive, leached of its salt, unwanted, undesirable, waste and disregarded', the latter two could hint at soils which have become infertile by bad agricultural practices. *Tlalzolli* is another word for bad soil, which will not support any growth, because it is worn out.

Unfortunately, we lack information as to soil tests in Aztec agriculture. They must have had some means of determining e.g. salty and bitter soil, but the extant sources do not give information on such tests apart from the knowledge of two grasses as indicator plants for *tequixquitlalli* and ashen soil (*nextlallili*), two infertile soil types to be avoided. While these soils were considered unfit for agriculture, they played a role in indigenous medicine.

Aztec farmers pulverized *tepetate* soils to make them fit for planting, a strenuous task, after which organic matter was added to increase fertility. Planting was sometimes done in holes, which served as sediment traps and collected rain or irrigation water, as Barbara Williams suggests. Glyphs exist also to describe fallowing and fertilizing the soil with a substance which could probably be nightsoil. The fertilizing agents comprised 'refuse', which cannot be identified, nightsoil, bat dung, organic ash and alluvium. Irrigation canals filled with mud over time, and this mud was brought back to the fields as another form of fertilization. One glyph suggests that soils were mixed; in particular a woodland soil with high organic matter content was mixed with clay, a procedure that makes agricultural sense, although it is not fertilization in the strict sense.

Aztex agriculturalists are most famous for the chinampas, ridges constructed in swampy areas and shallow lakes from the mud on the bottom mixed with aquatic vegetation, not unlike the Chinese floating fields. These anthrosol garden beds were highly fertile and supported the population concentrated in the Tenochititlan area before Europeans conquered their state. Chinampas could be used for continuous cropping, with seed beds (similar to those used in rice agriculture) saving space and time. In contrast to these wetlands, drier soils had to be irrigated, and a multitude of words exist in Nahuatl to describe irrigation related activities, implements and constructions.

Salinization of soils was a problem in the northern part of the Basin of Mexico, where salt production took place on the shores of saline lakes. Problems with salinization increased after conquest. Farmers apparently had figured out that saline soil is unproductive, but that the salt can be dissolved with freshwater. The farmers of the early 19th century are reported to have dumped soil which had become saline due to chinampas cultivation into Lake Texcoco, digging out fresh soil from another spot. They would come back to the dumping spot after a while, when the lake water had dissolved the unwanted sodium chloride, again ladling up the cleaned soil for their gardens.

Taxation seems to have been dependent on soil quality, so soil knowledge was not confined to farmers, but had to be possessed also by administrators. Like in all other agricultural societies, soil knowledge was an integral part of everyday life, and had cultic, medicinal and technical aspects. Only a very small part of this knowledge system can be reconstructed from the scarce evidence left after conquest.

The Aztecs, like so many other civilizations, built irrigation networks and agricultural terraces, thereby influencing pedogenesis, erosion and, of course, fertility.

Terrace building is not confined to the Aztec peoples. In Middle America alone, three common types of terraces can be distinguished. Firstly, one finds sloping-field terraces, which span gentle hillsides and do not change much the angle of the slope. They are built to retain soil behind the terrace wall, making it deeper, and to collect moisture. These terraces also retain nutrients by retaining topsoil, which might be washed away by rains. The second type of terraces are bench terraces, which are made to produce leveled platforms along the contour. In Middle America, these terraces were usually irrigated. Cross-channel terraces were built by placing check dams or weirs across narrow valleys or ephemeral brooks to collect sediments for cultivation, using erosion to build up new agricultural land.[41] Terrace building entailed a large investment of labor and material, and created long-lasting changes to landscapes and their soils. Which kind of changes terracing brings to the soil depends on many factors such as terrace and soil type, cultivation regime and fertilization techniques. Positive and negative effects on soil quality have been described in a worldwide survey of terraces by Jonathan Sandor.[42] Terracing in the Americas is an old practice. For Puerto Rico, terraces can be dated to at least 1300 BP (ca. 650 CE). Human induced or enhanced erosion is at least of the same antiquity in the region. Evidence of soil erosion appears in the sediment record around the same time on Hispaniola.

3.3.2.3 Amazonia

The lush vegetation of tropical rainforests mislead European conquerors to assume that the soils underneath the tropical paradise must be very fertile. It soon became clear that rainforest soils can only be cultivated sustainably by long fallow swidden techniques, as almost all nutrients are contained in the biomass. The soils are very poor and prone to degradation and accelerated erosion once the forest is cleared. So Amazonia became to be considered a vast tract of infertile land, until scientists at the end of the 19th century discovered patches of dark earth underneath the forests. *Terra Preta de Indio* or Amazonian Dark Earth is a local name for the soil of these patches.[43]

Nowadays it is known that the dark earths occur in several countries in South America and probably beyond. They were most likely created by pre-Columbian Indians between 1000 years BCE and 1500 years CE and abandoned after the invasion of Europeans. Many questions about their origin, distribution, and properties remain unanswered, but it is clear that they are a product of indigenous soil use. Whether they were intentionally created for soil improvement or whether they are a by-product of habitation is not clear at present. As the notion comprises earths of varied features throughout the Amazon Basin (*terra preta* and *terra mulata* being frequently distinguished); they might have different histories. *Terra mulata* is thought to be the product of intentional anthropogenic activities, based on intensive swiddens or patch cultivation, with long-lasting agricultural activity, involving recurrent clearing of vegetation and incomplete combustion of organic material.

Amazonian Dark Earths contain about five times as much carbon as the surrounding soils, and the enriched horizons are twice to three times as thick as the surrounding soil, most often about half a meter. Carbon is life's building block, and usually circulates relatively quickly through soils. But the organic matter in the dark

---

41 Whitmore and Turner (2001).

42 Sandor (2006).

43 Lehmann et al. (2003).

earths persists hundreds of years after they were abandoned, because it consists of black carbon, as in soot or charcoal, coming from burnt biomass. In addition to carbon they contain more phosphorus, nitrogen and calcium than the surrounding soils. Fallows on Amazonian Dark Earths can be as short as 6 months, whereas fallow periods on the surrounding forest soils need to be eight to 10 years long. The size of the patches varies, ranging from small patches of less than one hectare scattered within the normal upland soils, to areas of several km^2 stretched along river bluffs and interfluves. *Terra preta* often contains debris such as animal bone fragments, turtle carapaces, shells, excrement, potsherds and remains of plants used for houses, but the reason for their fertility are likely changes in the microbial community due to the changes in substrate. They are currently investigated by soil microbiologists.

Looking at the soil, it becomes clear that Amazonia's rain forest is not a pristine wilderness, but the location of probably the oldest ceramic finds in the Americas, and presumably, the world, dating to 5000 BCE–3500 BCE. Nearly continuous habitation has been proven for a site near the mouth of the Amazon river, which was occupied as early as 3500 BCE. Occupation of the Amazon, which has been widespread and is documented in a multitude of known archaeological sites, ended with the European conquest, or shortly thereafter. Agriculturalists who had by luck discovered how to transform the infertile forest soils into useful ones, and then probably experimented to create larger patches of this precious resource, vanished almost without trace. The record of the Spanish explorer Francisco de Orellana, of his voyage down the Amazon in 1542, where he provisioned himself by raiding villages along the river and reported to have seen very large settlements standing on a slope on the northern shore of the river near the Rio Madeira, has probably been wrongly dismissed.

The soils of the vanished people persist to date. *Terra preta* can be mined and is sold by truckload or pot for gardening and agriculture. Scientists try to create dark, fertile soils by inoculating rainforest soils with charcoal, with promising results. The high carbon storage capacity of these soils makes them an interesting experimental field for greenhouse gas policies.

3.3.2.4 Other Regions

The environmental histories of soils in South Asia, Africa, or Australia are less well researched than those of China or Europe, and are often based solely on archaeological evidence. Nevertheless, some observations on soil classifications, soil testing methods and diligent use of soils have been made and are available to the non-specialist through overviews. Before turning to two such examples, the case of Easter Island should briefly be mentioned. The remote Pacific island is a place where phases of sustainable and unsustainable land-use are visible in the archaeological and pedological record.[44] Easter Island is one of the sites of successful

[44] Bork and Mieth (2006).

cultivation using lithic mulching. This technique—adding a layer of stones on top of fertile soil—is used to curb erosion and limit evaporation. There are many instances of lithic mulching documented around the world.[45]

India

South Asia is the home of some of the most ancient human settlements. The Harappan Civilization in the Indus valley dates to 3200 BCE and thrived in Mohenjodaro from agriculture, grazing and trade until ca 1900 BCE.[46] The archaeological record of terraced fields allows us to assume knowledge about water and soil retention having been available at that time. In the Ganges valley, the situation is similar. Archaeological evidence shows that wild rice was being used as early as 9000–8000 BCE at sites near Allahabad, but reliable evidence from pollen of settled agriculture for most of the valley does not appear until about 6000 BP (~4000 BCE).

The testimony of the Vedas, religious texts of uncertain age, offers a coarse terminology for soils. Land was classified as waste (*ushara*), especially when saline, pasture (*gochara/vraja/ghoshtha*) and cultivated (*karsha*). Cultivated land could be furrowed (*sitya*) and plowed (*halya*). The Vedic texts show that ancient Indians acknowledged the positive effect of alluvium deposition by rivers, and distinguished two types of riverbed erosion. One text, the *Vishnu Purana* (1st century CE), offers distinctions of soils by color (black, white or yellowish, red or blue, and golden) and texture or morphology, namely gravelly and hilly or bouldery. Soil quality is connected to water quality in another of these texts, the *Brihat Samhita*, which also discerns soil and rock type by depth. The *Manusmriti*, a compilation of Hindu law composed in its final form in the 2nd or 3rd century CE, advises against the use of iron-tipped plows because they injure the earth and its creatures. Another piece of evidence for soil knowledge in India is contained in the *Arthashastra*, a treatise on how the well-being of a people should be organized, plausibly dated to 3000 BCE. In the chapter on the duties of the superintendent of agriculture (XXIV), the existence of systematic agricultural knowledge is presumed. The superintendent must either possess such knowledge himself or be assisted by those who do. The text advises on suitable lands for particular crops, without giving a classification—it has referred readers to the agricultural knowledge beforehand—but informs e.g. that lands that are frequently overflown by water (*parivahanta*) are good for long pepper, grapes and sugar-cane. The text does also give some indication of manuring practices, stating that water-pits at the root of trees are to be manured with the bones and dung of cows on proper occasions. Likewise, sprouts of seeds are to be manured with a fresh haul of minute fishes and

45 Lightfoot (1997).

46 Wasson (2006).

irrigated with the milk of *Euphorbia Antiquorum*, a succulent native to parts of India which has a latex of pungent odor.

A contemporary source (the *Krishi*) offers a difference between rain-fed and river-fed land, and discusses the cultivation of fresh alluvium, where vegetables can be grown with the aid of cow manure, which, the writer suggests, could also be used dried and pulverized in paddies. The difficulties of cultivation on hard, iron-black soil are acknowledged, soil was plowed often, as many as five times before sowing, and remaining clods were to be broken by a harrow. The *Kashyapiyakrishisukti* was written mostly in the 7th or 8th century CE. It is an agricultural treatise, giving a detailed account of methods of cultivation, the identification of suitable land and the construction of irrigation works. According to this text, good land should be devoid of stones and bones; a pliant (plastic) clay, very unctuous (greasy) with reddish and black hue, and glossy with water; neither too deep nor too shallow; conducive to speedy seedling emergence; easy to plow and cultivate; water absorbent, and replete with beneficial organisms such as earthworms; and devoid of thorns and cow dung, thickly set and compact, and heavy when it was lifted. There is some evidence of soil testing in the text: A hole is to be dug and the effects of the digging observed repeatedly. Several characteristics of the soil are to be observed, among them color and uniformity of color, taste, fluidity, and stickiness. Cow and goat excrement, compost and tendrils of creepers are mentioned as manuring substances.

An agricultural text from Mughal India, which was transcribed in 1693, the *Nushka Dar Fanni-Falahat*, is of particular interest for the diligence employed in preparation of planting sites and the wide range of manures mentioned: Pits dug for olive trees should be left open for a year or burnt, presumably to kill pathogens. Burnt cow bones and dung are recommended as fertilizers for trees. Apart from dung and bones, salt and nitre (saltpetre), vine sap, eggs, olive leaf-sap, pig dung, human urine, night soil and sheep blood are mentioned. The text also offers a description of suitable soils for a list of cultivars, similar to all other extant agricultural manuals.

The native classification of soils and the uses soils were put to during the Mughal reign were recorded by an East India Company employee in the 18th century, rendering four soil types and eight types of land, each was characterized by its situation with respect to elevation and vicinity to watercourses. Like in other agricultural civilizations, where revenue depended largely on land taxes, the distinction between classes of land was made by fertility.

For later times, 'native' classifications and methods are known mainly through the reports of colonial officials or travelers. In one of these accounts, dating to 1820, the pit-digging soil test which was used both in China and Europe, is described as being used by the farmers of the Malabar province. By using the test, these farmers distinguished sandy and clays soils, preferring clays.

Africa

The huge continent of Africa cannot be subsumed into one story. Its soils are varied and their distribution very patchy, depending on slope, bedrock, and rainfall. Two case studies aim at showing the diversity. Places as different as the Nile valley and the Kalahari desert, to name but two of the iconic landscapes of the continent, do exhibit very different histories with regard to their soils. Kate Showers has pointed out how greatly the perception of African soils has been shaped and influenced by colonial, European, mindsets and expectations. Many African regions offer no historical sources pre-dating colonial impacts; this scarcity of evidence and lack of research makes it difficult to compare African agricultural societies with those of other continents. We have to rely on ethnopedology for case studies except for the Nile valley.[47]

Deirdre Birmingham has done ethnological research to investigate the soil knowledge of the Bété and Senufro peoples in Cote d'Ivoire. The Bété people, who live in the equatorial forest zone, distinguished from 10 to 12 mutually exclusive soil types, on the basis of gravel, texture, and color. The Bété used their senses—touch, sight, smell, and even hearing (for testing soil grittiness)—to determine soil properties. While the determining characteristic was absolute, other properties varied. The range for nondetermining properties, such as that of colors, was often described. Although names for the same soil may vary due to dialectical differences among villages, descriptions of each soil type were essentially the same. Animals, earthworms, and termites that carry soils to the surface help the Bété identify soil types, particularly subsoils. The Senufo, living in the guinea-savanna zone, used an entirely different system. Land types include land that floods annually (*fa'a* or *fadoulgou*), land that does not flood annually (*shopegay* or *shofigay*), and land that is barren and rocky (*yandalga*). Each valley bottom may carry its own name, sometimes given as the soil or land type. While the Senufo know that gravelly soils are found on the crests and upper slopes, with softer soils on the midslopes to valley bottoms, they do not classify soils. They know, however, variations in their plots and prefer certain soil qualities. The features they use to distinguish among soils (gravel content, texture, and color) influence the soil properties of importance to them: these are primarily the rates of water infiltration and retention, and soil workability. The main difference to the Bété system is the integration of the distinction between soil types into those of land: Each place is characterized by a specific set of qualities, and soil is not analytically taken apart from these.[48]

Systems of soil description and sensually aided determination of soil qualities are widespread among agricultural people. As the two African examples show, the systems can differ markedly, but all agricultural peoples do possess traditional ecological soil knowledge acquired through experience and transmitted orally as part of their life-world knowledge. Oral traditions are often broken due to people

---

[47] Cooper (1977).

[48] Birmingham (2003).

removed from their ancestral land, and many of the soil classification and testing systems of Africa are probably lost with the people who had possessed them.

Written sources, by contrast, allow access to the soil history of peoples long gone. The Nile Valley is one of the most ancient agricultural landscapes, and the site of an elaborate civilization. It has been studied in particular by economic historians, who have also shown some interest in soils.[49] The basis of cultivation, the Nile floods, have been decisive for cultivation since the onset of agriculture in the valley. While it would be a fallacy to assume that agriculture did not change substantially from Pharaonic Egypt onwards, the overview of Egyptian agriculture from 640 to 1800 CE, in which Richard Cooper details how soils were used, provides a good overall description. The cultivable land is clayey, and of black color (which gave rise to the Ancient Egyptian name for the Nile valley, Kemet, meaning black). Like all clays, the land is difficult to plow, and can dry to hardpans, if worked at the wrong condition. While the annual Nile floods deposited a layer of mud from weathered sandstone and granite upstream before the Aswan dam was built, this fertile silt did not contain all necessary nutrients. Potassium and phosphorus were supplied, but the low nitrogen content had to be supplanted by the cultivation of legumes, in particular lucerne, which was used for grazing and then plowed under.

Ruins and trash-heaps were used to supplant nitrogen in the Islamic period, as was a nitrate-bearing clay available in some regions. The Egyptian farmers had a name for salt-contaminated land. They used land in the vicinity of such spots for manuring flax. The practice of keeping dove-cots was widespread and resulted in precious dove droppings available as manure. Keeping pigeons meant that fields needed protection from them after sowing, seeds had to be covered with earth. Therefore, the land was flattened with a tree trunk pulled across the fields after plowing and seeding. Cinders and the stalks of harvested plants were used as fertilizers, and the practice of fallowing was also known. When European implements were introduced, it turned out that local plows, which did not turn the soil, proved better adapted to the cultivation of Egyptian lands than the imported moldboard plows.

Through a fascinating set of sources on a large estate in the Fayum during the Roman period (30 BCE–640 CE) we can follow the reclamation of land for agricultural purposes by means of an elaborate system of dykes and canals.

The Graeco-Roman town of Philadelphia was situated on the eastern edge of the oasis of Fayum's cultivated land, in the north-western part of Egypt. It was a Roman garrison and its founder Apollonius, had a large estate laid out for him in the 1st century CE. His lands were partially sandy desert, partially marshland overgrown with brushwood and reeds, only some of which had been previously watered with irrigation works or drained. This land was transformed into good arable land for cereals, vineyards and orchards through the well-planned construction of dykes and the digging of canals and drainage ditches, after cutting wood and reeds.

---

49 Bowman and Rogan (1999).

The operation of the estate was not an easy task even after the irrigation and drainage system had been finished. We know of a complaint from a person named Psentaes, one of the peasants to whom land was given, that his whole plot had never before been plowed, and therefore was full of gullies.[50]

Experts were needed to plan irrigation and drainage works. Petechon, an engineer and contractor, was working on the major parts of the irrigation and drainage system, in particular on the great canal of Kleon, which irrigated sandy land, and on a complicated drainage system for the recovery of marshy land by means of ditches. In his correspondence we also find that land which was salty was considered hopelessly unproductive. The detailed sources we have on the construction of the irrigation and drainage system testify to the large investments of knowledge and labor that were undertaken in Antiquity to establish productive agricultural systems. We can safely assume that the amount of labor, material, knowledge and organization needed for the Egyptian system were comparable to those of equally complex systems found elsewhere in the world.

Colonial exploitation has scarred African landscapes, whose soils are often delicate and prone to degradation. A wealth of studies has confirmed the common pattern of ignorance on the part of the colonizers, coupled with their self-assuredness of knowing it better and their interest in supporting their own economies by often recklessly exploiting colonial environments as being destructive to soils.[51] African soils are nowadays part of a global economy based on unequal distribution which leads to high demands on soils in the poor countries of what used to be called the Third World. This does not aid to their sustainable use. A 'Doubly Green Revolution' has been proposed by agronomists in the 1990s to counteract the widespread unsustainable use of soils.

3.3.2.5 The Mediterranean

We possess abundant evidence about soils and their use for a large part of the human history of the Mediterranean.[52] Although landscapes are varied and an array of cultivation techniques has been developed, there are some common features in the climatic regime and soil types commonly encountered which allow us to discuss common features. One contrast to the northern parts of the European continent is the semi-arid character of its climate. The most important limiting factor for agriculture was water, whereas the shortness of the growing season and the danger of frost were limiting factors further north.

The history of agricultural knowledge in the Mediterranean starts with a work of which only blurred traces exist, Mago of Carthage's comprehensive treatise on agriculture, which is said to have comprised 28 volumes. It was rescued during the

[50] Rostovtzeff (1922).

[51] Fairhead and Leach (1996), Showers (2005).

[52] Winiwarter (2006b, c).

fall of Carthage (146 BCE) so must be older than that. The Roman Senate commissioned its translation. Many other works have been lost and are only known through references, but we have the works of Hesiod, Theophrast and Xenophon in Greek, and those of Cato the Elder, Varro, Virgil, Columella, Pliny the Elder and Palladius in Latin. Taken together they allow a detailed reconstruction of soil terminology and tests for soil quality. Sources on soil uses are likewise abundant. All authors represent members of the landed elite, so what we actually have information about is not peasant agriculture, but for the most part typical of the system of large estates. A 10th century CE Byzantine collection, the *Geoponika*, is the sole surviving manual for the Byzantine empire, itself based on works from the 5th and 6th centuries. As has already been noted for the Chinese tradition, knowledge in the Mediterranean realm has been compiled, excerpted and combined with personal observations and experiences over the centuries. Rodgers has produced a graphical rendering of the connections between texts on Agriculture in Arab, Latin, and Greek.[53]

Xenophon's dialogue *Oikonomikos* has some information on soil qualities, albeit not very differentiated: The basic idea is that soil and plant are an interactive system. To be a successful farmer one must first know the nature of the soil, for if one does not know what the soil is capable of growing, one cannot know what to plant or what to sow.[54]

Soils can be 'fat' or 'lean', dry or moist, and by means of the plants that grow ('indicator plants') on a piece of ground one can judge the quality of this ground. Xenophon is concerned about the right timing for plowing and sowing. Manuring is suggested as a means of bettering the soil. Xenophon's information on soil quality is the most basic knowledge one can expect in agriculture, and in this does not differ from knowledge transmitted from other agrarian civilizations.

The soil classification of the Roman writers is much more elaborate. In contrast to the Chinese taxonomy, which worked in a holistic way, Latin soil descriptions are based on an array of adjectives, which comprise all important aspects of soils in several classes per aspect. Of these aspects, grain size, density and structure, humidity and color correspond to modern categories, whereas fertility, 'taste', 'temperature' and some special properties do not. There are important conceptual differences between the authors. The practically minded, though very learned agriculturalist Columella set up a simple systematic soil classification system, which rests on dichotomies: soil can be dry or wet, dense or loose, and fat or lean. His interest is the overall soil quality for agriculture, for which these three qualities are surely encompassing and accurate. Varro, less pragmatic than Columella, uses a theory about soils which holds that the different types are generated by mixing of eleven kinds of (mineral) substances. Varro explains: "For there are many substances in the soil, varying in consistency and strength, such as rock, marble, rubble, sand, loam, clay, read ochre, dust, chalk, ash, carbuncle (that is, when the

[53] Rodgers (2002).

[54] Xenophon (1992).

ground becomes so hot from the sun that it chars the roots of plants); and soil if it is mixed with any part of the said substances, is e.g. called chalky, as well as according to other differences as mixed."[55]

Like all other terminological systems the Ancient Mediterranean ones were made to aid agricultural practice. They are therefore all concerned with combinations of soils and plants—there is, in fact, no bad soil, only a badly treated one, writes Columella in one instance, probably slightly overstating his case. Xenophon was the first to make clear that there are suitable soils, suitable for a specific plant or, more generally, under specific circumstances of cultivation.[56] In the centuries after him, authors tried to define what makes a soil good and/or suitable. Some authors refined the determination of soil qualities in order to be able to give suitability descriptions which are inter-subjective. Others took into account that plants will grow optimally at a particular site, but might still be grown elsewhere, thus developing a plant-relative quality measure.

The Roman authors are most explicit about the methods needed to test a soil for its quality. They describe the same test which we have encountered already in China and India, a test for structural stability of soils. After digging a hole into the ground, one tries to refill the hole. If the earth has increased in volume, leaving a small hill, it is fertile. If the volume has not changed and the hole can be filled evenly, the earth is of middle quality. The soil quality is meager if the soil volume has decreased due to digging and the earth leaves a trough.[57] According to Columella, this test is not applicable to black soils. Land cover of pristine land has been important in assessing soil suitability in several of the agricultural civilizations. Indicator species are named that allow one to distinguish between sweet and saline grounds, and especially between grounds fit for grain or not. Columella gives a long list of plants[58]; Pliny remarks generally on the possibility of plant cover as a soil quality indicator[59] and refers to a list of plants that Cato had already given. Columella, who is generally the most cautious author, warns against relying on plant cover as the sole indicator.

Like the Bété of Cote d'Ivoire, like Ancient Chinese and probably also like the Aztecs, Roman agriculturalists relied on their senses to determine soil quality. Nowadays it is known that the typical smell of earth is due to the activity of fungi, in particular Actinomyces. Fungi play a crucial role in soil biology, and a soil that exhibits the typical earthly odor is obviously healthy even by modern standards. Pliny[60] already mentions soil smell as an indicator of overall fecundity, which according to him is strongest when rain wettens a surface that has dried out, and also becomes stronger when the soil is worked. Taste was likewise employed. As it

---

55 Varro (1996–1997).

56 Xenophon (1992).

57 Columella (1982a), Vergilius [5](1987a).

58 Columella (1982b).

59 Plinius Secundus (1995).

60 Plinius Secundus (1994a).

is very unpleasant to taste the soil directly, and as the compounds which make a soil salty or bitter are water-soluble, soil-impregnated filtrates could be used for testing. One takes clear, fresh water and mixes it thoroughly with a soil sample. Once the mixture is filtered, through an unglazed earthenware or a sieve as used in wine-making (so Virgil)—details vary depending on the author—one can cautiously taste the water, which will have taken on the taste of the soil.[61] This test is said to be particularly important for wine-growing as the wine will pick up the taste of the soil. In modern oenology this concept has been resurrected as 'terroir'.

Water retention capacity and organic matter content are important factors determining the quality of a soil, in particular under semi-arid conditions. The Roman writers report a specific test for these qualities, which is still used by soil scientists in the field. One takes a small clod of soil, adds a few drops of water to moisturize it and then moulds it in the palm of the hand. If it becomes sticky and clings to the skin, the soil is of good quality (explained in terms of 'natural humidity' and 'richness').[62] Pliny remarks that, because potters clay, which is infertile, also shows this stickyness, the test is inconclusive.[63]

Interestingly, there was no hiatus or breach in the transmission of agricultural knowledge between pagan Antiquity and the Christian Middle Ages. A proof of this is a commentary to the Bible by the church father Jerome who lived ca. 350–420 CE. Jerome was thoroughly a part of late antique Roman culture, but his works helped transmit some of that to the Middle Ages. He was born in Dalmatia, had traveled the Middle East, and then lived in Trier and Rome, and spent his last years in a monastery he had founded. He wrote several commentaries to books of the bible. In the Commentary to Isaia 14, 22–23 (*I will cut off from Babylon her name and survivors, her offspring and descendants, declares the LORD. I will turn her into a place for owls and into swampland*) Jerome uses vocabulary contrasting swampland, which is muddy and silty, to a fertile irrigated field, writing: "… where there is not an irrigated field, which produces diverse fruit from seed, but infertile swamps, muddy and silty, in which nightly animals enjoy themselves." The vocabulary he uses in the commentary (*limosus, lutosus, irriguus*) is found nowhere in the Vulgate Bible. Jerome tries to explain what one is to understand by the biblical 'swampland' and its owls using terms he must have known from elsewhere, namely from the agricultural heritage of pagan Antiquity.

The Byzantine tradition is similar to that of the early Latin west. The beginning of the *Geoponika* discusses the quality of land: "The best land is that, the soil of which is of a black color, recommended above all, for it is proof against wet and drought. The next is that of a yellowish hue, and that which is thrown up by rivers, on which they bestow the epithet miry, and that which is sweet, and that which feels warm; for these kinds are known to be adapted to vines and trees, and to the propagation of corn. A deep soil is also recommended, especially if it is friable and not hard to work,

[61] Palladius (1976a), Vergilius ^{5}(1987b).

[62] Vergilius ^{5}(1987c), Columella (1982c), Palladius (1976b).

[63] Plinius Secundus (1994b).

and not calculated to the production of trees only; but a red mould is very good for other things, it is not however fit for the production of trees".[64] This short text contains the main indicators for soils in an agricultural context: Besides color and depth, sweetness (that is, the absence of salt and bitterness, both indicators of soils unfit for cultivation), and warmth are listed. Soil quality is seen as being dependent on the plants to be cultivated, not just as an intrinsic quality of the soil.

Learned medieval writers read Greek, and Latin remained the language of the learned in the western Mediterranean, it was even read by authors in Muslim Spain, who wrote in Arabic. The agricultural knowledge of Antiquity remained available mainly to monks, who had access to monastic libraries, but was not confined to them. New soil knowledge was mainly incorporated into encyclopedias in the Middle Ages. However, a learned North Italian citizen by the name of Petrus de Crescentiis, or Piero de Crescenzi who had worked as legal adviser to city councils for three decades before writing his book, wrote an agricultural manual. His *Ruralia Commoda* were finished around 1300. They offer a full-scale treatment of all matters rural. It enjoyed great popularity and was translated into several European languages, among them Italian, French and Polish. It was also printed at least six times in three languages before 1500, with a circulation of several hundred books for each printing. He developed soil terminology further, using 12 new adjectives not found in the manuals of Antiquity.

Petrus de Crescentiis devotes a chapter of his second book to the question how the fecundity of soil can be determined. He starts with a detailed list of qualities a good soil should not have, such as being gravelly, clayey, bitter or salty, and then goes on to describe indicator plants such as raspberry, cane or clover, which, if they grow of their own accord on unused land, can be taken as indicative of good soil. The color of the soil is not of high value in determining its quality; its fatness and sweetness are most important. The fatness can be tested by sprinkling a clod with water and kneading it, if it gets gluey and sticky, it is fat. The sweetness can be tested by placing a sample from the part of the field which looks least desirable into fresh water and taste it.[65] With the exception of the advice that sampling should take place at the worst looking part of the field, Petrus de Crescentiis offers old wisdom here: all his tests and considerations are already described in texts from Antiquity. But the practical advice to choose the sampling place well is a welcome addition to knowledge. Specific indicator plants are also given in the book on viticulture (IV, 6). His treatment of soils in the general chapter is not very detailed, but he names the suitable soil for many of the tree and herb species he lists in the compendium.

About a century earlier, Ibn al-'Awwâm's *Kitab al-filaha*, written in Muslim Spain, offered a complete treatment of all aspects of agriculture. We know next to nothing about the author, who lived probably in Seville and wrote his elaborate

---

[64] Owen (1806a).

[65] Petrus de (1995–2000).

treatise in the middle of the 12th century. He used Arabic, Greek and Latin sources and wrote in Arabic.

Ibn al-'Awwâm's book offers a very detailed treatment of soils, which is compiled in the first chapter. The main determinants of quality to him are the capacity of a soil to warm up and to hold water. Black soils (here color is used in indicator function) are usually able to absorb heat. Red and yellow soils have a smaller heat absorbing capacity. White soil is cold and has no affinity to warmth. But Ibn al-'Awwâm knows that color can be deceptive, and he gives other qualities as well. The most desirable texture of a soil is that which looks similar to old and ready-to-use manure (probably compost), a soil which is not at all compact. Alluvial soil, a mixture of alluvium and sand, is second to the compost-lookalike in humidity and freshness (coolness). The coarser a soil, the drier it is, and a soil which is composed of gravelly sand without any clay or alluvium hence is the driest. Cohesion and hardness of the clod are main indicators for the dryness of a soil; hardness can reach that of stone. Soils containing a lot of alluvium are good soils for growing vegetables, unless the texture of the soil surface is gummy and viscous.

Experiments to determine soil quality, of which Ibn al-'Awwâm offers several, are best conducted at the onset of winter. Smell and taste, as well as the visual indicators given above, are the main criteria. Indicator plants, although they sometimes can be misleading, play an important role, too. The lushness of natural vegetation indicates water holding capacity, a main criterion for quality. Soil testing is done by mixing soil with water and tasting the liquid. If it is salty, it can only be used for palm plantations. Smelling the water can add to the information, as a badly smelling earth is not fertile. Quoting a writer named Sidagos, Ibn al-'Awwâm even presents a parallel experimental setting for the tasting. The most astounding and unique experiment prescribed by the author is a test aimed at soil biota. A sample of soil is placed in a tightly sealed clay pot and reburied where it was taken. After 1 week and again after 2 weeks, if the pot is not yet covered with humidity after the first period, the soil is inspected. If the insects in the pot are black, purple or green, the soil quality is mediocre. If they are yellow, red or grey, or of a brownish or light green or white color, the soil is of good quality. The smell of the sample is also taken into consideration: foul smell indicates bad soil.

Soils were treated elaborately in the Mediterranean, terracing, irrigation, crop rotation and green manuring were known and practiced. Columella even measures the quality of a farmer by his ability to procure a given amount of manure from his stables.[66] The prescriptions on manuring are manifold and comprise not only organic manures but also mineral soil amendments, in particular, marl.[67] Like medicine and astrology in Antiquity, agricultural knowledge was rooted in the system of four humors (humorism). The theory, systematized by Ancient Greek thinkers around 400 BC, is based on analogies between the four body fluids black

---

66 Columella (1983) While this is the best edition and commentary, for quick reference of an English translation: Columella (1941/1948).

67 Winiwarter and Blum (2008).

bile, yellow bile, phlegm, and blood, the four elements water, fire, earth and air, but also seasons, organs, celestial bodies and more. Paired qualities (cold/hot and wet/dry) were associated with each humor. Imbalances between the qualities were conceived as the reason for illnesses and moodiness in humans, but also for the lack of fertility of a soil leading to dietary recommendations, cures of ailments and to prescriptions of manures aimed at re-establishing the balance in the soil. Therefore, many agricultural recipes talk about situations where something is too cold, too hot, or too dry or wet. While some of these qualities correspond with physically measureable parameters of soils, not all do. While this theory has long been discarded, it lead to an astoundingly systematic discussion of soils in the agricultural literature.

The lesser-known *Geoponika* can serve as an example of the typical knowledge about manures: Its twenty-first chapter concerns manure: "Manure makes good land better, and it will be of greater service to that which is bad; but that which is naturally good does not want much manure; that which is of a middling quality, a little more; and that which is thin and weak, a great deal".[68] We further read that manure should be dispersed, as an over-abundance causes scorching. Unmanured land becomes stiff. Compost should not be thrown directly on the roots, but packed between two layers of earth. This will prevent the burning of the roots and, by covering the compost, prevent its heat from evaporating. Heat, one of the qualities discerned in humor theory, can be interpreted in modern terms as the energy contained in the molecules of fertilizing agents such as nitrates, which would indeed be better preserved under a protective layer of earth. Bird droppings, with the exception of that of water birds, are useful, because of their moisture. Pigeon dung is superior; therefore it is sometimes mixed with the seed when planted, without further preparation. In addition to rendering an impotent soil more powerful, pigeon dung can also eradicate agrostis, a grassy weed, which is mentioned throughout the work. The high appreciation of pigeon dung is shared by Latin writers and the Ancient Egyptians, but without the explanation that its value is in the moisture it brings. In fact, bird droppings are a good fertilizer because of their relatively high phosphorus content.

Like most other agricultural manuals, the *Geoponika* ranks manures according to their quality. Next to pigeons, human feces is listed, because it is destructive to weeds. Human feces, so we are told, is prepared in Arabia by drying it, then macerating it in water, and drying it again. This manure is used particularly for vines. Third best is donkey's dung, very fertile and of good use for any plant. Goat dung, ranked thereafter, is more pungent than sheep's dung, which comes next, followed by ox dung. Hogs dung is superior in quality, but by its instant heat burns grain fields, and hence is of no use for them. Horse and mule dung is inferior, but can be made useful by mixing it with other sorts. The ranking differs from that of Columella in valuing goat's and sheep's droppings higher than cattle excreta. Compost (that is, manure made from dung) should never be used fresh, as this will create many noxious animals (worms). The author(s) think that 3–4 year old

---

[68] Owen (1806b).

compost is best. Finally, the *Geoponika* links the moon's cycle to agricultural operations, again in accordance with Latin writers: manure should never be spread during the waxing of the moon, as this would lead to more weeds.

From the literature as well as from the archaeological record it is quite evident that techniques like terracing, irrigation or manuring were developed to a high standard in the Mediterranean world of Antiquity and that much of this knowledge and practice was carried on into the Middle Ages. Erosion did happen and was a problem, a problem to which solutions were not abundant. Terracing was widespread, but was costly and did not solve all erosion problems. While soil could be collected downhill and carried back up, there was no way to prevent massive erosion of cultivated landscapes which could come with only very few, but very effective catastrophic torrential rainfalls spread over the centuries. The histories of climate and soils are linked.

3.3.2.6 Europe North of the Alps

The practice of marling was, as Pliny the Elder reports in his encyclopedia, brought to the Roman Empire from the north of Europe. Peasant societies had evolved to some standard of fertilization techniques by the 1st century CE, manuring was, as detailed above, common from the Bronze age onwards, and so was the generation of anthrosols from adding grass sods to sandy soils, a technique leading to fertile 'Plaggen' soils. For a long period, until the 14th century, knowledge about soil management was not collected into the kind of manuals we know from Antiquity. One has to turn to the various medieval encyclopedias about nature to find such information. But besides orally transmitted local experiences the manuals of antiquity, in particular Palladius, were relatively widespread, and translations of the Latin texts began to appear as soon as the vernaculars became more important. Agricultural manuals exist in all major European languages. One of the best was written in German in the late 16th century, and it shall serve as the sole example of a library's worth of books on the subject. Johannes Coler, son of a Lutheran minister and a minister himself, was one of many churchmen who engaged in the development of agriculture. He was born in Silesia, brought up in Berlin and later had his own parish in Parchim in Brandenburg, in north-easter Germany. A learned man, well versed in the languages of the Mediterranean agricultural treatises, Coler was one of many early modern writers who adapted the knowledge produced for condition of the Mediterranean to the different soils and climates of Europe north of the Alps.[69]

The first part of Coler's Oeconomia (an agricultural calendar) was published in 1591, and its last parts were printed in 1606. During the 17th century alone this multi-volume compendium enjoyed 14 reprints. Coler often notes differences

[69] McDonald (1908/1968).

between his sources and his own observations due to different climate and natural conditions.

Coler talks about the suitability of land for grain and vines, referring to Columella and Virgil, and he recounts the entire list of soil quality tests already known in Antiquity: smell, digging of a hole and refilling it, mixing soil with water to test its stickiness, and taste by filtering a suspension. Clover and the existence of strong trees are Coler's vegetation indicators. As to color, he suggests to differentiate: Although color has been dismissed by authorities (he quotes Palladius, the most popular work from Antiquity), black color does indicate a fertile soil, unless it is boggy. A boggy soil is cold and tenacious, but the addition of horse-dung can improve it.[70]

The entire book seven is devoted to the cultivation of fields. There he gives several remedies for fields of a 'bad complexion', as he puts it, rooting him firmly in Galenic humoral pathology derived from the four-quality concept. But a growing variety of refuse from proto-industrial operations, workshops of button- and combmakers, tanners, spinners of wool, and soap boilers was added to the repertoire of soil amendments over time. To give but one example, in Coler's 7th book, the 29th chapter is devoted to the treatment of cold and wet fields. Marl, horse dung, fine sand, pulverized limestone, quicklime, saw mill waste, and coal-dust as found on places where charcoal piles are made, added and plowed under, make the field mellow and fat (*geil*). Workshops, waters, forests, meadows, waste- and other marginal lands were connected by way of animal excreta and the animal's draft power with the arable, whose sustained fertility was the prime concern. All agriculture was adapted to specific circumstances: fertilizers such as seaweed, small fish and shells added marine resources to the repertoire where the situation allowed and/or demanded.

Experts from the Netherlands travelled through late medieval and early modern Europe whenever large drainage works were planned, such as in the Oderbruch, near Berlin, to turn large marshes into fertile soils for peasants. The inhabitants of the Netherlands had learned to drain peats by the 11th century and had created fertile soils for agriculture. Crisscrossing the landscape with drainage ditches and draining peats with their high amount of organic matter had unintended side-effects. With waterlogging ceased, the organic parts of the soil would oxidize and mineralize and consequentially, shrink. Sinking peat bogs in a flat country close to the sea meant a growing danger of inundation, and dykes had to be built. As the land surface sank further below the level of the sea, needing constant pumping to stay useful, poldered landscapes with their drainage fuelled by the wind and later, fossil energy resulted. A second, very important use of peats was as fuel for a region with few woodlands (and those now cut down) and a growing economy with more and more urban population. This meant a new demand on soils. As everywhere in Europe, urban populations disconnected the nutrient flows which had in the agricultural countryside remained fairly tightly knit. Human excrement was often lost to

[70] Coler (1591).

soil cycling in urban areas. This was not only a profound intervention into nutrient cycles, but created a hygienic menace as well. Human feces, when discarded into open water used for drinking, can lead to transmission of several diseases such as typhoid fever. In Japan or China, nightsoil collection worked also in urbanized areas, leading to better hygiene and less nutrient loss. But with much more livestock on the land in Europe, human feces were commonly less economical there.

Concern about lost nutrients accompanied urbanization and predates soil science. The alchemist Johann Rudolf Glauber (1704) had already described a nitrate cycle. He based his reasoning on observations that saltpeter (nitre) could be obtained from the clearing of cattle sheds, and hence had to be the product of the digestion of vegetable matter, as it could not be a product of the animals themselves.

While in a process of trial and error, guesswork and experiment, with many wrong alleys taken but many advances, too, an understanding of chemical elements, their cycles and the intricacies of nutrient uptake by plants developed from the 17th century onwards, having a first culmination in Liebig's agricultural chemistry. Land-use, however, stayed traditional: Cultivation practices had to be adapted to the soil, in an agro-economy which was at the same time an agro-ecology.[71]

Erosion and measures against it were a concern on slopes and valleys in Northern Europe, too. Terracing, especially for vineyards, continues to shape some of the most renowned cultural landscapes. But when one third of the total erosion of the past 1500 years in the middle part of Germany happened within one week in July, 1342, due to excessive rainfall and subsequent floods, when gullies were cut into landscapes, roads and infrastructure buried under precious topsoil washed away by the current, medieval Germans had no remedies.[72] Subsequent centuries brought an increase in floodwater protection measures, but floods and the erosion they bring will always be part of human life, which has been concentrated in the rich alluvial landscapes along rivers and coasts. Humans are, where soil is, and although they do not appreciate the dynamic of riverine ecosystems, their life does in part depend on the fresh soils brought by the rivers.

3.3.3 Soils Over the Long Course of Human History: Sustainability and Ecological Inheritance

Rather than discussing soils in industrialized societies as a series of case studies, a long-term view will place the role of soils in industrialized societies in perspective. Whoever visits the rice paddies of China, the terraces of the Rhone valley vineyards, the poldered fields of the Netherlands or the 'Danger, No Entrance' signs around a patch of polluted soil, faces an 'ecological heritage'. This concept has been elaborated by evolutionary biologists Laland, Odling-Smee, and Feldman in

71 Krausmann (2004).

72 Bork et al. (1998).

1999. The ways of living typical for a species can include lasting changes to the environment. Such changes are effective longer than the lifespan of the generation undertaking the changes. An ecological inheritance is the result. Dens and burrows are a good example for the heritable parts of a niche. Such constructed niches can be quite permanent structures, used (and changed) by several generations of inhabitants. Progeny inherit a changed lifeworld, which then acts as a selective force for future generations. Not only the environmental conditions as such, but also an ecological inheritance (e.g. the burrow) are a means of natural selection.

Humans construct their ecological niche by building their version of dens, houses, and by permanently and semi-permanently altering natural systems. The lasting changes they make act as means of natural selection on them. Soils left from one generation of humans to the next can also bear lasting, discernible marks of previous cultivation, leaving a particular ecological inheritance.

The first such change is due to the widespread and extensive use of fire to alter vegetation, because vegetation cover is one of the main forces relevant in pedogenesis. Soil formation was changed when fires in the post-glacial steppes were set at times when they might not have occurred naturally, and at human-induced frequencies.

The second and perhaps most outstanding change is the development of agriculture, with smaller distinctions to be made for the shift first from hoe to plow, and then to the moldboard plow. The most important factor changing pedogenesis in agriculture is again the change of vegetation, in particular the invention of monoculture. The uppermost layer of soils was also changed mechanically by plows. In addition to changing the mechanical properties of soils, influencing root growth and water holding capacity, plowing changed the micro-habitats of soil organisms. The use of fallowing is somewhat similar to plowing in its effects. Fallowing in addition changes the water balance of soils, as does the change from fallowing to a rotation system, which does also impact soil biota.

Manuring is yet another important intervention into soil ecosystems, bringing in food for soil organisms. The organisms destroy the larger remains of vegetation and residues of animal metabolism into smaller organic and inorganic compounds, ideally into a chemical form (as ions) which can be taken up by plant roots as nutrients. Soil organism turnover is speeded up by the extra food available, and the ecological niche of the manured soil is made up by organisms in different relative abundances.

Marling or liming changes the acidity of soils and therefore again intervenes into soil biota communities, shifting the balance from bacteria to fungi. Finally, irrigation techniques need to be mentioned. Besides the fundamental change in water availability through irrigation, the chemical balance of the soil is changed. Traditional surface and ground water sources contain dissolved minerals that remain at the soil surface when the water evaporates or is transpired by plants, salinization results. The innovation of flooding irrigated fields to cleanse the soil of accumulated salts marks an early change, as does the 20th century invention of drip irrigation which reduces the amount of water needed by placing it directly at the plant.

Vegetation cover change, mechanical changes in the uppermost soil layer, changes in water retention capacity and changes in soil biota due to practices such as manuring and marling taken together are the most profound change in soil architectures which happened in the Holocene. This is linked to reverberations in human societies: sedentary lifestyle, higher population densities, and the development of stratified societies with uneven distribution of surplus all are in one way or other connected to soil changes.

Interventions into the nitrogen balance of soils are of particular importance, because they change the linkage between soils and atmosphere. Earlier nitrogen sources were limited to naturally existing oxidized (as nitrate) or at least chemically bound (as ammonia) forms. Airborne nitrogen became available only through the deposition of nitrous oxides produced by lightning in thunderstorms in minute quantities. The first such intervention was the use of leguminous plants (nitrogen fixing species) as domesticated plants; their ability to bind nitrogen from the air, thus greatly enlarges the nitrogen pool for cycling. Legumes provide nutrient-rich food for humans and their livestock; beans, peas and lentils are common around the world. A major innovation was the much larger influx of oxidized airborne nitrogen produced by the deliberate use of nitrogen fixing species as green manure, by plowing under the entire plant.

To give but one example of the historical significance of this technique, the cultivation of nitrogen-fixing clover in England dates from the early 17th century. Although data on the chronology and location of their introduction is patchy we know for example that by the 1740s about half the farmers in Norfolk and Suffolk were growing turnips and about a quarter had clover on their farms.[73] It seems likely that it was not until after the mid-eighteenth century that these crops were having much effect on cereal yields. They were important in enabling some light lands, like the chalk downlands of southern England and parts of Norfolk, to be brought under the plow for the first time. A third of the increase in arable productivity in northern Europe between 1750 and 1850 has been attributed to legumes such as clover.[74]

Nitrates were created in the soil by nitrogen-fixing bacteria living in symbiosis with leguminous roots. The largest reservoir of nitrogen, the air, could however not be tapped directly before the invention of the catalytic reaction of nitrogen and hydrogen gas, developed into an industrial process called the Haber-Bosch-Synthesis, patented in 1910. If we look at human interventions into the nitrogen balance in soils, the invention of legumes as domesticated plants, the invention of green manure and the invention of an air-based synthesis of nitrogen are the main turning points.

As to the two other key nutrients necessary for plant growth, the restoration of potassium by means of wood or other plant ashes was developed in antiquity, whereas the mining of potassium-containing minerals as well as the mining of

[73] Overton (1984).

[74] Chorley (1981).

phosphates is a late 19th–early 20th century invention, only possible by means of fossil energy.

Anthrosols, termed so for the reason that they bear more than just the mark of human intervention, because they are essentially man-made, can be found in gardens (hortisols) and fields (e.g. plaggen soils). The most astounding example of an anthrosol is that of the terra preta and terra mulata patches in the Amazon basin.

Aside from fertilizer, available through mining and the Haber-Bosch synthesis, the invention of potent pesticides marks a decisive event in the environmental history of soils. Their massive impact on soil biota long went largely unnoticed because many soil organisms have never been identified, let alone their changes due to chemical interventions monitored.

The widespread changes of agricultural practice according to the theory of mineral plant nutrition as elaborated by Liebig in the 19th century are probably the most important contribution of chemistry to the use of soils. Plants which do not respond well to the availability of mineral nutrients in large quantities are no longer used as cultivars and breeding was and is done to enhance nutrient uptake. One can argue that agricultural practice was changed in order to match the simplified theory that inorganic nutrients are the single most important factor in soil fertility.

All the described changes correspond with an increase in the intensity of human impact on soils, fueled by the abundance of fossil energy. Due to the abundance of energy, humanity can currently afford not to adapt their cultivation to the soil, but rather adapt soils to their necessities, with consequences such as farm-machinery induced compaction, pollution and erosion. Hydroponics, the cultivation of plants in greenhouses with nutrient solutions rather than on soils, finally detaches plants from the soil, but at high energy cost. Massive colonizing interventions into plant life are counterbalanced by a decrease in impact on soils. As so often in environmental history, this is a mixed blessing at best.

3.4 Concluding Remarks

In environmental history and beyond, monocausal explanations are more often wrong than right. By identifying soil degradation as the culprit for the collapse of agricultural empires, one tends to oversimplify stories in which sustainable use of soils plays a very important part. But not all agriculture was sustainable. In particular colonization of new territories, which brought with it the loss of acquired knowledge about soils, could bring dramatic consequences. Pioneers were often harbingers of soil destruction.

The abrupt failure of the control system for the Yellow River in 1855 and catastrophic displacement of its mouth by 400 km,[75] the Dust Bowl phenomenon of the U.S. Great Plains during the 1930s, with dust storms blackening the skies and

[75] Dodgen (1991).

tens of thousands of people as environmental refugees, and the dramatic erosion in Haiti today are just some examples of recent and ongoing unsustainable use of soils which has been documented in many soil science and agronomy papers and monographs over the past decades.

Feeding 9 billion people is a challenge for the soils of the world, although much of this challenge is politically rather than naturally induced. Poverty and lack of education are damaging to soils, as is a high-input lifestyle based on the excessive use of fossil energy.

The recent (2008) attempt to write a blueprint for the development of sustainable agriculture (International Assessment of Agricultural Knowledge, Science and Technology for Development[76]) calls for sustainable use of agriculture's natural resources, most importantly of soils and water, but the ideas how to reach this goal differ widely, from more implementation of traditional knowledge to increased use and development of GM crops. Some of the stakeholders left the process because visions about a sustainable future could not be reconciled. Such a vision would benefit from inclusion of historical data.

What Peter Lindert wrote about China and Indonesia in 2000 still holds true: "The study of trends in soil quality needs the kind of quantitative history that has so enriched climatology and geology. Virtually all estimates of current soil trends, are in fact, not data. Rather they are expert's predictions, derived by combining data on slope, climate, and land use with what happens to such soils under experimental conditions. Sometimes they are refined into "expert opinion" as in the GLASOD map, but they are still not based on any observation before the mid-1980s".[77] Unlike other fields of environmental history, the environmental history of soils is still in its infancy. Providing long-term data on sustainable and unsustainable use of soils in the past is a daunting task for environmental historians for the next years and decades.

Acknowledgments The empirical work for this contribution was compiled during an APART fellowship of the Austrian Academy of Sciences (2003–2006). Herwig Weigl and Richard Hoffmann helped improve this text with their critical remarks. Anita Hipfinger assisted with formatting and references, Andrea Bottanova produced the newly formatted footnotes.

References

Altieri MA (1999) The ecological role of biodiversity in agroecosystems. Agric Ecosyst Environ 74(19–31):22

Arnold RW, Szabolcs I, Targulian VO (1990) Global soil change. Institute for Applied Systems Analysis, Laxenburg

Bakels CC (1997) The beginnings of manuring in western Europe. Antiquity 71:442–445

[76] International Assessment of Agricultural Knowledge, Science and Technology for Development (IAASTD) (2014).

[77] Lindert (2000b).

Beach T, Luzzadder-Beach S, Dunning N (2006) Soils and history in Mesoamerica and the Caribbean. In: McNeill JR, Winiwarter V (eds) Soils and societies: perspectives from environmental history. The White Horse Press, Strond, pp 51–90

Bennett HH, Chapline WR (1928) Soil erosion a national menace. (U.S. Department of Agriculture Circular No. 33). Government Printing Office, Washington DC

Berthelin J, Babel U, Toutain F (2006) History of soil biology. In: Benno PW (ed) Footprints in the soil: people and ideas in soil history. Elsevier B.V, The Netherlands, pp 279–306

Birmingham DM (2003) Local knowledge of soils: the case of contrast in Côte d'Ivoire. Geoderma 111(3–4):481–502. http://www.sciencedirect.com/science/article/B6V67-4778DCC-1/2/d91b7d603b6e36f26fe046ab9fb4073d. Accessed 17 Jan 2014

Blaser P (2004) Field techniques: soil systems. In: Achim S (ed) Environmental systems. In: Encyclopedia of Life Support Systems (EOLSS). Eolss Publishers, Oxford. http://www.eolss.net. Accessed 17 Jan 2014

Blum WEH (2008) Threats to soil quality in europe. In: Gergely T, Luca M, Ezio R (eds) JRC scientific and technical reports. Publications Office of the European Union, Luxembourg, pp 5–10:5f. http://eusoils.jrc.ec.europa.eu/ESDB_Archive/eusoils_docs/other/EUR23438.pdf. Accessed 18 Jan 2014

Boardman J (2006) Soil erosion science: reflections on the limitations of current approaches. CATENA 68(2–3):73–86

Bork H-R (2006) Landschaften der Erde unter dem Einfluss des Menschen. Wissenschaftliche Buchgesellschaft, Darmstadt

Bork H-R, Bork H, Dalchow C et al (1998) Landschaftsentwicklung in Mitteleuropa: Wirkungen des Menschen auf Landschaften. Klett-Perthes, Gotha

Bork H-R, Mieth A (2006) The dynamics of soil, landscape and culture on Easter Island (Chile). In: McNeill JR, Winiwarter V (eds) Soils and societies: perspectives from environmental history. The White Horse Press, Strond, pp 273–321

Bowman AK, Rogan E (1999) In: Proceedings of the British Academy 96: Agriculture in Egypt, From Pharaonic to Modern Times. Oxford University Press, Oxford

Bray F, Needham J (1984) Science and civilisation in China. Agriculture 6 (2):724 (Cambridge University Press,Cambridge)

Chorley P (1981) The agricultural revolution in northern europe, 1750–1880: nitrogen, legumes, and crop productivity. Econ Hist Rev 34(1):71–93

Cogo NP, Levien R (2006) Erosion and productivity: effects on human life. In: Encyclopedia of soil science, vol 1. Taylor and Francis, Florida, pp 540–543

Coleman DC, Crossley DA, Hendrix PF ^{2}(2004) Fundamentals of soil ecology. Academic Press, Burlington, p 11

Coler J (1591) Calendarium oeconomicum and perpetuum: vor d. Haußwirt, Ackersleut, Apotecker und andere gemeine Handwercksleut, Kauffleut, Wanderssleut, Weinherrn, Gertner und alle diejenige so mit Wirtschafft umbgehen. Wittenberg

Columella LIM (1941/1948) On agriculture (trans: Boyd AH) Harvard University Press/The Loeb Classical Library, Cambridge, MA

Columella LIM (1982a) De re rustica. In: Richter W (ed) Zwölf Bücher über Landwirtschaft 2. Artemis, München, pp 18–19

Columella LIM (1982b) De re rustica. In: Richter W (ed) Zwölf Bücher über Landwirtschaft 2. Artemis, München, p 20

Columella, LIM (1982c) De re rustica. In: Richter W (ed) Zwölf Bücher über Landwirtschaft 2. Artemis, München, p 18

Columella LIM (1983) De re rustica. In: Richter W (ed) Zwölf Bücher über Landwirtschaft 3 II, 14, 4. Artemis, München

Cooper RC (1977) Agriculture in Egypt, 640–1800. In: Spuler B (ed) Handbuch der Orientalistik: Geschichte der Islamischen Länder 6: Wirtschaftsgeschichte des vorderen Orients in islamischer Zeit. Brill, Leiden, Köln, pp 188–204

Darwin C (1883) The formation of vegetable mould, through the action of worms. John Murray, London, p 316

De Deyn GB, Van Ruijven J (2005) The role of above- and belowground linkages in ecosystem functioning. In: Wilhelm B, Eduard KL, Stefan P (eds) Biodiversity: structure and function. In: Encyclopedia of Life Support Systems (EOLSS). Eolss Publishers, Oxford. http://www.eolss.net. Accessed 17 Jan 2014

Dodgen RA (1991) Hydraulic evolution and dynastic decline: The Yellow River conservancy, 1796–1855. Late Imperial China 12(2):36–63

Einwögerer T (2005) Die Auffindung einer jungpaläolithischen Säuglings-Doppelbestattung im Zuge neuerer Ausgrabungen am Wachtberg in Krems, NÖ. Das Waldviertel 54(4):399–404 (Elsevier B.V, The Netherlands)

Eldor PA [3](2007) Soil microbiology, ecology and biochemistry. Academic Press, Amsterdam, pp 5–9

Evtuhov C (2006) The roots of Dokuchaev's scientific contributions: cadastral soil mapping and agro-environmental issues. In: Benno PW (ed) Footprints in the soil: people and ideas in soil history. Elsevier B.V, The Netherlands, pp 125–148

Fairhead J, Leach M (1996) Misreading the African landscape: society and ecology in a Forest-Savanna Mosaic (African Studies Series 90). Cambridge University Press, Cambridge

Feller C, Blanchart E, Yaalon D (2006) Early scientists describing soil profiles. In: In: Benno PW (ed) Footprints in the soil: people and ideas in soil history. Elsevier B.V, The Netherlands, pp 85–105

Flanagan DC (2006) Erosion. In: Encyclopedia of soil science, vol 1. Taylor and Francis, Florida, pp 523–526

Gisi U, Schenker R, Schulin R, Stadelman FX, Sticher H (1997) Bodenoekologie. Thieme, New York, p 237

Glauber JR (1704) Deß Teutschlandts-Wolfahrt. Caspar Wussin, Prag

International Assessment of Agricultural Knowledge, Science and Technology for Development (IAASTD) (2014) Executive summary of the synthesis report. Island Press, Washington, DC. http://www.agassessment.org/docs/IAASTD_EXEC_SUMMARY_JAN_2008.pdf. Accessed 17 Jan 2014

Jenny Hans (1941) Factors of soil formation. McGraw Hill, New York

Juergens N (2006) Combating degradation in arid systems. In: Barthlott W, Linsenmair E, Porembski S (eds) Biodiversity: structure and function. In: Encyclopedia of Life Support Systems (EOLSS). Eolss Publishers, Oxford. http://www.eolss.net. Accessed 18 Jan 2014

Koehler HH (2005) Application of ecological knowledge to habitat Restoration. In: Wilhelm BK, Eduard L, Stefan P (eds) Biodiversity: structure and function. In: Encyclopedia of Life Support Systems (EOLSS). Eolss Publishers, Oxford. http://www.eolss.net. Accessed 18 Jan 2014

Krausmann F (2004) Milk, manure, muscle power. livestock and the transformation of preindustrial agriculture in Central Europe. Hum Ecol 32:735–772

Kuntze H, Roeschmann G, Schwerdtfeger G (1994) Bodenkunde. Ulmer, Stuttgart, p 202

Lehmann J, Dirse CK, Bruno G, William WI (2003) Amazonian dark earths: origin, properties, management. Kluwer Academic Publishers, Dordrecht

Lightfoot D (1997) The nature, history and distribution of lithic mulch agriculture: an ancient technique of dryland agriculture. Agric Hist Rev 44(2):206–222

Lindert PH (2000a) Shifting ground: the changing agricultural soils of China and Indonesia. MIT Press, Cambridge, MA

Lindert PH (2000b) Shifting ground: the changing agricultural soils of China and Indonesia, 43 f. MIT Press, Cambridge, MA

Lysenko EG (2004) Interactions: food and agricultures/environment. In: Lysenko EG (ed) Environmental and ecological sciences, engineering and technology resources. In: Encyclopedia of Life Support Systems (EOLSS). Eolss Publishers, Oxford. http://www.eolss.net. Accessed 18 Jan 2014

McDonald D (1908/1968) Agricultural Writers, from Sir Walter of Henley to Arthur Young, 1200–1800. Originally published in London. Burt, Reprinted 1968 in New York (for the English tradition)

McNeill JR, Winiwarter V (2004) Breaking the sod: humankind history and soil. Science 304:1627–1629

McNeill JR, Winiwarter V (2006) Soils and societies: perspectives from environmental history. The White Horse Press, Strond

Montgomery DR (2007) Dirt. The Erosion of Civilizations. University of California Press, Berkeley

Muoghalu JI (2003) Priority parameters: abiotic and biotic components. In: Hilary II, John LD (eds) Environmental monitoring. In: Encyclopedia of Life Support Systems (EOLSS). Eolss Publishers, Oxford. http://www.eolss.net. Accessed 17 Jan 2014

Muscolino MS (2008) The yellow croaker war: fishery disputes between China and Japan, 1925–1935. Environ Hist 13:306–324

Needham J, Lu G-D, Tsien T-H et al (1984) Science and civilisation in China. Botany 6(1):47–117 (Cambridge University Press, Cambridge)

Needham J, Sivin N, Lu G-D (1986) Science and civilization in China. Biology and biological technology 6(6):102–103 (Cambridge University Press, Cambridge)

Nortcliff S (2009) The soil: nature, sustainable use, management, and protection—an overview. GAIA 18(1):58–68

Orgel LE (1998) The origin of life—a review of facts and speculations. Trends Biochem Sci 23 (12):491–495, 493

Overton M (1984) Agricultural revolution? Development of the agrarian economy in early modern England. In: Baker A, Gregory D (eds) Explorations in historical geography: some interpretive essays. Cambridge University Press, Cambridge, pp 118–139

Owen T (Trans.) (1806a) Geoponika. Agricultural Pursuits 1. White, London, p 54

Owen T (Trans.) (1806b) Geoponika. Agricultural Pursuits 1. White, London, p 67

Palladius RTA (1976a) Opus Agriculturae I, II. In: Martin R (ed) I, V, 3. Paris

Palladius RTA (1976b) Opus Agriculturae I. In: Martin R (ed) 3. Paris

Patzel N, Sticher H, Karlen D (2000) Soil fertility—phenomenon and concept. J Plant Nutr Soil ScI 163:129–142

Petrus de C (1995–2000) Ruralia Commoda: das Wissen des vollkommenen Landwirts um 1300. In: Richter W (ed) Editiones Heidelbergensis II. C. Winter, Heidelberg, p 26

Pidwirny M, Heimsath H (2008) Soil. In: Cutler JC (ed) Encyclopedia of earth, Washington, DC, Environmental Information Coalition, National Council for Science and the Environment. http://www.eoearth.org/article/Soil. Accessed 17 Jan 2014

Pimentel D, Lach L, Zuniga R, Morrison D (1995) Environmental and economic costs of soil erosion and conservation benefits. Science 267:1117–1123

Plinius Secundus (the elder) G (1994a) Naturalis Historiae, Liber XVII (ed and trans: König R, Hopp J) Artemis and Winkler, München, p 39

Plinius Secundus (the elder) G (1994b) Naturalis Historiae, Liber XVII (ed and trans: König R, Hopp J) Artemis and Winkler, München, p 27

Plinius Secundus (the elder) G (1995) Naturalis Historiae, Liber XVIII (ed and trans: König R, Hopp J, Glöckner W). Artemis and Winkler, München, p 34

Reardon-Anderson J (2005) Reluctant pioneers: China's expansion Northward, 1644–1937. Stanford University Press, Stanford

Reij C, Scoones I, Toulmin C (1996) Sustaining the soil: indigenous soil and water conservation in Africa. Earthscan Publications Ltd, London, pp 1–27

Reintam L, Lang V (1999) The progress of pedogenesis within areas of prehistoric agriculture. Pact 57(3):415–431

Rodgers R (2002) Kepopoiia: garden making and garden culture in the Geoponica. In: Littlewood A, Maguire H, Wolschke-Bulmahn J (eds) Byzantine garden culture. Dumbarton Oaks Research Library, Washington. http://www.doaks.org/ByzGarden/ByzGarch8.pdf. Accessed 17 Jan 2014

Rostovtzeff MI (1922) A large estate in Egypt in the third century B.C. a study in economic history. University of Wisconsin Studies, Madison

Sandor JA (2006) Ancient agricultural terraces and soils. In: Warkentin BP (ed) Footprints in the soil: people and ideas in soil history. Elsevier B.V, The Netherlands, pp 505–534
Showers KB (2005) Imperial gullies: soil erosion and conservation in Lesotho. Ohio University Press, Athens
Showers KB (2006) Soil erosion and conservation: an international history and a cautionary tale. In: Benno PW (ed) Footprints in the soil: people and ideas in soil history. Elsevier B.V, The Netherlands, pp 369–406
Varro MT (1996–1997) (ed) Res Rusticae (trans: Dieter Flach) (Gespräche über die Landwirtschaft I, 9), pp 2–3. Wissenschaftliche Buchgesellschaft, Darmstadt
Vergilius MP [5](1987a) Georgica. In: Götte J, Götte M (eds) Gespräche über die Landwirtschaft II. Tusculum, Munich, pp 226–237
Vergilius MP [5](1987b) Georgica. In: Götte J, Götte M (eds) Gespräche über die Landwirtschaft II. Tusculum, Munich, pp 238–247
Vergilius MP [5](1987c) Georgica. In: Götte J, Götte M (eds) Gespräche über die Landwirtschaft II. Tusculum, Munich, p 250
Wasson RJ (2006) Human interaction with soil-sediment systems in Australia. In: McNeill JR, Winiwarter, V (eds) Soils and societies: perspectives from environmental history. The White Horse Press, Strond, 243–272
Whitmore TM, Turner BL (2001) Cultivated landscapes of middle America on the eve of conquest. Oxford University Press, New York
Williams BJ (2006) Aztec soil knowledge: classes, management, and ecology. In: Warkentin BP (ed) Footprints in the soil: people and ideas in soil history. Elsevier B.V, The Netherlands, pp 17–41
Winiwarter V (2006a) Prähistorischer Umgang mit den Böden. In: Blume H-P , Felix-Henningsen P, Fischer WR et al (eds) Handbuch der Bodenkunde. Loseblattwerk in 3 Ordnern. Ecomed, Heidelberg, München, Berlin, pp 1–6
Winiwarter V (2006b) Prolegomena to a history of soil knowledge in Europe. In: McNeill JR, Winiwarter V (eds) Soils and societies: perspectives from environmental history. The White Horse Press, Strond, pp 177–215
Winiwarter V (2006c) Medieval and early modern soil indicators. Mitteilungen der Österreichischen Bodenkundlichen Gesellschaft 73:21–30
Winiwarter V, Blum WEH (2008) From Marl to Rock Powder: on the history of soil fertility management by rock materials. J Plant Nutr Soil Sci 171(3):316–324
Winiwarter V, Blum WEH (2009) Religious aspects of soil-human relationships. In: Rattan L et al Encyclopedia of soil science. http://www.tandfonline.com/doi/abs/10.1081/E-ESS-120044721#.UtgObPvZfAk. Accessed 16 Jan 2014
Winiwarter V, Gerzabek M, Baumgarten A et al (2012) The challenge of sustaining soils: Natural and social ramifications of biomass production in a changing world. (Interdisciplinary Perspectives 1). Verlag der Österreichischen Akademie der Wissenschaften, Wien, pp 12–47
Xenophon (1992) Oikonomikós. In: Ökonomische schriften (trans: Gert Audring) (Schriften und Quellen der Alten Welt 16, 2). Akademie-Verlag, Berlin

Chapter 4
Environmental History of Water Resources

Stéphane Frioux

Abstract Water resources, indispensable to life, are relatively limited with respect to the Earth surface and unequally distributed in the humanized spaces. They have been targeted by many different human actions through the centuries, which makes environmental history of water a well represented approach, among environmental history in general. In addition to its use for drinking and cleaning, water has contributed to the development of an increasingly productive agriculture with irrigation; and for centuries, rivers and canals have also played a major role in facilitating the traffic of people and merchandise and have structured the economics of many cities. Historians have emphasized in their researches the changes occurred during the Industrial Era (ca. 1780–1960 in the West). This period was crucial for the use of water resources because of the competition between different activities needing water: washing, factories needing pure water, factories using rivers as a receptacle for their wastes, professional or leisure fishing, etc. With the decrease of industrial discharges and boat traffic, in the last decades of the twentieth century (and mainly in the Western countries), water has been integrated in many urban development programs as a source of leisure and of aesthetic value. Restoration projects allowed for the return of salmon and other fish species, but history raises the question of how to restore. After having played an important role in culture and leisure, water resources remain an object of public policies that nowadays defend a very different goal from the post-WW2 modernist ideologies which led to vast dam and irrigation schemes. Nevertheless, great disparities still exist on a global scale. Climate change, making "wars for water" a potential threat for the next century, can be used by environmental historians to find lessons from the past which could be provided to policy-makers.

S. Frioux (✉)
Laboratoire de Recherche Historique Rhône-Alpes (LARHRA),
Université de Lyon, Lyon, France
e-mail: stephane.frioux@univ-lyon2.fr

M. Agnoletti and S. Neri Serneri (eds.), *The Basic Environmental History*,
Environmental History 4, DOI 10.1007/978-3-319-09180-8_4

4.1 Introduction

No water, no life. Nearly every environmental historian encounters water-related issues, at one stage or another of his or her research. Such issues have even been at the core of proto-environmental history essays.[1] The statistics indeed aim at being aware that water is almost everywhere on Earth, the "Blue Planet": 70 % of the surface of our planet is covered by water, of which 97 % are in oceans and 2 % in ice (this last proportion has been decreasing for the last few decades). The concept of "Water resources" implies focusing on the relationships between water and the human societies and questioning the various uses of the different types of resources (sea, streams, ponds, groundwater, ice). Of course, one has to keep in mind that people do not have the same relationships with water, if they live in Greenland, in the United States or in Sahara. For environmental historians, the relationships between societies and their "waterscapes" are crucial not only because water is a necessity for life, but also because the dramatic demographic, cultural, economic, social, spatial and political changes since the late eighteenth century had many consequences on water resources.

The uses of water resources throughout history can be classified into five major categories: *humankind vital needs* and comfort (including personal hygiene and cleanliness of the daily environment); *agricultural use* for cattle and for irrigation; *industrial use* as a component of the fabrication process or as a source of energy; *means of transportation* of goods and people; lastly, the *aesthetic or cultural dimension* of waterscapes. Besides, we need to take into account the potential hazards carried by water or climate (epidemics, flooding, landslides, and drought). Even if, during the last century, humankind has altered the biosphere as never before, it is quite evident that the great water manipulation efforts pursued by various societies, engineers and political leaders have produced significant imbalances in the relationships between local societies and their waterscapes. These production of new socio-technical environments, reshaped by vital, political, economic is at the core of water resources' environmental history.

Water has always been a topic of interest in environmental history, whether approached by the disappearance of fish in rivers, the history of resources conservation or the pollution by urban wastewater and industrial effluents.[2] It is now recognized as a major research subject, as illustrated by the founding of an International Water History Association (2001),[3] and very recently by a new scientific and interdisciplinary journal, *Water History* (July 2009).

The present essay has no claim to be exhaustive about such a broad topic, around which many fields of study meet. Rather than trying to cover all aspects of water

---

[1] Reclus (1869). About the "The Rivers of America" series: see Mink (2006).

[2] See respectively: Carson (1962), Hays (1959) and Tarr (1996). For reviews about the urban environment: Dieter (2004). A case-study about Italy: Neri Serneri (2007).

[3] Review articles of one of its international conferences can be found in a special issue of *Environment and History,* May 2010.

environmental history, it will provide some elements of synthesis, and some highlights carefully chosen to exemplify a number of issues. I will explore the various facets of the humanity/water resources relationships, by studying on a very long term the main usages of water, even if my attention will be drawn upon the nineteenth and twentieth centuries, in which a hugely increasing demand and use of water transformed the daily life and the environment of millions of people.

4.2 Narratives of Water Resources' "Benefits"

4.2.1 Water Between Environment and Health

Human communities—and especially cities—had always been dependent on basic vital needs, and the problem of food and water supply is one of the most crucial for urban authorities. Getting enough water for a growing population was a major challenge, inspiring sometimes ingenious solutions. For ancient Middle East, archaeology and environmental sciences discussed the so-called Mesopotamian model and some case studies emphasized the idea of an initial development of local hydraulic communities, able to produce a social regulation system of water resources, even without a centralised state and bureaucracy.[4] During the Roman period, cities acquired most public facilities related to water control and management: the aqueduct/fountain couple became a standard pattern of the Roman city-planning.[5] Water towers (*castella*), networks of pipes (*fistulae*), public taps (*salientes*) represented expensive equipment that was firstly financed by public funds, then by rich people like Agrippa, the son-in-law of Augustus. The Augustean period marks the generalization of a water management policy: beyond technical progress and the diffusion of hydraulic monuments, new administrative structures were set up, and the distribution of water carried by aqueducts to domestic buildings broadened from aristocratic houses to "middle-class" *domus*.

For centuries, water was generally not delivered to urban houses (or provided to a very small percentage of buildings), while the locations of rural settlings were often chosen because of the existence of a spring, a pond, or the digging of a well, indispensable to satisfy the vital needs of the human beings as well as domestic animals and plants. A few cities had aqueducts, the best example being Republican and Imperial Rome; in other cases, urban dwellers relied upon hundreds of thousands of private wells or sometimes cisterns gathering rainwater, or a combination of those systems, like in Jerusalem. The richest citizens could afford the service of water haulers. It is only very recently, compared to the long-standing nature of the urban phenomenon, that people have been able to turn on a tap to get the precious

[4] Wittfogel (1957). For a recent discussion, Special Issue: Ancient Near East and Americas 2010. Water History 2/2.

[5] Bruun (1991).

liquid. Often, it has been possible through a change in the supply system, involving the use of surface water from lakes or rivers, or groundwater related to the river. In a country like France, after the Second World War, an inquiry stated that in 38,014 "*communes*" (the smallest administrative division), only approximately one third had functioning water supply systems.[6] In many rural districts, villages and scattered farms were linked to a public water system only in the late 1960s. Before that time, each farm relied on harnessed springs or on its own well. When studying the history of the relationship between humanity and water resources for the last 150 years, it is necessary to deal with issues of scarcity and water allocation. The modern period, with its demographic increase (from less than 2 billion people by 1910 to approximately 6.8 billion one century later) brought a twofold problem: the inadequate provision of water and the purification of water once it had been used.

To quench thirst, water has always been in competition with other—tasty—beverages, since the Ancient times: beer, wine, various types of alcohol, and nowadays sodas. Was it only a question of taste, or a question of accessibility and/or conservation of safe water? Waterborne diseases, like cholera or typhoid fever, struck Western cities until the first half of the twentieth century (for instance, the population of Hamburg was decimated by a cholera epidemics in 1892; Russian and Italian cities suffered from the same disease as late as in 1910–1911). Nowadays, waterborne illnesses are still considered as the second highest cause of mortality. Thus, the World Health Organization continues to foster the fight against the consumption of polluted water. Today, approximately two billion people still do not have any domestic access to public water (and much more cannot drink bacteriological safe water). While some Western suburban neighborhoods consume more than 400 l per capita per day, millions of African or Asian inhabitants live with less than 10 l per capita per day. In Latin America, a continent where "Mega-Cities" have developed since the 1980s, a substantial proportion of the citizens of Mexico City and São Paulo do not have access to the public water supply network. Some inter basin water transfers, initially planned to satisfy agricultural irrigation have now become a vital source of water for agglomerations like Los Angeles and Phoenix (Colorado watershed) or Johannesburg and its surrounding area in South Africa.

Water resources are not only useful for the consumption of food and beverages; at the other end of the domestic cycle, they serve to carry away all wastes produced by human bodies and by cooking and housekeeping. Environmental history is concerned with was Joel Tarr has called the "search for the ultimate sink".[7] At least since the Roman Times, urban authorities had to think about a water-carriage system to deal with sewage problems. But for centuries, the problem has been difficult to solve in many cities: either the canalizations were too small, or water was insufficient to flush the sewers. Sometimes, there were even no underground pipes, but only small gutters or narrow drains between buildings; wastes stagnated

6 "Distribution d'eau et assainissement",*L'Eau*, January 1946, p. 4.

7 Tarr, Joel. *The search for the ultimate sink*, op.cit.

until a rainfall flushed them, and in summer time, the accumulation of organic wastes produced foul smells to which some urban dwellers attributed diseases. Jorgensen studied sanitation practices in late medieval England.[8] Rivers were long seen as a convenient public good to dump wastes at will; even if some regulations were established, like the forbidding of throwing wastes during the daytime, nightly discharges did not suppress water pollution.

4.2.2 Water Resources, Agriculture, and Food

Mapping the first great civilizations rapidly provides an interesting element of their geography and of their hydraulic environment. They generally correspond to river basins: Indus River and Ganges in India (Indus Valley Civilization: 2500–1900 BC), Tiger and Euphrates in Mesopotamia, Nile in Egypt. The seasonal floods of these big rivers offered agriculture the opportunity to take advantage of very fertile soils, thanks to the thin silt layer dropped on the river banks. In the twentieth century, human societies, encouraged by their political leaders and by engineers, have tried to gain independence from natural cycles. Dams—which will be treated later on—were conceived for this purpose and allowed two or three annual crops. But their disadvantages consisted in a frequent salinization of soils, enhancing a loss of productivity. What was gained on the one hand was lost on the other after a few years or a few decades. Other environmental damages have been produced by the overexploitation of water resources for agricultural purposes: huge groundwater aquifers have been pumped—or are still pumped—with much intensity in Midwest America, in Libya, and in Saudi Arabia.

Through ponds, lakes, streams or coastal areas, water is the natural habitat of fish, an important component of human food. Professional fishermen were already active and powerful in Middle Ages. During the same period, fish farming in natural ponds or artificial basins was an important resource to feed cities and was one of the many activities developed by monasteries (let us not forget that religious precepts about food consumption were strictly observed).[9] Fish were also present in thousands of natural ponds located next to rivers and in side channels, functioning as sorts of "hydraulic annexes" regulating water in case of flood or drought. The surface occupied by those ponds decreased drastically since at least the eighteenth century. The urban and industrial growth entailed the drainage of many ponds and subsequent ecological changes, both in terms of species present in the environment and of water system regulation. In the nineteenth century took place an "aquaculture revolution", consisting in the artificial breeding of fish species in basins placed near streams (or near coasts, when one aimed at reproducing mussels and oysters). In the Second Empire France, Victor Coste was the man who fostered the development of

[8] Jorgensen (2008).

[9] Hoffmann (1996).

what he called *aquiculture*, with the institutional support of the regime. In southern Alsace, near the Swiss border, and with the financial aid of Napoleon III, Coste settled a fish breeding farm, from which thousands of eggs and young fish were send throughout France and to many countries (as far as Brazil and USA). The objective of many local fishing societies founded during the late nineteenth century was to repopulate the streams.[10]

From the second half of the eighteenth century, an important transformation simultaneously affected cities and countryside: the increasingly negative view concerning wet zones. The link between occasional fevers and swamps is very ancient. However, the drainage effort strengthened and was renewed at the end of the eighteenth century, for economic reasons as well as medical ones: the hostile speech against the commons and the agricultural revolution urged elites to look for new medical justifications in the transformations of the territory. Urban transformations also contributed to this medical debate. The wet zones were less and less perceived as defensive or productive areas—which they often were in Northwestern Europe since the late Middle Ages[11]—to become places to be cleaned up and to fit out in the interest of the city-dwellers. In another geographical context, since the late nineteenth century, European colonization of Asia and Africa led to medical research about malaria, which was endemic in many areas. But the wetlands were also places with great biodiversity; human efforts to adapt wetlands to their new needs and visions led to the disappearance of many species. The last century or so has experienced a huge movement of land reclamation in coastal wetlands, leading to profound physical transformations of thousands square kilometers. Because of these changes, even if public authorities did not always support this environmental movement as enthusiastically as the goal of reclaiming wetlands, various species belonging to these ecosystems (birds, flowers, and so forth) became central in biodiversity conservation programs elaborated by ecologists.

In many large-scale hydraulic projects, integrated water management was not secured. One famous example lies in the region of the Aral Sea. Located in the former Soviet Union, the Aral Sea has almost disappeared because of intensive irrigation from rivers Syr Daria and Amu Daria for cotton growing in Uzbekistan and Kazakhstan, a non traditional usage planned by Soviet engineers in the 1950s. Such a large amount of water was diverted that the Sea level started to fall in the following decade, provoking many environmental changes in the region (including climatic ones). The major ecological problem touched the social framework of the former seashore. Fisheries disappeared in less than 30 years. Everything, from the groundwater to the air and the soil, was salinized. The coastal area is now far removed from the Aral waters, the seashore receded by many kilometers and professional fishermen have been forced to abandon their livelihood. Projects are currently in progress to return the sea water as close as possible to its former shoreline.

---

[10] Malange (2009).

[11] Soens (2011) and Morera (2011).

Many major water projects have been the occasion of a lot of water waste because of the lack of ecological knowledge or a will to ignore the regional conditions: in Egypt like in Central Asia, about half the water of reservoirs and canals evaporated or seeped into the earth before being usable for people. Some estimations state that by the 1990s, salinization seriously affected about 10 % of the world's irrigated lands. In Southeast Asia, and other places, the drainage of wetlands for rice growing and shrimp farming went along with the disappearance of the mangrove, and a subsequent loss of biodiversity.

Integrating social and environmental history leads to studying conflicts provoked by lack of water resources, or big scale water projects. Various scales of analysis can be used, from the local one (inside a community, for example), to the catchment area (between upstream and downstream communities), the State or even the international level.[12] Categories like "public"/"private", "rights of use" are discussed by various scholars. Barca has argued that in the Italian Liri Valley the cultural and political changes from the Enlightenment led to the remaking of a river as enclosed property, used by a rural bourgeoisie with no consideration of social and environmental costs.[13] In addition this new economy of water disrupted the river ecosystem; as early as the mid-nineteenth century, fish species were disappearing from the river due to both excessive canalization and industrial pollution. Riverine environments have been very inspirating for environmental historians, because of the complex entanglement between social, natural and cultural factors of change.

4.3 The River's Attraction to Historiography

Environmental history of water resources has developed, during the past decades, a particular focus on rivers. Studies have moved from the narrative of the capitalistic exploitation of nature to the assessment of socio-natural evolutions, showing how natural forces and human actions are entangled.[14] Some scholars have thus proposed to see rivers as "envirotechnical" landscapes.[15] Science and technology studies (STS) on the one hand, interdisciplinary projects on the other hand, inspired an increasing amount of essays emphasizing the hybridity of rivers—as other waterscapes. At the local scale, well before the invention of the "white coal", the water mills illustrate the human reshaping of rivers for economic and social purpose. Playing the role of mediators between human societies and natural resources, they were central to the agricultural and industrial economy and also important in

[12] For a case study about the USA: Paavola (2006).

[13] Barca (2010).

[14] To measure the evolution between some past and present environmental history approaches: Worster (1985), White (1995) and Cioc (2002).

[15] Pritchard (2011).

social and political relationships, in feudal Europe as well as in Muslim world.[16] They were indeed instruments of power for their owners (individuals as well as abbeys or collectivities) who collected taxes by users: peasants had to use them to transform cereals into flour to produce bread or beer, to grind olives into oil, etc.

4.3.1 Water and Transportation

For hundreds of years, water has been the best way to ship goods. Political powers, like Louis XIV in France or the Chinese emperors, took into account the economic importance of digging canals.[17] Many cities became prosperous because they functioned as a harbour (either on riversides or on the seaside). Some of them were places of off-loading, as a resource for the transportation of goods and people. Since Ancient Times, water has been a crucial factor of economic development and of employment. Water-related activities shaped the urban space, by creating specific areas: in Imperial Rome, near the Tiber River, the emporium produced a hill (the "Testaccio") just with broken amphorae and terra cotta utensils. When its harbour silted up, Bruges lost its predominance among Flemish cities at the end of Middle Ages. Rivers and canals were major transportation arteries until the development of roads and the railway. In the early Middle Ages, Vikings penetrated into many kingdoms thanks to their ships, well-fitted for shallow rivers.

The shape of many European and North American rivers has changed dramatically over the past two centuries because of the rise of their use for transportation in the nineteenth century and the construction of roads, houses and factories along the river banks. In the first decades of the Industrial Revolution, canals demonstrated their importance, such as the Erie Canal, linking New York harbor to the Great Lakes region: hundreds of kilometers of canals were constructed in France, the United Kingdom, Germany and the United States.[18] In this country, even at the end of the nineteenth century, waterways were still considered as useful means of shipping by merchants and businessmen, for instance in the Midwest. After the increase of railroad shipping prices, they turned to the hope of a new development of the Mississippi valley, in order to facilitate trade with the southern States and even with Latin America. The artificialization of waterways has often revealed itself as hazardous, as the flooding in New Orleans, consecutive to the hurricane Katrina (2005), dramatically recalled. In 1927, the city had been preserved from a huge flooding only by dynamiting levees upstream, a decision which led to the spread out of the Mississippi river over rural areas. If the channelization of many rivers by the Army Corps of Engineers eased navigation, it also permitted the settlement of more people near the banks, in former wetlands, and hence increase the human and

---

16 The tide mill was taken as a symbol by the first issue of *Water History* (Tempelhoff et al. 2009).

17 Mukerji (2009).

18 Jones (2010).

economic damages of exceptional floodings. Since the second half of the nineteenth century, various specialists and practitioners, who joined in the "conservationist movement", in the United States as well as in other countries—like France—inquired into the relation of forest cover to stream flow and even landslide.

Because of their role in promoting traffic and trade, rivers often provided the foundation for urban growth and its spatial expression. Progressively, especially from the end of the eighteenth century to the beginning of the twentieth century, rivers have been the subject of various plans, designed by local authorities or by central States, to overcome obstacles to navigation (fluctuations in river flow throughout the seasons, exceptional droughts and floods, sandbars, rapids, etc.). Because canals came to be seen as the ideal means of transporting goods, rivers were increasingly engineered, e.g. standardized in width and depth, to resemble these artificial waterways. At that time, canals were indeed the key means to transport coal. As early as 1815, the Rhine Commission, an international body, was created to oversee the river and mold it into what would amount to a commercial canal. In the 1830s, the United States federal government began to reshape the Mississippi river, viewed as "nature's highway to market".

The rise of railroad sealed the decline of waterways. In the second half of the 20th century, only big channels (deep enough to allow the transit of several thousand-ton barges and boats) remained in activity. From this time onward, some of the smaller channels were converted into recreational boating facilities. The seas remain the biggest waterway for shipping. They are regularly hit by environmental catastrophes, such as the major oil slicks consecutive to the tanker spills of the Torrey (1967), Amoco (1978) and the Exxon (1989). As early as 1969 a conference was held in London on the subject of sea pollution by hydrocarbons. Environmental history has perhaps a little less explored history of seas than history of inland water, except for history of fishing and marine animal populations which is still a very active field.[19]

4.3.2 The Dam Building: An Ambiguous Balance

Environmental historians have paid attention to the controversies surrounding some domestic water improvement schemes, leading to dam buildings to set up reservoirs for instance. At the beginning of the 20th century, the Hetch Hetchy Valley project, to furnish water to San Francisco was a famous case. It entailed national protests from the preservationist movement and its prominent figures, like John Muir. A more positive aspect in the same Bay Area region was that thanks to their land reserves around the springs and pumpings, some water districts and companies

[19] Danish environmental historians are very engaged in Marine Environmental History and belong to various transdisciplinary networks, such as the International Council for the Exploration of the Sea.

provided open spaces available for hiking.[20] Throughout this century, reservoirs became popular for outdoor recreation, swimming and boating, for example. The multiplication of dams and the grand change of scale in water manipulation schemes count among the landmarks of the last century's environmental history, embodying the self-confidence in technical progress and the human mastership of art wilderness, after several millennia of traditional modes of dam building (with earth and rocks). In 2000, the World Commission on Dams counted more than 45,000 large dams on the globe; the scope and scale of dam building in the twentieth century was unprecedented.

Two purposes mainly motivated the physical reorganization of the environment they embody: firstly, the need to provide more water than the natural supply of streams and rivers could offer, especially when the climate led to great seasonal variations in rainfall; secondly, in other regions the creation of artificial reservoirs was linked to power-consuming industrialisation, functioning as a water supply source for hydroelectricity plants. Both grew dramatically after 1945.[21] The world's total irrigated surface rose fivefold, from 50 million ha at the eve of the twentieth century to nearly 250 million in 1995, with the fastest expansion taking place after the Second World War. At the beginning in the Western world, and then in colonies becoming newly independent nations, a number of rivers have merely become a series of slack-water ponds, losing many of the physical and ecological characteristics of a stream.

In most cases, dam building could be related to a certain form of bureaucratization of water management. The process involved vertical schemes of information diffusion and deprived local people of their traditional way of life, provoking the upheaval of water usage in affected environments. For social and political histories, big dam projects can often be associated with a "top-down" decision making. This kind of central planning and manipulation of resources often paved the way for popular protests.

In Spain, the use of water during the first half of the twentieth century was devoted mainly to agriculture. The Franquist regime set up urban and industrial use as a priority (1950s–1970s). A centralized governance of water resource imposed the industrialist development model on the entire territory. During the second half of the twentieth century, after the boom in dam building launched by the Hoover Dam, outside Las Vegas, "pharaonic" projects consisted of big dams and inter basin rural-to-urban water transfers. They were carried out without any compensation to rural inhabitants. Social and environmental costs were not taken into account. This was made following a national central planning. These projects also received financial support and expertise from international agencies like the World Bank and Western countries such as West Germany or the United States. It is true that economic development benefited the general standard of quality of life in Spain. However, the industrial and urban model of water allocation created social and

[20] Walker (2007).

[21] McNeill (2001).

environmental inequalities, because this model of centralized and planned economic growth was made at the expense of rural livelihoods. For that reason, in the final period of dictatorship's regime and after the death of Franco, rural and peripheral regions and inhabitants protested and called for democratic water governance, including participation in decision-making processes.[22] In South Africa, until 1990, most of the projects designed were integrated into the apartheid framework, and into a process of land appropriation by White people. The environmental history of South African rivers cannot be separated from the general history of this country.[23] The last century has proved that any political regime can be harmful to environmental resources: grass-roots green and "blue" movements can easily develop in democracies, but "technocratic" management remains powerful and potentially dangerous.

Presented as triumphs of engineering—this view prevailed until late twentieth century—the dams materialized both the idea of nature's domination by humankind and the economic development and modernization goals claimed by governments of all kinds. Progressively, other voices emerged which tried to be heard: they stated that these huge concrete works provoked ecological and social problems. For instance, they underlined that dams were opposed to natural behaviors of animals, like salmon in rivers. But dams not only constituted an impediment to the life of ecosystems, they also changed the course of life for millions of people. Dams have indeed provoked the displacement of people whose towns were inundated as part of the water projects: sometimes, only a few families in what seemed urban/rural conflicts in a period of depopulation of rural areas; in other occasions, several thousands of people in India, or even more than one million in China, for the Three Gorges Dam (for which structural work was finished in 2006 and which produced a lake of several-hundred-kilometers length), three decades after a rapid decline in dam building in the Western countries. It is difficult to keep precise figures, but in 2000, the World Commission on Dams estimated that between 40 million and 80 million people had been displaced by the settling of reservoirs.

4.3.3 Cities and Rivers: Intertwined Histories

Recent monographs (not only in environmental history) have explored the relationship between rivers and cities, which still stands as a fruitful topic for doctoral dissertations.[24] In the mid-twentieth century, geographers were already aware of the importance of water resources in the birth and growth of cities. Researchers focused either on major cities and major rivers,[25] or on small rivers which have been used as

[22] Swingedouw (2007) and Corral Broto (2012).

[23] Blanchon (2009).

[24] For a review of some recent theses, Frioux (2012).

[25] Dagenais (2011).

open sewers and then often covered and "forgotten" by urbanites in their everyday life.[26] Rivers, often described through anthropomorphizing discourses, have always been essential to the life of cities—and indeed, most cities are riverine cities.[27] They could provide drinkable water if there was no available spring and serve the "sanitary city" ideal.[28] They also offered industrial water, as a source of energy or a component of industrial processes.[29] They were indispensable to boat traffic, to float timber or to ship merchandise to and from the agglomeration.

After the Middle Ages, the diversification of water use increased. At the same time, in the Western world, a form of expertise about water management was developed. In the eighteenth century, the French monarchy created the first corps of engineers partly specialized in waterworks (the "Ponts et Chaussées"). In cities, scientists assessed water quality based on physical criteria (taste, odor, appearance…). They did not know yet that a great upheaval was about to happen in many cities, with the huge increase of industries (mining, iron and steel mills, textile factories…) which materialised the advent of a new age of urban water resources. In Europe, the purity problem had long been raised (in the Middle Ages, for instance, some crafts needed clean water for their processes), but only from the nineteenth century onwards did specific sciences devoted to this issue develop. In part this was due to the cholera epidemics which struck industrialising nations and their capital cities (like Paris and London) several times: 1832, 1854, 1884, 1892 and 1910. Providing safe water became a public health issue and a duty for public authorities after different investigations demonstrated the link between bad water and outbreaks of these diseases.[30]

From the early nineteenth century onwards, the problem of supplying city dwellers with sufficient quantities of water to meet their basic needs (drinking, washing, cooking) and the issue of meeting the contradictory demands from industrialists (some of them needing pure water, others using rivers as receptacles of their wastes), became increasingly acute, throughout Europe and North America. In most cases, the quest for an expanded water supply system arose indeed, mainly from industry. Urban environmental historians can establish links, in that case, with business history. The water-supply service became a real industrial venture: eight companies shared the water market in London, the most populated city in the world at that time. In 1852, French capitalists created the *Compagnie Générale des Eaux* which took over water-supply of Lyon, the second largest city of the country; a few years later, it also obtained a contract with the Parisian administration, under the auspices of its ruler, the Baron Haussmann. This new age of "conquest of water" used sophisticated devices and breakthrough technological innovations: powerful steam pumps, big reservoirs, long canals (like the *Canal de Provence* for

[26] Castonguay and Evenden (2012).

[27] Blache (1959).

[28] Melosi (2000).

[29] Steinberg (1991).

[30] Hamlin (1990).

Marseilles), water towers (like in Chicago). Infrastructure which modified the urban environment and its peripheries had counterparts in the cultural domain (see below). In the case of London, social and political factors accounted in the late establishment of a publicly operation of water supply. Broich argues that politics and environment intertwined to drive the physical reorganization of the environment—of resource collection and delivery—and the changes in the administration of the environment: in London, at the turn of the twentieth century, the eight companies competing on the water market were bought by the new Metropolitan Water Board.[31] Studying history of the provision of safe water to urban dwellers from the end of the nineteenth century also highlights environmental inequities on various scales, between the haves and have-nots. People receiving purified water and benefiting from sanitation—above all, people who are able to pay for these services—suffer less from waterborne diseases than people lacking connections to the public water supply and sewer system. In 1995, this was the case of more than 80 % of the residents of important Asian capitals (Manila, Dhaka, Karachi).

4.4 Floodings, Pollution, Depletion: The Dark Side of Water Resources History

History of water resources during the Industrial Era can neither be reduced to a series of failures, nor to substantives like "pollution" and "deterioration". The setting up of safe water supply systems allowed for the urban expansion and even a part of the demographic growth of the twentieth century. From the middle of the nineteenth century onwards, especially in Great Britain where worked a national Rivers Pollution Commission, riverine dwellers and users became aware of the damages produced by the rapid expansion of industrialization and urbanization. British streams experienced first the tremendous increase of chemical waste disposal and pollution by textile industry, and were severely affected because of their relatively low flow rates. Different responses could be proposed to the challenge of pollution: legislative answers were not sufficient and had to be completed by more practical and technical solutions. These technological fixes were offered by the rapid progress made in microbiological theory and in bacteriological analysis, which means by the urban science and technology.

Nevertheless, efforts of public health officials and sanitary engineers proved inefficient to prevent what became the major source of water alteration: industrial discharges, whose chemical compounds and effects on living organisms remained hardly known for a long time. In almost every part of the Earth, from the Columbia River (USA) to the Danube River (Europe), since the late 18th century and increasingly from around 1850, rivers have been the receptacle for toxic loads from mining, chemical production or the food-processing industry. For a long time, the

[31] Broich (2013).

chemical pollution of water resources was not considered as a big problem, the fecal pollution retaining the attention of public health specialists (especially during the nineteenth century). After the Second World War, the development of scientific ecology and the rise of modern environmentalist movements highlighted other forms of water resources degradation.

As illustrated by a French 1829 law, the conservation of water resources emerged relatively early and not later than policies to preserve drinkable water quality. But hygienic measures encouraged many people engaged in commercial and recreational fishing to fight situations they perceived as ongoing "water crises". The Izaak Walton League, created in 1922 (named after a famous British writer and angler of the seventeenth century, author of *The Complete Angler*), acted to promote the protection and restoration of American waters, before broadening its range of interests to topics related to the preservation of "Outdoor America" air pollution, clean energy, and sustainable agriculture. In 1951 appeared in West Germany the *Vereinigung Deutscher Gewasserschutz* to arouse public interest in water resources issues. Fishers and scientific *amateurs* were forerunners of ecologists and launched many operations long before the 1970s and the popularization of environmental protection topics. During the 1950s, the Amsterdam waterworks played a major role in forging international links between other waterworks along the Rhine River.[32] A new cause of environmental degradation has been "thermal" pollution from nuclear reactors, whose wastewater can raise the temperature of rivers high enough to kill all organisms in the water. In the 1970s for instance, several protests took place in West Germany to attempt to block nuclear reactor projects along the Rhine. In 1986, the catastrophe provoked by a fire at a chemical site near Basel (Switzerland), showed how downstream users of river water (and living organisms of the river ecosystem) have to face upstream pollution. In some cases, a short event can have consequences for many years on the water resources, hundreds of kilometers downstream.

Another change upon water resources has been the fantastic increase in chemical fertilizers and pesticides. As early as 1962, the American ecologist Rachel Carson wrote that synthetic pesticides "have been recovered from most of the major river systems and even from streams of groundwater flowing unseen through the earth" and even found "in fish in remote mountain lakes". Since her pioneering work, pesticides have been largely incriminated. However they are not the only threat to water resources brought by the twentieth century farming modernization. The use of fertilizers on agricultural lands indirectly produced algal blooms and lakes eutrophication, especially in the 1930s–1970s period. This phenomenon also happened in enclosed seas like the Baltic Sea, the Adriatic, the Red Sea and numerous inland lakes. The excess nutrients provoked a proliferation of algae, but not the same expansion for fish stocks.

[32] Disco (2007).

Most of the works analyzed for this chapter deal with the place held by water in "conquest of nature" (Blackbourn) over time.[33] Another trend focuses on rivers as source of hazards and on the impact of major floods. Some research projects have been funded in order to apply historical knowledge to current and future flood-risk management. They can reveal how the social and political processes set up after flood events already contained measures to reduce vulnerability of flood-threatened areas. In south-west Germany for instance, after an extreme flood in the Neckar River catchment in October 1824, a letter from the Royal Government of Württemberg stated that houses that were destroyed or severely damaged should not be rebuilt on their original flood-exposed sites but elsewhere, on higher ground and further away from the river, where risk of repeated flooding was lower.[34] Vulnerability can also be historically constructed in an unequal way. During the nineteenth century, in the outskirts of London, thousands of new homes and hundreds of factories supplanted the marshes of the Lower Lea that once absorbed floods. They increased the potential for damage when water flowed into these former wetlands. In 1888, 1897, 1904, and 1928 the poorest people in West Ham, who lived in the suburb's oldest wetland neighborhoods, suffered as water rushed into their homes.[35]

From events that occurred in the last century, as well as from small-scale, almost microscopic, changes that anyone can notice during regular visits to the same waterscape, it is possible to state that both water-related ecosystems and human societies are dynamic forces rather than static entities clashing with one another. History and geography have thus much to teach environmental managers and policy-makers. In Europe, the EU's Water Framework Directive requires stream restoration, to ensure that a good ecological status is achieved in all water bodies in the EU by 2015. In Sweden, the first step before restoration necessitates knowledge about the history of timber floating and its physical effects on river channels. Environmental history can work besides natural science, in order to use historical archives providing data about floatway structures, their purposes, their localizations and their effectives. Historians are also useful to raise the question of cultural remnants of the timber floating age (1850s–1980s). Stone piers reflected the socioeconomic reality at that time and the need for wages for the working people along the rivers in northern Sweden. For Erik Törnlund, "one important part of restoration work should be to consider both ecological *and* cultural aspects".[36]

---

[33] Blackbourn (2006).

[34] Seidel et al. (2011).

[35] Several water-related cases can be found in Massard-Guilbaud and Rodger (2011). Example from Clifford (2011).

[36] Erik Törnlund, "From Natural to Modified Rivers and Back? Timber Floating in Northern Sweden in 1850–1980 and the Use of Historical Knowledge in Today's Ecological Stream Restoration", in *Thinking through the Environment. Green Approaches to Global History, op. cit.*, 241–267. About restoration, see Hall (2005).

4.5 Water Resources, Political Power and Culture

Environmental history cannot ignore the cultural dimension of water resources. Throughout several millennia, the representations of water drastically changed, evolving from a vision in which water was like a treasure, or a tool, given by divinities to men, to an increasingly anxious point of view, conscious of the multiple risks associated with water pollution and, more recently of the non-renewable character of many sources of water and of the value of the biodiversity in water landscapes.

Water has been a decorative element in cities since Ancient Rome: for centuries, monumental sculpted fountains in the heart of the urban centers were the norm of local government. The tradition has been prolonged to our times, like in Atlanta near the Olympic Park. But the combination of decoration and usefulness could also hide —and still does some time—some political goals.[37] Therefore, environmental history cannot forget the cultural dimension of water: the Roman civilization was famous for the importance given to the public baths (*thermae*): water was an object for evergetism and tourists can still visit Caracalla's *thermae* in Rome, as well as dozens of public baths in the former "Roman world". The Imperial regime in Ancient Rome was accompanied by the birth of a "water evergetism", serving the hygienic needs of the citizens at the same time as being a means of publicity for the local aristocrats in charge of the water service at their own expense. Use of water for cleaning bodies—and souls—remained a core value in other civilizations, like in the Muslim world; while popular baths in Christian Middle Ages cities were progressively denounced by clerks as places of immorality. During the Middle Ages, control of water resources and infrastructures related to water—such as bridges, mills, water supply systems, was a stake for rival authorities, for instance ecclesiastical institutions and communal councils. On crucial rivers for transportation of people and goods, boatmen developed particular social and cultural characteristics.[38]

A few centuries later, like the Roman emperors, the French king Louis XIV demonstrated his absolute power partly thanks to the hydraulic system of his gardens at Versailles. The power was expressed through fountains, ornamental pools, crisscrossing jets and so forth, themselves created by diverting water from a river and from the surrounding marshes, by installing a complex network of pipes and large reservoirs. In the same country, in southern towns of the late nineteenth-century, the municipal councils showed particular attention to the presence of republican symbols on monumental public fountains. Water participated both in the beautification of the city and in the expression of political messages. At the same time, the control of water resources for domestic usage came into public hands in many British and German cities: this was the age of the "municipal socialism". Western countries were rather an exception at the turn of the twentieth century: few

[37] For a collection of case-studies outside the Western World, in an environmental politics perspective, see Baviskar (2007).
[38] See Rossiaud (2007).

cities in Southern Europe had an efficient water supply and existing systems were subject to poor operating conditions. Moreover, this public involvement was not always put in practice in France or in the United States, and nowadays questions of public management of water supply systems are still a topic of local political debates. The problem moved, after 1945, in the postcolonial world and in Latin America. Lastly, water was seen in the twentieth century as a resource which had to be exploited to provide the human beings a better future: water-diversion projects appealed to political authorities. Environmental history of former authoritarian regimes, where archives recently became available (former Communist European countries) reveal how hydraulic projects, embodied by gigantic dams and reservoirs, were used to affirm the prestige of the State. Control of nature was thus an expression of political power. It also created networks of specialists (engineers, scientists), who contributed to first environmental criticisms, in the absence of any grassroots movement.

Legends, storytelling, songs and many other cultural facts were attached to some waterscapes, mainly wetlands and rivers. The Ganges continues to be important in Hinduism. In the Book of Genesis, in Greek Mythology, rivers structure the landscape and many myths and stories explore the healing or evil dimensions of water. Rivers have often been idealized in literature and the visual arts, for instance in Russia with references to "Mother Volga" and in the USA to "Old Man River" (Mississippi).[39] The Rhine was also important for German conservationists and nationalists; a Nazi official referred to it in 1941 as the "pulsating life's vessel" of the nation. Water resources protection in the Rhine valley became also a grounding issue for the modern ecological movement. The folklore surrounding water is extremely rich, from the huge rivers just mentioned above, to little ponds or fountains that were linked with faeries, witches or healing waters. In the nineteenth century, hundreds of artists painted panoramas of rivers, ponds, or seashores. Riverscapes presented Romantic features such as waterfalls and gorges (especially the Middle Rhine). What is interesting to be noticed is that in the twentieth century, stereotypes about water resources, like rivers, were used both by conservationists and developers: in Germany, each side drew on Romantic imagery of German rivers and on the tourist potential of the river they had in mind ("natural environment" vs artificial lakes created for hydroelectricity) to support their respective positions.

Within the different civilizations, the perception of water evolved across centuries. In the Western world, during the Early Modern Period, cleanliness was provided by the change of clothing and the whiteness of clothes, rather than by bathing and using water to clean the body. Then, the eighteenth and nineteenth centuries drastically changed the human relationship to water resources: hygienists pointed out the necessity to provide pure water to human bodies in order to avoid diseases; the rise of mandatory schooling, home economics classes and enrollment

[39] Some cultural studies can be read in Mauch and Zeller (2009).

in the army taught young people the importance of showering and washing hands.[40] Water resources have indirectly been a factor of the birth of a new culture of leisure and new places of sociability, first in the West, and now spreading out to the rest of the world. From the eighteenth century onwards, two sorts of new residential and seasonal resorts appeared, firstly for the European aristocracy: seaside towns and spas. The first category is linked with a change of perception towards the coastal areas: once perceived as dangerous, used only by local populations either for fishing or shells, kelp or algae picking, they became both aesthetic landscapes and a form of treatment (cold baths) prescribed by physicians.[41] Dozens of seaside resorts experienced the birth of tourism thanks to their "water resources". Inland, the development of hydrotherapy and the progress of transportation (railroad) led to the same phenomenon in "spas" (from the Belgian town of this name). The proximity of water, either from sea or coming out of the earth (hot springs), gave rise to a new type of urbanism and the settling of urban amenities even in very small resorts. Environmental history of touristic development has yet to be developed. Since the 1930s, with the development of public swimming pools and swimming lessons in cities, and after the Second World War, the sea—associated with sun—became a major popular attraction. People also experienced the sand, whereas the elites had once been separated from it by boardwalk promenades. Environmental history of beaches is a promising field of research, necessitating cultural history as well as geographical investigation.[42] Islands, once seen as the paradise of mosquitoes and considered as under-developed areas, are now oriented towards tourism and activities like boating and diving (particularly in the Mediterranean and Caribbean Seas, the Indian Ocean). Inland, artificial lakes created upstream from dams and hydroelectricity plants were also used for bathing, fishing or boating purposes, as well as for new sports like windsurfing or water skiing bringing seasonal activities to rural regions confronted with human desertification. Other places with water or landscapes sculpted by water have attracted millions of tourists: the Niagara Falls and the Grand Canyon are well-known for this aspect.

4.6 Conclusion

As was written in the first issue of the new environmental history journal *Water History*, "water has been and is likely to continue to be one of the most pressing environmental resource concerns". Water resources appeal both to archaeology and to contemporary history, because first settlements, first cities, were developed thanks to first hydraulic works, and because the struggle towards universal access to water entailed fierce competition during the nineteenth and twentieth centuries: to

---

[40] Goubert (1989).

[41] Corbin (1994).

[42] Devienne (2011).

the traditional requirements of agriculture was added a hugely increased urban demand. In the case of surface water (streams, rivers, lakes), one can draw a three phase scheme from many case-studies. Most rivers of the Western world have now entered the third sequence. During the first period, pre-industrial rivers were essential to feed the riverine population, thanks to their fish; they also could provide benefits to agriculture, through the fertile lands in the floodplains; when hydrological conditions were favorable for navigation, they served as cheap and important means of transportation. After a second phase of intense exploitation of water resources and multi-scale environmental changes, new reasons have added value to the place of water in cities, for aesthetic, ecological as well as touristic purposes. Meanwhile, environmental history of water resources has grown as a fertile topic, bridging various disciplinary approaches. The numerous case studies written about the nineteenth and twentieth-century Western world can be useful both for water managers dealing with restoration projects and for historians working on water issues outside Europe and North America and globalizing environmental history.

References

Barca S (2010) Enclosing water: nature and political economy in a Mediterranean valley, 1796–1916. The White Horse Press, Cambridge

Baviskar A (ed) (2007) Waterscapes: the cultural politics of a natural resource. Permanent Black, Ranikhet

Blache J (1959) Sites urbains et rivières françaises. Revue de Géographie de Lyon 34(1):17–55

Blackbourn D (2006) The conquest of nature: water, landscape, and the making of modern Germany. Norton, New York

Blanchon D (2009) L'espace hydraulique sud-africain. Le partage des eaux. Karthala, Paris

Broich J (2013) London: water and the making of the modern city. Pittsburgh University Press, Pittsburgh

Bruun C (1991) The water supply of Ancient Rome: a study of Roman imperial administration. Societas Scientiarum Fennica, Helsinki

Carson R (1962) Silent spring. Riverside Press, Cambridge

Castonguay S, Evenden M (eds) (2012) Urban rivers: re-making rivers, cities and space in Europe and North America. University of Pittsburgh Press, Pittsburgh

Cioc M (2002). The Rhine: an eco-biography, 1815–2000. University of Washington Press, Seattle

Clifford J (2011) A Wetland suburb on the edge of London: a social and environmental history of West Ham and the River Lea, 1855–1914. Ph.D. dissertation, York University, Toronto

Corbin A (1994) The lure of the sea: discovery of the seaside in the Western World 1750–1840. University of California Press, Berkeley

Corral Broto P (2012) De la plainte légale à la subversion environnementale: l'aménagement des rivières dans l'Espagne franquiste (Aragon 1945–1979). Vingtième siècle. Revue d'histoire 113:95–105

Dagenais M (2011) Montréal et l'eau. Une histoire environnementale. Boréal, Montréal

Devienne E (2011) Beaches in the city. The quest for the ideal urban beach in Postwar Los Angeles, La Vie des idées. http://www.booksandideas.net/Beaches-in-the-City.html

Dieter S (2004) Urban environmental history: what lessons are there to be learn. Boreal Environ Res 9:519–528

Disco C (2007) Accepting father Rhine? Technological fixes, vigilance, and transnational lobbies as 'European' strategies of Dutch municipal water supplies, 1900–1975. Environment History 13:381–411

Frioux S (2012) At a green crossroads: recent theses in urban environmental history in Europe and North America. Urban Hist 39(3):529–539

Goubert J-P (1989) The conquest of water: The advent of health in the industrial age. Princeton University Press, Princeton

Hall M (ed) (2005) Restoration and history. The search for a usable environmental restoration. University of Virginia Press, Charlottesville

Hamlin C (1990) A science of impurity. Water analysis in nineteenth century Britain. University of California Press, Berkeley

Hays S (1959) Conservation and the gospel of efficiency: the progressive conservation movement, 1890–1920. Harvard University Press, Cambridge

Hoffmann R (1996) Economic development and aquatic ecosystems in medieval Europe. Am Hist Rev 101:631–669

Jones C (2010) A landscape of energy abundance: anthracite coal canals and the roots of American fossil fuel dependence, 1820–1860. Environmental History 15(3):449–484

Jorgensen D (2008) Cooperative sanitation: managing streets and gutters in late medieval England. Technol Culture 49(3):547–567

Malange J-F (2009) Pêcheurs, pisciculteurs, science et Etat français face au "sauvage" aquatique de 1842 à 1908. In: Frioux S, Pepy E-A (eds) L'animal sauvage entre nuisance et patrimoine. ENS Editions, Lyon, pp 149–164

Massard-Guilbaud G, Rodger R (eds) (2011) Environmental and social justice in the city: historical perspectives. The White Horse Press, Cambridge

Mauch C, Zeller T (eds) (2009) Rivers in history. Perspectives on waterways in Europe and North America. University of Pittsburgh Press, Pittsburgh

McNeill JR (2001) Something new under the Sun. An environmental history of the twentieth-century world. Norton, New York

Melosi M (2000) The Sanitary City: urban infrastructure in America from Colonial times to the present. The Johns Hopkins University Press, Baltimore

Mink N (2006) A narrative for nature's nation: constance lindsay skinner and the making of rivers of America. Environmental History 11(4):751–774

Morera R (2011) L'assèchement des marais en France au XVIIe siècle. Presses Universitaires de Rennes, Rennes

Mukerji C (2009) Impossible engineering : technology and territoriality on the Canal du Midi. Princeton University Press, Princeton

Neri Serneri S (2007) The construction of the modern city and the management of water resources in Italy, 1880–1920. J Urban Hist 33:813–827

Paavola J (2006) Interstate water pollution problems and elusive federal water pollution policy in the United States, 1900–1948. Environment History 12(4):435–465

Pritchard S (2011) Confluence: the nature of technology and the remaking of the Rhône. Harvard University Press, Cambridge

Reclus E (1869) Histoire d'un ruisseau. Hetzel et cie, Paris

Rossiaud J (2007) Le Rhône au Moyen Age. Aubier, Paris

Seidel J et al (2011) Reconstruction and analysis of the flood Catastrophe along the River Neckar (South-West Germany) in October 1824. In: Myllyntaus T (ed) Thinking through the environment. Green approaches to global history. The White Horse Press, Cambridge, pp 201–217

Soens T (2011) Floods and money. Funding drainage and flood control in coastal Flanders (13th–16th centuries). Continuity Change 26(3):333–365

Special issue: dealing with fluvial dynamics: a long-term, interdisciplinary study of Vienna and the Danube. Water History, 5/2 July 2013

Steinberg T (1991) Nature Incorporated. Industrialization and the Waters of New England. Cambridge University Press, New York, Cambridge

Swingedouw E (2007) Technonatural revolution: the scalar politics of Franco's hydro-social dream for Spain, 1939–1975. Trans Inst Br Geogr 32(1):9–28
Tarr J (1996) The search for the ultimate sink: urban pollution in historical perspective. Ohio State University Press, Akron
Tempelhoff J et al (2009) Where has the water come from. Water Hist 1:1–8
Walker R (2007) The country in the city : the greening of the San Francisco Bay Area. The University of Washington Press, Seattle
White R (1995) The organic machine: the remaking of the Columbia River. Hill and Wang, New York
Wittfogel K (1957) Oriental despotism: a comparative study of total power. Yale University Press, New Haven
Worster D (1985) Rivers of Empire: water, aridity and the growth of the American West. Oxford University Press, New York

Chapter 5
Environmental History of Air Pollution and Protection

Stephen Mosley

Abstract Concerns about air pollution have a long and complex history. Complaints about its effects on human health and the urban environment were first voiced by the inhabitants of ancient Athens and Rome. But urban air quality worsened considerably during the Industrial Revolution, as the widespread use of coal in factories in Britain, Germany, the United States and other nations ushered in an 'age of smoke'. Despite the tangible nature of this form of air pollution, early laws to control it were generally weak and ineffective—regardless of its high socio-environmental costs—reflecting the importance of coal-fuelled steam power to economic growth. Not until the mid-twentieth century, after major air pollution episodes such as London's 'Great Smog' had demonstrated beyond doubt that polluted air was as harmful to the public's health as polluted water supplies, were stringent national laws to abate smoke finally introduced to clear the skies over the cities of the first industrial nations. However, while the citizens of the developed world now breathe cleaner air, smoke pollution is still a significant environmental problem in many industrial cities of developing countries today. In terms of their scale, the effects of coal smoke in the nineteenth and early twentieth centuries were largely local and regional. But after the Second World War a number of invisible threats began to emerge—acid rain, photochemical smog, ozone depletion and climate change—that were transnational and global in character. It often required the cooperation of scientific experts across academic and political borders, as well as new techniques such as computer modelling, to make these new threats 'visible' to the public. Global environmental problems also required collective political and legislative action on the part of nations if solutions were to be found. The success of the Montreal Protocol in phasing out the use of ozone-depleting CFCs stands as a successful example of international environmental governance. However, it will need a strong commitment to international cooperation if an effective agreement to reduce greenhouse gas emissions is to be reached, particularly as global warming is a concept that the public (and many politicians) still find difficult to grasp.

S. Mosley (✉)
School of Cultural Studies, Leeds Metropolitan University, Leeds, UK
e-mail: s.mosley@leedsmet.ac.uk

M. Agnoletti and S. Neri Serneri (eds.), *The Basic Environmental History*,
Environmental History 4, DOI 10.1007/978-3-319-09180-8_5

5.1 Introduction

Air pollution is a major environmental problem and it comes in a variety of forms, from visible particles of soot or smoke to invisible gases such as sulphur dioxide and carbon monoxide, and it can be created indoors and outdoors. Although some sources of atmospheric pollution are emitted naturally, from volcanoes or forest fires, most are the result of human activity in the home or workplace. This chapter aims to provide an overview of anthropogenic air pollution problems, and attempts to solve them, focusing in particular on the past two centuries. Atmospheric pollution, however, has a longer history. It was also a significant issue for pre-modern societies, with the burning of biomass and fossil fuels damaging both human health and the local environment. But the rise of modern urban-industrialism—and the shift from fuel-wood to coal and then to oil—extended the scale and scope of air pollution problems dramatically. Indeed, some harmful airborne contaminants like chlorofluorocarbons (CFCs), chemicals used in aerosol cans, refrigerators and air-conditioning systems that deplete the ozone layer, did not exist before the twentieth century.

The chapter is set out in three parts. Firstly, it examines early examples of air pollution and its effects, especially in northern Europe and the Mediterranean basin where archaeological records and a variety of written sources provide clear evidence of environmental change. Secondly, it will explore the development of air pollution problems between 1780 and 1950, as the availability of cheap wood supplies declined and coal became the chief source of energy in the rapidly industrialising world. Lastly, it will discuss transboundary pollution caused by acid rain, high levels of emissions from gasoline-fuelled cars, and the emergence of new global threats from invisible air pollutants after 1950, such as CFCs and human-induced climate change, as well as the various international measures put in place to tackle them.

5.2 Preindustrial Air Pollution

Indoor air pollution caused by cooking and heating with open fires in poorly ventilated dwellings was a significant cause of ill-health from the earliest times. Scientific studies of samples of mummified lung tissues from Egypt, Peru, Britain and elsewhere have revealed that ancient societies suffered from anthracosis, (blackening of the lungs), from long exposure to the acrid smoke of domestic fires.[1] Smoke was most likely tolerated indoors because it helped to keep mosquitoes and other insect pests at bay. But poor domestic air quality—with concentrations of harmful particulates high in cramped conditions—undoubtedly increased the risk of illness and death from chronic respiratory diseases. The testimonies of Aretaeus of Cappadocia, Aulus Cornelius Celsus, Pliny the elder and other medical writers

[1] Brimblecombe (1988, 2008) and Colbeck (2007).

indicate that diseases of the lungs were widespread in the classical civilisations of the Mediterranean basin. However, deaths from 'normal' diseases like bronchitis have to date attracted little sustained attention from medical historians of the classical period.[2] In the developing world, where heating and cooking with smoky biomass fuels such as wood and animal dung remains commonplace, in 2013 around 2 million deaths were linked to indoor air pollution.[3]

Outdoor air pollution only became a major issue with the rise of cities. Early cities were very different in many respects from their modern counterparts. They were, for example, compact 'walking cities', with the marketplace, religious and public buildings all being easily accessible on foot. The influential Hippocratic treatise *Airs, Waters, Places*, written c.400 BCE, stressed the importance of good air quality, as well as pure water and a salubrious setting, in choosing settlement sites. But where large numbers of people crowded into urban centres, smoke and other noxious fumes from households and small manufacturing works soon became a cause for concern. Air pollution was an everyday part of life for the inhabitants of cities like Athens (population c.200,000 in 430 BCE) and Rome (population c.1 million in 150CE), where the emissions from homes, smelting furnaces, potteries and other preindustrial workshops darkened the skies.[4]

The residents of ancient Rome referred to their city's smoke cloud as *gravioris caeli* ('heavy heaven') and *infamis aer* ('infamous air'), and several complaints about its effects can be found in classical writings.[5] The poet Horace (65 BCE–8 CE), for instance, lamented the blackening of Rome's marble buildings by countless wood-burning fires, while the statesman and philosopher Seneca (4 BCE–65 CE) wrote in a letter to a friend:

> I expect you're keen to hear what effect it had on my health, this decision of mine to leave [Rome]. No sooner had I left behind the oppressive atmosphere of the city and that reek of smoking cookers which pour out, along with clouds of ashes, all the poisonous fumes they've accumulated in their interiors whenever they're started up, than I noticed the change in my condition at once. You can imagine how much stronger I felt after reaching my vineyards.[6]

Some 2,000 years ago civil claims over smoke pollution were heard before Roman courts, and in 535CE the emperor Justinian promulgated the *Institutes* which included a section that acknowledged the importance of clean air to breathe (and pure water to drink) as a birthright: 'By the law of nature these things are common to mankind—the air, running water, [and] the sea'. Earlier Babylonian and Assyrian laws dealt with similar issues, and around 200CE the Hebrew Mishnah sought to control sources of air pollution in Jerusalem.[7] Atmospheric pollution in

---

2 Shaw (1996), Sallares (1991) and Schiedel (2001).

3 World Health Organisation (2013).

4 Hughes (1996) and Mosley (2010).

5 Hughes (1996).

6 Colbeck (2007) p. 375.

7 Brimblecombe (2008), Colbeck (2007) and Mamane (1987).

the ancient world, then, was recognised as damaging to both human health and the built environment, and it was in early cities that the first legislative steps were taken to abate it—and to protect the air as common property—albeit with limited success.

Domestic smoke problems from wood and charcoal burning were mainly confined to a limited area in and around urban centres. The noxious emissions from smelting and mining metals, however, had more serious and far-reaching consequences. The leading sources of metallic pollutants were lead and copper production, which had environmental impacts on a regional and hemispheric level long before the Industrial Revolution. First smelted in Anatolia and Mesopotamia around 5000 BCE, the production of lead increased sharply during the Greco-Roman period (peaking at around 80,000 tons per annum). Indeed, it was central to the Roman's daily lives, and they used it extensively for everything from domestic water pipes and roofing to kitchen utensils and coinage—and even as a sweetener of wine (lead arsenate). The adverse health effects of long-term exposure to lead (impaired fertility and neurological damage), which disproportionately affected the Roman aristocracy and upper classes, have been linked to the end of empire. Lead extraction and smelting also posed a serious health hazard for workers (often forced labour) in Roman mining operations in the Iberian Peninsula, England, Gaul, Greece and elsewhere, as well as leaving behind hill-sized mounds of black slag that transformed the landscape. The Greek geographer and historian Strabo (c.64 BCE–c.23 CE) described how toxic metallic emissions from smelter furnaces were discharged into the air from 'high chimneys'; and small-sized particles were transported on the prevailing winds to pollute large regions of the northern hemisphere. Analysis of Arctic ice-core studies has shown that imperial Rome increased the release of lead into the environment by a factor of ten, mainly due to inefficient smelting in open furnaces.[8]

The Roman period also saw a marked rise in copper production, which—often alloyed with tin to form the harder metal bronze—was utilised to make tools, weapons and coins. Reaching a peak of over 15,000 tons per annum approximately 2,000 years ago, Roman copper supplies were sourced mainly from Spain, Cyprus and central Europe. The widespread use of copper coinage in medieval China under the Sung Dynasty contributed to a second preindustrial boom in production (rising to about 13,000 tons per annum at its peak; a scale comparable to that of Roman times). But the primitive technologies and techniques employed by Roman and Chinese metallurgists resulted in around 15 % of all smelted copper being expelled into the atmosphere. The data from Arctic ice-cores shows that the cumulative deposition of copper pollution in the northern hemisphere was much greater before the Industrial Revolution than afterwards, when smelting technologies had improved.[9] It is important to note that because smelters and other smoky trades (such as brick-making) made disagreeable neighbours, they were often situated well beyond the boundaries of a city. Pragmatically, they were generally located in the

[8] Hong et al. (1994), Hughes (1996) and McMichael (2001).

[9] Hong et al. (1996) and Colbeck (2007).

countryside close to forest and woodland areas where fuel was abundant, its transportation costs were low, and where few people would be troubled by air pollution problems.

A shortage of fuelwood and charcoal in sixteenth-century London, as its growing population placed unsustainable demands on surrounding woodlands, led to the increasing use of coal as a substitute—a harbinger of things to come. One of the largest cities in Europe, London's population is estimated to have more than doubled from 75,000 in 1550 to 200,000 in 1600, with an almost threefold increase to 575,000 in 1700. As wood became scarce its price increased dramatically, by some 780 % between 1540 and 1640, forcing large numbers of Londoners to switch to cheaper supplies of 'sea-coal' as an alternative source of fuel for their homes and businesses. Imported into London via coastal shipping routes and the Thames, mainly from north-eastern England, the growing consumption of 'sea-coal'—up from c.10,000 tons in 1580 to c.360,000 tons in 1680—brought increasing complaints about smoke emissions.[10]

Coal had been shipped to London since medieval times, where it was burned mainly by tradesmen in small workshops during fuelwood shortages. The dense smoke billowing from smiths' forges, breweries and lime kilns soon attracted criticism from its citizens, who were worried about deteriorating air quality in the city. In the late thirteenth century, for example, two royal commissions were appointed to inquire into complaints about pollution caused by coal-fired lime kilns operating in London. And Edward I issued a royal proclamation in 1307 to prohibit the use of smoky 'sea-coal' in the city's kilns because of the 'annoyance' caused to its inhabitants and concern over 'the injury of their bodily health'. In 1578 Queen Elizabeth I objected to the 'taste and smoke' of sea-coal issuing from brewing houses sited near the Palace of Westminster. By the seventeenth century the city's smoke-cloud had thickened, as more and more Londoners made the transition from wood and charcoal to fossil fuel consumption.[11] This provoked some influential figures of the day to protest about the state of London's atmosphere, best exemplified by the publication of John Evelyn's pamphlet *Fumifugium* in 1661, in which he spoke indignantly of:

> … that Hellish and dismall Cloud of SEA-COAL …perpetually imminent over her head … mixed with the otherwise wholesome and excellent *Aer*, that her *Inhabitants* breathe nothing but an impure and thick Mist accompanied with a fuliginous and filthy vapour, which renders them obnoxious to a thousand inconveniences, corrupting the *Lungs*, and disordering the entire habits of their Bodies; so that *Catharrs*, *Phthisicks*, *Coughs* and *Consumptions* rage more in this one City than in the whole Earth besides.[12]

In addition, he described how the smoke-cloud damaged the city's architecture and green spaces, as well as Londoner's clothes and possessions. Smoke emissions on such a scale would not be tolerated in the other great cities of Europe, Evelyn

---

[10] Te Brake (1975), Brimblecombe (1988) and Jenner (1995).

[11] Te Brake (1975), Brimblecombe (1988) and Sieferle (2001).

[12] Evelyn (1976) p. 5.

argued, and he recommended that polluting industries be relocated outside of London. But from the end of the eighteenth century, rapid urban-industrial growth would see air pollution from the burning of fossil fuels become a major environmental problem throughout Europe and the wider world.

5.3 The Age of Smoke, c.1780–1950

Coal was essential to the rise of industrial towns and cities, first in Britain, followed by northern Europe, the United States and other parts of the world. World coal output was around 10 million tons in 1800, with about 80 % of the total mined in Britain. By 1900, global coal output had increased enormously to about 780 million tons, with both Germany and the United States now major producers. There were few big cities before coal was used to power machines (in 1800 just six cities in the world had over 500,000 inhabitants), their size and number held in check by the low productivity of their hinterlands. The shift from a solar to a fossil energy system removed old constraints on urban and economic expansion (by 1900 there were 43 cities that exceeded half a million in population, including 16 of more than one million); but at a high socio-environmental cost.[13] Air pollution rose to unprecedented levels as coal replaced wood and charcoal for industrial and home energy uses, impacting negatively on nature and human health in new manufacturing centres like Birmingham, Leeds, Manchester, Chicago, Pittsburgh, St. Louis and Germany's Ruhrgebiet.

The concentration of industry was made possible by improved transport infrastructure—canals and railways—that could get massive volumes of coal into towns and cities quickly and cheaply. While each place had its own experience, due to differences in topography, climate, population density and economic base (such as textiles, steel or chemicals), the spread of coal-fired industrialisation also brought air pollution problems that were common to all. The tall smokestacks of factories and furnaces dominated nineteenth and early-twentieth century cityscapes, and their emissions permanently filled the air. On arrival, visitors to new industrial towns found that the acrid smoke stung their eyes and inhibited their breathing, while the falling soot soiled their clothes and skin. Smoke pollution, denser in winter than in summer, seriously damaged architecture in urban-industrial areas. As early as 1854 Charles Dickens, in his lesser-known role as journalist, complained that many of Britain's grand public edifices appeared to be 'built of coal' rather than of brick or stone. Monumental public buildings, such as new town halls that expressed civic pride, were soon defaced by soot and grime. In addition, the stonework of historic buildings, such as the Houses of Parliament, St. Paul's Cathedral and York Minster, began to erode under the effects of acid rain.[14] Writing on 'The Air of Towns' in 1859, the Manchester-based scientist Robert Angus Smith first coined the term:

---

[13] Smil (1994), Nye (1998) and Sieferle (2001).

[14] Mosley (2008).

> It has often been observed that the stones and bricks of buildings, especially under projecting parts, crumble more readily in large towns where much coal is burnt than elsewhere… I was led to attribute this effect to the slow but constant action of the acid rain… it is not to be expected that calcareous substances will resist it long, and one of the greatest evils in old buildings in Manchester is the deterioration of the mortar. It generally swells out, becomes very porous, and falls to pieces on the slightest touch.[15]

Using Smith's extensive chemical analyses of free acids in Manchester's rainwater, Dietrich Schwela has converted his figures to estimate that the pH value of the city's rainfall over a century ago was a very low 3.5.[16] This is a far more acidic value than today's measurements for acid rain in Europe and the United States. In the USA and Germany smoke, soot and acid rain had similar deleterious effects on the built environment. Chicago's dazzling 'White City', an exhibit of elegant buildings created for the 1893 World's Fair by some of America's best architects, quickly darkened and decayed. While in industrialising Germany famous buildings such as Cologne Cathedral were seriously corroded by air pollution.[17]

The omnipresent smoke that blanketed fast-growing industrial areas was also linked to the destruction of urban nature. Green spaces and blue skies had largely become 'meaningless terms' for poor city-dwellers during what some contemporaries called 'the age of smoke'. In Manchester, the world's first real industrial city, damage to local vegetation was devastating, as Robert Holland, consulting botanist to the North Lancashire Agricultural Society, noted:

> Some years ago I had the honour of making an inspection of all the public parks of Manchester on behalf of the Corporation… I scarcely need say that, going as I did from the fresh green country, I was horrified to see the havoc that was being made. Fine open spaces… which ought to have been beautiful, and would have been picturesque if well covered with trees, and which should have supplied pleasant recreation grounds for a population that sees far too little of country life and breathes far too little of fresh country air – rendered hideous by the blackness of everything with[in] them – trees stunted, dying, flowers struggling to bloom and sometimes their species scarcely recognisable. It is no exaggeration; and as long as the surrounding chimneys send out volumes of sulphurous acid and of carbon there can be no improvement …[18]

The effects of smoke on flora in Manchester's parks were representative of the conditions in and around coal-fuelled industrial cities more generally. In the 1920s, for example, the authorities of Missouri Botanical Gardens in St. Louis, Brooklyn Botanical Gardens and New York's Central Park all lamented the loss of trees, shrubs and plants from smoke-related problems. At roughly the same time, official reports into air pollution emanating from the Ruhr in Germany revealed damage to cropland and woodland on a regional scale.[19] While some vegetation could withstand the unrelenting assault from air pollution (such as ash, elder, hawthorn, poplar

[15] Smith (1859) p. 232.

[16] Schwela (1983).

[17] Stradling (1999) and Frenzel (1985).

[18] Mosley (2008) p. 41.

[19] Obermeyer (1933) and Brüggemeier (1994).

trees, privet hedges, rhododendrons and willows), there was a considerable loss of diversity in urban-industrial areas.

Smoke pollution—by absorbing and scattering light—was recognised to lower sunshine levels significantly for city-dwellers. Early scientific investigations in Leeds and Manchester suggest that they lost as much as 50 % of available sunlight and daylight. Fogs formed readily in industrial towns and cities as the particulate pollution (smoke and soot) provided in abundance the necessary nuclei for condensation and the formation of water droplets, especially at low temperatures. As a result of the presence of the 'smoke nuisance', fogs grew denser and became more frequent, particularly during the winter months. An investigation undertaken on behalf of the Meteorological Council in 1901–1903 concluded that 20 % of London's fogs were smoke induced. In cold, calm atmospheric conditions fog often went hand-in-hand with a temperature inversion; when the air overhead was warmer than at ground level it acted like a lid, trapping polluted air in the streets for long periods. Moreover, coal smoke formed a sticky film around the water droplets, which meant that town fogs evaporated far less easily than country fogs. When thick fogs occurred it was not unusual for the inhabitants of British and American industrial cities to experience 'night at noon', which seriously disrupted business, transport and urban life as visibility diminished to just a few metres.[20] Reduced sunlight, dirty air and the increase in the frequency and persistence of urban fogs were also a threat to human health on both sides of the Atlantic.

Rickets, a disease of childhood which resulted in softened, calcium-deficient bones, was caused by poor diet and insufficient exposure to sunlight. It was endemic in the gloomy manufacturing towns and mining areas of northern Europe and the United States until the 1940s. Respiratory diseases were rife where air quality was poor, and by the turn of the twentieth century bronchitis was the biggest single killer in Britain's factory towns. Smoky industrial centres in Germany and the United States also suffered inflated death rates from diseases of the lungs. The doors and windows of people's homes were routinely kept closed to exclude soot and smoke, which meant that urban populations—who spent a good deal of their time indoors—were overexposed to tuberculosis and other infectious diseases. Contemporary observers also noted that the same factors that influenced the high incidence of rickets and respiratory diseases in large towns and cities could also have serious psychological effects on their inhabitants. In 1913, for example, research conducted by Dr. J.E. Wallace Wallin, as part of an investigation undertaken by the Mellon Institute at the University of Pittsburgh, linked smoke pollution to a wide variety of psychological problems, including: 'chronic ennui'; 'morbid emotions'; 'instability of attention'; 'irritability'; and 'lessened self-control'. And as well as affecting the physical and mental well-being of city-dwellers, air pollution was also closely connected to a perceived decline in moral standards. Drunkenness and criminal activity were both thought to increase beneath the thickening smoke-cloud, while many housewives simply gave up the arduous struggle to keep their

[20] Luckin (2003), Mosley (2007, 2008), Tarr (1996) and Stradling (1999).

homes and children clean. In Britain, the middle classes relocated to semi-rural suburbs beyond the reach of the smoke, leaving the working classes behind with few appropriate 'role models' to emulate.[21]

Common concerns about the detrimental effects of air pollution on industrial cities and their communities led to the establishment of numerous smoke abatement societies in Britain and the United States, but not in Germany where citizen's associations generally played less of a role in municipal politics.[22] Britain's first anti-smoke groups were founded in 1842, including a Committee for the Consumption of Smoke at Leeds and the Manchester Association for the Prevention of Smoke. Eager to secure clean air for the 'health, comfort and well-being' of city-dwellers, these early reformers tended to focus on technical solutions to the smoke problem. For example, meetings of the Manchester Association for the Prevention of Smoke were held in the lecture room of the Royal Victoria Gallery for the Encouragement of Practical Science, and they were attended by some of the city's foremost scientists, technologists, and industrialists, including William Sturgeon, Peter Clare, William Fairbairn, and Henry Houldsworth. Its main goal was to persuade manufacturers that preventing pollution was good business, as the installation of the latest abatement devices and efficient boiler-furnaces would reduce fuel bills as well as smoke emissions. Education was thought to be the key to making progress in smoke abatement, and later societies such as the London-based Smoke Abatement Committee and the National Smoke Abatement Society organised major exhibitions of smoke abatement technologies to encourage large numbers of industrialists and householders to adopt cleaner, fuel-efficient appliances (such as automatic stokers and closed stoves).[23] Anti-smoke activism did not begin in earnest in the United States until the late 1880s and early 1890s, when reformers in Chicago, Cleveland, St. Louis and Pittsburgh organised to address the problem.[24]

Unlike Britain, middle-class women, rather than professional men, were in the vanguard of the smoke abatement movement in the United States. Reflecting the fact that women, as housewives and mothers, bore a heavy burden in keeping homes clean and healthy in the face of a constant barrage of soot and grime, influential women's groups such as the Ladies Health Protection Association in Pittsburgh and the Wednesday Club in St. Louis spearheaded the anti-smoke campaign. But as questions of technology, economy and efficiency began to come to the fore, by the early 1910s male engineering experts were playing a central role in the American smoke abatement movement. There was increasing cooperation between British and American anti-smoke activists with, for example, delegates from Chicago, Pittsburgh and other US cities attending the 1912 International Smoke Abatement Exhibition in London. However, as bituminous coal was both plentiful and inexpensive, and smoke-consuming appliances were expensive to buy

---

21 Mosley (2003, 2008), Andersen (1994) and Stradling (1999).

22 Mosley (2008), Stradling and Thorsheim (1999) and Uekötter (1999).

23 Mosley (2007, 2008) and Ranlett (1981).

24 Stradling (1999).

and often unreliable to run, at this time many industrialists and most householders chose to ignore calls to burn fuel more efficiently. Anti-smoke activists often spoke of the 'curse of cheap coal'. Although the educational approach had reduced the amount of smoke in some places, air pollution remained a serious cause for concern in most industrial towns and cities.[25] By the turn of the twentieth century, frustrated reformers—especially those campaigning in Britain—believed that tougher anti-smoke legislation needed to be introduced by national governments if the movement for clean air was to be successful.

Despite economic and cultural differences, a number of detailed studies have clearly shown that the British, American and German legislatures all embarked on a 'process of compromise' with the Industrial Revolution so as not to unduly hinder urban-industrial growth. As the first industrial nation, Britain was a pioneer where legislation to control smoke pollution was concerned. Up until the end of the eighteenth century, its Common Law courts had accepted that there was a natural right for an individual to enjoy clean air (and pure water) on their own property. It was no defence for a businessman causing nuisances to claim that his operations brought jobs and prosperity. Thereafter, a growing reliance on coal-fuelled steam power in new industrial towns provided the impetus for a shift in the focus of traditional nuisance law. As smoke and other forms of industrial pollution worsened in the manufacturing districts of Britain, the dilemma faced by nuisance law judges was, 'How best to reconcile the often conflicting goals of environmental quality and economic development'?[26] The increasing use of the doctrine of 'social-cost balancing' by early nineteenth century judges weakened the plaintiff's right to protection, and a rigid interpretation of liability was abandoned in industrial areas. Indeed, Victorian judges often stated that 'life in factory towns required more forbearance than life elsewhere'.[27] The utilitarian concept of 'social-cost balancing' allowed the courts to weigh the costs of imposing sanctions on a polluter against the benefits of abating the pollution. In new factory towns like Manchester the benefits of abating smoke were thought to be more than outweighed by the possible negative repercussions for the economy of obtaining injunctions to stop polluting businesses. Thus, traditional restrictions on economic enterprise that had previously shielded people and the environment from the injurious emissions of industry were relaxed. The Common Law became ineffectual because the new industrial society had made a pragmatic trade-off: dirty air in return for economic success, employment, and consumer goods. As the number of smoking chimneys rapidly multiplied in Britain's industrial towns, relatively few air pollution cases were brought before the Common Law courts. In the 90-year period after 1770 there were, on average, only one or two actions in England every 10 years.[28]

---

[25] Gugliotta (2000), Stradling (1999), Stradling and Thorsheim (1999), Mosley (2008) and Thorsheim (2006).

[26] Rosen (1993).

[27] Brenner (1974).

[28] Rosen (1993), Brenner (1974) and McLaren (1983).

Parliament appointed several committees in the nineteenth and early twentieth centuries to investigate the 'nuisances' arising from smoke pollution, including: the Taylor Committee on Steam Engines and Furnaces, 1819–1820; the Mackinnon Committees on Smoke Prevention, 1843 and 1845; the Select Committee on Smoke Nuisance Abatement, 1887; and the Newton Committee on Smoke and Noxious Vapours, 1914 and 1920–1921. And both municipal and national government enacted a raft of legislation to deal with the 'smoke nuisance', such as: the anti-smoke clauses in Local Acts, passed in Derby 1825, Leeds 1842, Manchester 1844, and elsewhere from the 1840s onwards; the Smoke Nuisance Abatement (Metropolis) Act, 1853; the anti-smoke clause in the Public Health Act, 1875; and the Public Health (Smoke Abatement) Act, 1926. However, as smoke was closely associated with jobs and economic well-being, and because it was not possible for contemporary physicians to prove conclusively that air pollution caused ill-health, there was little political or public support for tough action to tackle the problem. The evidence of anti-smoke 'experts' to parliamentary committees largely went unheeded; and successive anti-smoke laws were poorly drafted and weakly enforced.[29] They all contained the following flaws:

- An ambiguous 'best practicable means' clause. The 'best practicable means' of smoke abatement did not mean 'the best available means' using the latest technology. Rather, it meant the abatement apparatus that industrialists felt they could install at a cost they believed reasonable. Businessmen who argued in court that they had done their utmost to curtail excessive air pollution from their plant by employing technological measures that were commensurate with their financial means, however rudimentary, were usually assured of a sympathetic hearing.
- Insignificant fines. Legislation introduced to combat smoke—both local and national measures—only imposed low fines on offenders, effectively providing Britain's industrialists with a 'license to pollute'.
- Exemptions. The failure to regulate emissions from domestic fireplaces, which were major polluters of city air, when it was possible to reduce smoke significantly by burning fossil fuels in closed stoves. No government of the day was willing to incur the public's displeasure by taking action to interfere with their freedom to enjoy the warmth of a traditional open coal fire. Metal, brick, glass, pottery and other trades that used heating power intermittently were also exempt from early anti-smoke legislation.

Moreover, in Britain's factory towns many of the magistrates who dealt with smoke pollution cases were local industrialists, or they had close connections to local businessmen. Before the Second World War, weak and poorly enforced legislation had little effect in cutting smoke pollution and improving air quality in British cities.[30]

---

[29] Ashby and Anderson (1981) and Mosley (2007, 2008).

[30] Mosley (2007, 2008) and Ashby and Anderson (1981).

Concern for industrial interests also played a significant role in undermining legislative attempts to control smoke in the United States and Germany. As in Britain, few US citizens attempted to enforce the Common Law principle that harmful 'smoke nuisances' must be abated, especially as it was so difficult to link any damage caused to property to a particular polluter. And the commonly accepted notion that smoke meant employment and prosperity similarly prevented the tough municipal regulation of polluting businesses. In the United States, the first 'smoke ordinances' were not passed until the early 1880s, with Chicago, Cincinnati, Cleveland, Pittsburgh and St. Louis leading the way. But these anti-smoke laws were never rigorously implemented, for fear that 'over-regulation' might hamper economic growth in industrialising America. Despite the calls of some anti-pollution activists for the authorities to 'get tough' regarding smoke control, few lawsuits were brought against manufacturers. Instead, in the US the emphasis came to be placed on 'smoke inspection' by well-trained engineers, who advised businessmen about the latest technological advances for reducing air pollution from their works.[31] In Germany, laws to control smoke were passed during the 1870s and 1880s in cities such as Breslau, Nuremburg, Stuttgart and Freiburg. By the early twentieth century, local laws in Hamburg, Dresden and Munich were even being used systematically in the fight against smoke. But before the Second World War, German municipal government overall was unsuccessful in protecting townspeople and the environment from the deleterious impacts of smoke pollution. Indeed, it has been argued that the *Ruhrgebiet*—Germany's main urban-industrial region—was a place where fledgling industries were 'consciously protected' rather than nature.[32]

Smoke abatement made little progress until the interwar years, when the transition to cleaner energy systems, based on gas, oil and electricity, began in earnest. In the United States, cleaner sources of energy—oil and natural gas—were abundant and inexpensive, and their rapid development challenged the dominance of coal (although 'King Coal' remained sovereign in many northeastern and midwestern cities until the late 1940s and 1950s). In 1920, coal still accounted for almost 70 % of American energy production. But it only provided around a quarter of the nation's energy in 1955, replaced by oil and natural gas and which by this time held 41 and 26 % of the market share respectively.[33] In Britain, where coal was used to produce both gas and electricity, rates of growth were slower and progress disrupted by the outbreak of the Second World War ('town gas' was used extensively in the interwar home for lighting and cooking, but it remained uncompetitive for heating purposes). Moreover, gas works and power stations soon became major polluters in their own right.[34] But the growing availability of affordable 'smokeless' forms of energy saw anti-pollution campaigners on both sides of the Atlantic increase the pressure on governments to eradicate smoke,

31 Stradling (1999) and Stradling and Thorsheim (1999).

32 Uekötter (2009) and Brüggemeier (1994).

33 Nye (1998), Stradling (1999) and Tarr (1996).

34 Luckin (1990), Sheail (1991), Thorsheim (2002) and Mosley (2009).

which could no longer be considered as an 'inevitable' consequence of urban-industrial life.

The outbreak of war in 1939, however, and the need for rapid industrial production, saw public and political interest in smoke abatement diminish. It was London's 'Great Smog' of 1952, which according to official figures caused some 4,000 deaths from respiratory and cardiovascular disorders (and perhaps as many as 12,000 deaths according to recent research), that proved to be the catalyst for the introduction of stringent legislation to control air pollution in Britain. Following this tragedy, which demonstrated conclusively that polluted air could be just as deadly as polluted water supplies, the public and the press supported the passage of the Clean Air Act of 1956, which for the first time regulated both domestic and industrial smoke emissions. Widely considered by historians to be an important milestone in environmental protection, the legislation included powers to establish 'smokeless zones' in towns and cities, and householders and industrialists were required to burn cleaner fuel to meet their energy needs.[35] Local ordinances in American cities such as St. Louis and Pittsburgh had given the authorities the right to control the type of fuel consumed by industry prior to London's 'killer smog', reducing smoke significantly. Following this well-publicised disaster, and the Donora smog of October 1948 which had killed 20 people, concerns about health saw federal air pollution control acts passed in the United States in 1955, 1963 and 1967. But not until the Clean Air Act of 1970 were national air quality standards set for the first time. In post-war Germany, lacking influential citizen's associations and needing to rebuild its shattered industries, the government subsidised Ruhr coal production in order to boost economic growth. The first national German legislation to combat air pollution had to wait until the Federal Emissions Control Act of 1974, which sought reductions in emissions through the use of *Stand der Technik* or the 'best practicable means' of abatement. As was the case in the nineteenth century, this meant using abatement technology installed at a cost that took the economic circumstances of a particular firm into account.[36] Nonetheless, by the late 1960s and 1970s, the skies were clearing over the cities of the first industrial nations; but it had taken more than a century to solve the smoke problem.

In early twentieth-century Japan, the industrial Hanshin region between Kyoto, Kobe and Osaka experienced roughly the same smoke pollution levels as western European and American manufacturing centres. But after the passage of national smoke control legislation in 1962, with regulations being further tightened in 1970, city air in Japan also improved in quality. But smoke pollution continued to be a major problem in other parts of the world, particularly in the cities of the Soviet Union, eastern Europe and China where industrialisation was still heavily dependent on coal. Before the collapse of the USSR (1989–1991), the commitment of communist countries to economic growth—to match or even out-produce the

[35] Davis (2002) and Thorsheim (2004).

[36] Tarr and Zimring (1997), Stradling (1999), Davis (2002) and Morag-Levine (2003).

capitalist West—resulted in 'polluted skies as never before'.[37] Monopolistic state control over heavy industries such as coal mines, chemical works, and iron and steel foundries meant that there was little pressure to introduce abatement technologies or environmental protection legislation that might impede economic expansion. While smoke emissions were coming under control in the West, they were rising rapidly behind both the Iron and Bamboo Curtains. In Nizhni Tagil, an industrial city 700 miles east of Moscow, urbanites regularly experienced 'night at noon' because the smog was so thick. Chinese industrial cities, such as Shenyang and Lanzhou, were also characterised by high levels of smoke pollution. China, which now has over 120 cities with more than one million inhabitants, currently burns in excess of two billion tons of coal per annum (and it is likely to remain its dominant fuel for decades to come). In 1998, of the ten most polluted cities in the world, nine were to be found in China. Lanzhou's inhabitants had to breathe air with average levels of pollution that were more than 100 times the World Health Organisation's (WHO) guidelines. In 2013, the WHO estimated that over 1.3 million people around the world were killed by respiratory and other diseases associated with outdoor air pollution, the great majority in developing countries.[38]

China, which manufactures low-priced goods for export to the West, is now the world's biggest producer and consumer of coal (many developed nations have essentially 'outsourced' their pollution). Industrialising India (where smoke emissions were strictly regulated under the British Raj), is also among the top five nations for coal production, with cities such as Delhi, Kolkata and Mumbai all suffering from severe air pollution. The air quality in many Chinese and Indian manufacturing centres today is as poor as it was more than a century ago in Manchester or Pittsburgh (with emissions from car exhausts adding to the problem —see below). It is worth noting too that while coal now provides less than a third of the world's energy, between 1900 and 2000 global production continued to rise sharply from 780 million tons to 3.5 billion tons per annum.[39] And smoke due to biomass burning and large-scale forest fires as more land is cleared for agriculture in South Asia, creating the so-called Asian Brown Cloud, is now causing health and environmental problems across borders.[40]

5.4 The Era of Invisible Threats, c.1950: Present

In the nineteenth and early twentieth centuries, smoke and winter 'smog'—a term coined in 1905 by Dr. Harold Des Voeux of the London-based Coal Smoke Abatement Society to describe the fusion of smoke and fog—were the most obvious

---

[37] McNeill (2000).

[38] Davis (2002), Mosley (2010)and World Health Organisation (2013).

[39] McNeill (2000), Anderson (1995) and Dorf (2001).

[40] United Nations Environment Programme (2002) and Chakrabarti (2007).

air pollution problems caused by coal-burning and other activities in industrial towns and cities. The impacts of less visible atmospheric pollutants, however, were also becoming conspicuous to some contemporaries. Indeed, the British government had enjoyed some early success in regulating hydrochloric acid gas emissions from the alkali industry, based mainly in Merseyside, Tyneside and Glasgow, which produced soda ash and caustic soda for use in soaps, detergents, dyes, bleaches, and glass and paper making. Landowners living in close proximity to alkali works complained that the environmental damage they caused was severe, with acid deposition destroying both agricultural crops and woodland. The Alkali Act of 1863, overseen by the country's first Alkali Inspector, the aforementioned Robert Angus Smith, compelled manufacturers of soda ash to reduce their acidic emissions by 95 % using a simple and inexpensive condensing technique.[41] But the act did not control other gaseous pollutants from factories, most notably the sulphur dioxide—a key component of acid rain—that was released when fossil fuels were burned.

Acid rain had first been identified as a deleterious influence on the environment in and around urban-industrial centres during the mid-nineteenth century (see previous section). But it was not until the 1960s that acid rain began to attract significant public and political attention. While smoke pollution had declined in the developed world from the mid-twentieth century, the problem of acid rain persisted and spread. Legislation designed to control visible coal smoke did little to curb invisible emissions of sulphur dioxide, largely because the economic costs of fitting preventive flue-gas scrubbing systems were high. In Britain, for example, long-standing ideas about employing the 'best practicable means' of pollution abatement meant that solutions still had to be both technically and economically feasible. Worried that low-level concentrations of sulphur dioxide posed a threat to human health, regulators instead insisted on another 'technical fix'—raising the height of industrial chimneys—to better disperse and dilute this harmful pollutant. Coal-fired power stations, providing 'clean energy' in the form of electricity to homes and industry, produced much of the sulphur dioxide that reacted with moisture in the atmosphere to form acid rain (solving one environmental problem can often exacerbate another). By 1960, Britain had built more than 60 new power stations and greatly extended the generating capacity of many older installations, with their chimneys reaching heights in excess of 135 m. These tall chimneystacks, intended to reduce local air pollution, transported sulphur emissions over hundreds and even thousands of kilometres. By the end of the 1960s, Scandinavian scientists had shown that enormous flows of air pollution from Britain, carried by the prevailing winds, were causing lakes and rivers to acidify in Norway and Sweden. The ecological consequences also included the widespread decline of forests (although acid rain is just one of a number of cumulative stresses that can cause die-offs).[42] For a long time the impacts of acid rain were mainly local or regional, but new scientific research gradually revealed that it was causing environmental damage on an international scale.

---

[41] Ashby and Anderson (1981), Dingle (1982), Hawes (1995) and Garwood (2004).

[42] Osborn (2004), Sheail (1991), Brimblecombe (2008) and Lundgren (1998).

Industrial sulphur dioxide emissions travelled unhindered across national borders, and this type of transboundary pollution also became an issue elsewhere in Europe and the wider world. South-westerly winds carried air pollution from Germany, Czechoslovakia and Poland to Scandinavia. In addition, by the mid-1980s about half of Canada's annual sulphur deposition was found to originate in the United States, (although emissions from Ontario also drifted the other side of the border). And from the 1990s, Japan was regularly showered by acidic rainfall from China and South Korea, where the demand for electricity had increased dramatically due to rapid urban-industrial growth.[43] International air pollution problems required international cooperation if solutions were to be found. An early example of ground-breaking environmental diplomacy over cross-border air pollution is the Trail smelter dispute (1927–1941) between the United States and Canada, which set an international precedent by establishing the 'polluter pays' principle.[44] After almost 15 years of wrangling between negotiators, a decision was reached that stated:

> Under the principles of international law, as well as the law of the United States, no state has the right to use or permit the use of its territory in such a manner as to cause injury by fumes in or to the territory of another or the properties or persons therein, when the case is of serious consequence and the injury is established by clear and convincing evidence.[45]

The Trail smelter tribunal ruled that US farmers should receive compensation for crops and farmland damaged by sulphurous smoke emissions, and the principle that the 'polluter pays' became one of the fundamental underpinnings of international environmental law. This precedent clearly influenced later political debates over acid rain in northern Europe, North America and East Asia.

In June 1972, the United Nations Conference on the Human Environment convened in Stockholm, bringing world leaders together for the first time to talk about the state of the earth. The Swedish government took this opportunity to highlight how acid rain from Britain and other European countries was damaging Scandinavian lakes and forests. Despite most industrial countries having already accepted the 'polluter pays' principle, this had previously applied to emissions from easily identifiable sources such as the Trail smelter, located close to a shared border. As far as the larger-scale acid rain problem was concerned, with the aggregated emissions from an industrial region in one country harming the environment of another hundreds or even thousands of kilometres away, coming to an agreement was still to prove difficult. In this case, it was no easy matter to demonstrate conclusively that environmental damage in one place was caused by pollution emissions originating in another.[46] But two important principles proclaimed at the Stockholm conference, which echoed the Trail decision, helped to provide the political impetus necessary to reach an agreement to tackle acid rain. Principle 21 stated that:

---

[43] McNeill (2000), Schmandt et al. (1988), McCormick (1997) and Brimblecombe (2008).

[44] Wirth (2000).

[45] Elsom (1992) p. 309.

[46] Sheail (1991), McCormick (1997) and Lundgren (1998).

> States have, in accordance with the Charter of the United Nations and the principles of international law, the sovereign right to exploit their own resources pursuant to their own environmental policies, and the responsibility to ensure that activities within their jurisdiction or control do not cause damage to the environment of other States or of areas beyond the limits of national jurisdiction.

And Principle 22 stated that:

> States shall cooperate to develop further the international law regarding liability and compensation for the victims of pollution and other environmental damage caused by activities within the jurisdiction or control of such States to areas beyond their jurisdiction.[47]

Following the Stockholm conference, pressure began to build for international political action to be taken to reduce and prevent acid rain.

In 1979, 34 governments and the European Community signed up to the Convention on Long-Range Transboundary Pollution (including the United States and Canada), which had been drafted by the United Nations Economic Commission for Europe. Signatories to the Convention set long-term targets to cut sulphur dioxide emissions, and for the past three decades they have cooperated on various research and monitoring programmes concerning the movement of pollutants across borders. The Convention's initial provisions have been extended eight times since it was implemented to include the control of pollutants such as nitrogen oxides (1988), volatile organic compounds (1991), heavy metals (1998), and the abatement of eutrophication and ground level ozone (1999). As a result of this cooperative approach to pollution control, in Europe the environment is now beginning to recover from the effects of acid rain.[48] In 1980, the National Acid Precipitation Assessment Program (NAPAP) was authorised in the United States to coordinate long-term monitoring and research into acid rain. A cooperative federal project, re-authorised through the 1990 Clean Air Act Amendments, it also assessed the effectiveness of an emissions trading system that utilised economic incentives to cut releases of sulphur dioxide into the atmosphere. As with Europe, efforts to control acid rain in the US, especially from coal-fired power stations, have resulted in the gradual regeneration of some—but not all—of its lakes and forests.[49] In 1993, a meeting organised by the Environment Agency of Japan at Toyama, attended by senior officials from China, Indonesia, South Korea, Malaysia, Mongolia, the Philippines, Singapore, Thailand and Russia, saw the establishment of the Acid Precipitation Monitoring Network in East Asia (EANET) to assess the ecological impacts of transboundary air pollution.[50] But despite this cooperative initiative, which now involves thirteen participating countries, acid rain still remains a serious

---

47 United Nations Environment Programme (1972).

48 United Nations Economic Commission for Europe (2004) and Colbeck (2007).

49 National Acid Precipitation Assessment Program (2005) and Brimblecombe (2008).

50 Acid Precipitation Monitoring Network in East Asia (EANET) (2013) and Brimblecombe (2008).

problem in this fast-growing region of the world, with China's coal-burning power plants a major source of sulphur dioxide emissions.

After the Second World War, air pollution problems associated with mass car ownership began to accelerate, first in the developed and then in the developing world. Emissions of nitrogen oxides from automobile exhausts are a significant source of acid rain. And until recently, vehicle exhausts spewed millions of tons of toxic lead into the atmosphere (tetraethyl lead was used as an additive in petrol to help the engine run smoothly), impairing the normal intellectual development of children in urban areas. Other vehicle pollutants known to have direct or indirect adverse effects on human health include carbon monoxide, sulphur dioxide, fine airborne particulates and volatile organic compounds (VOCs) such as benzene, which contribute to the increase or severity of cardio-respiratory conditions and even pose a cancer risk to city dwellers.[51] Studies have shown that children and the elderly are particularly at risk from exposure to these pollutants. Somewhat ironically, from the mid-1950s smoke control initiatives in Western cities allowed more sunlight to penetrate to the streets, where it reacted with pollutants such as VOCs and nitrogen oxides emitted from vehicle exhaust pipes to form dangerous ozone-laden photochemical smog. Excessive ozone in the air can cause breathing problems and trigger asthma attacks. But laws that controlled smoke in urban areas did not cover less visible emissions from cars (initially thought to be a minor issue in comparison with industrial sources of pollution). The rapid rise of automobile-centred transport systems—with high levels of individual car use—led to an air pollution problem that was difficult to control.[52]

Concerns about the effects of air pollution from automobiles on human health first emerged as early as the 1940s in Los Angeles—a city built for cars—as its residents began to complain of smarting eyes and a wide range of respiratory ailments. Suggested measures to limit exhaust emissions proposed during the 1950s, as vehicle ownership in California topped 7 million, included encouraging car pools, prohibiting car use at certain times in certain places, and imposing a 'smog' tax on drivers.[53] However, 'solutions' that infringed on people's personal freedoms, and that increased the cost of motoring, unsurprisingly proved to be unpopular with the public—then and now. By the 1970s, traffic pollution problems were emerging elsewhere in the world, with, for example, photochemical smog episodes in Athens, Greece, (called *nephos* locally), rivalling those of Los Angeles and damaging both human health and its historic buildings. At the turn of the twenty-first century, there were around 600 million cars registered worldwide, 200 million of them in the United States alone. But as car ownership has risen sharply in the developing world, places like Bangkok, Buenos Aires, Mexico City and Mumbai now rank among the most heavily polluted 'smog cities'. Globally, there has been a tendency for older, more polluting cars to be exported from the

---

51 Read (1994) and McMichael (2001).

52 Read (1994), Elsom (1992), Dupuis (2004) and McNeill (2000).

53 Elsom (1992) and Rajan (2004).

affluent West to developing nations, from the United States to Mexico for instance, where used vehicles are more affordable.[54]

Governments in the developed world have begun to include performance and fuel efficiency standards for automobiles in Clean Air Acts to obtain breathable air for their citizens, and to monitor air quality in urban centres in an attempt to meet internationally agreed standards, such as those published by the World Health Organisation.[55] However, there has been a much slower adoption of controls on exhaust emissions in the highly congested cities of the developing world. During the early 1990s, for example, national air quality standards were violated in Mexico City over 300 days every year (despite regular monitoring). Like Los Angeles, hemmed in by mountains Mexico City's photochemical smog was difficult to disperse. A pollution reduction scheme which prohibited cars with odd and even-numbered licence plates being driven in the city on particular days was ineffective, as many people switched vehicles or even plates. More recently, vehicle inspections twice a year and tax incentives to use cleaner fuels and new technologies, such as catalytic converters, have begun to reduce Mexico City's ozone levels.[56] While a great deal of progress has been made, its 4 million cars—especially older vehicles—are still responsible for most of the city's atmospheric pollution, and peak ozone levels still regularly exceed recommended limits.

In most parts of the world vehicle pollution has mainly been seen as a scientific or technological problem, to be solved by the development of cleaner fuels, more efficient engines and catalytic converters to reduce harmful exhaust emissions, rather than a social and legal one. The World Health Organisation has recently warned that vehicle exhausts are a major contributor to more than million premature deaths worldwide every year from urban outdoor air pollution. At present there are around 1 billion motor vehicles on the world's roads, and by 2030 this figure may double as a result of China's and India's rapid economic growth. The benefits of reducing emissions using cleaner technologies may be greatly diminished by sheer weight of numbers, as low-cost cars such as the Indian Tata motor company's Nano are changing the way millions of people travel in the developing world.[57]

At ground level, then, ozone is a pollutant that adversely affects human health and the environment. However, about 90 % of all ozone is found high in the stratosphere (between 10 and 50 km above the earth's surface), where it shields life on the planet from harmful ultraviolet (UV) radiation. Since 1930, when the DuPont company began to manufacture the first of the family of chlorofluorocarbons (CFCs) under the trade-name Freon, the stratospheric ozone layer has been under invisible assault. Chlorofluorocarbons like Freon made the widespread use of

---

[54] Papaioannou and Sapounaki-Drakaki (2001), McNeill (2000) and United Nations Environment Programme and Organisation for Economic Cooperation and Development (1999).

[55] Robinson (2005).

[56] Jacobson (2002), Lezema (2004) and United Nations Environment Programme and Organisation for Economic Cooperation and Development (1999).

[57] Elsom (1992), World Health Organisation, Health and Environment Linkages Initiative (HELI) (2013) and Sperling and Gordon (2010).

refrigerators and air-conditioning systems possible, and were later utilised in industrial solvents, insulated packaging and as propellants in aerosol cans for deodorants, hairsprays, paints, furniture polishes, and numerous other household products. Up until the 1920s, most natural refrigerants had been extremely hazardous, such as ammonia (both a flammable and toxic gas). Freon and other CFCs were stable, nonflammable, nontoxic, and inexpensive to produce. Chemically, they were stable and unreactive until they reached the stratosphere, a slow journey that took several years, where UV radiation broke them down and released the ozone depleting substance chlorine. However, because demand for CFCs was modest before the Second World War, and the process of ozone depletion was slow, it was not until the 1970s that concerns about the thinning of the ozone layer emerged.[58]

As late as 1950, emissions of CFC compounds were still moderate at around 20,000 tons per annum. But by the 1970s, as DuPont's original patents expired and the market became crowded with new producers—and consumers, as demand for refrigerators and air-conditioning units rose—CFC emissions increased sharply to some 750,000 tons annually.[59] During the same decade, scientific studies of CFCs in the stratosphere indicated that they had the potential to seriously deplete the ozone layer. In 1985, the depletion of the Earth's protective ozone layer was finally confirmed by scientists with the British Antarctic Survey, and was soon made 'visible' to the public by the processing and dissemination of data from NASA's Nimbus-7 satellite. Numerous studies since then have revealed a global decrease in what is sometimes called 'good ozone'. Inert at ambient temperatures, human-made chlorofluorocarbons reacted very aggressively with ozone in the extreme cold of the polar stratosphere in the months of early spring, when sunlight returned. The destruction of ozone by the chlorine in CFCs resulted in 'ozone holes' appearing over both the Antarctic and Arctic regions (the latter being less pronounced). The ozone layer also thinned significantly above Australia and Chile, and slightly above much of the Northern Hemisphere (but not over the tropics). The increase in exposure to ultraviolet radiation can cause skin cancer and eye disorders (especially cataracts and 'snow blindness') in humans, as well as impairing photosynthesis in plants and reproduction in fish and phytoplankton—adversely affecting terrestrial and aquatic ecosystems.[60] The threats to human health and the environment were clear and attracted considerable media attention, which led to widespread public support for action to combat ozone depletion.

Governments lost no time in responding to the threat posed by CFCs, in no small part because competitively priced substitutes were easy to manufacture, such as hydrofluorocarbons (HFCs) which contain no ozone-depleting chlorine. In 1985, the United Nations Environment Programme organised the Vienna Convention for Protection of the Ozone Layer, which encouraged international cooperation on research into ozone depletion and its effects. It was followed in 1987 by the

[58] Jacobson (2002) and McNeill (2000).

[59] Ackermann (2002) and McNeill (2000).

[60] Lambright (2005), Davis (2002), McMichael (2001) and McNeill (2000).

negotiation of the Montreal Protocol on Substances that Deplete the Ozone Layer, an agreement that aimed to substantially reduce the production and consumption of CFCs. Since entering into force in January 1989, it has been ratified by 196 states, including all members of the United Nations (and adjusted and amended on ten occasions). The first treaty to achieve universal ratification, the Montreal Protocol stands as the most successful example of international action to tackle a global air pollution problem, with 95 % of ozone depleting substances now having been phased out. But because the overall lifetimes of CFCs are between 50 and 100 years, the stratospheric ozone layer is not expected to recover to its original levels, nor the 'ozone holes' to finally close, until well into the second half of the twenty-first century.[61] The replenishment of stratospheric ozone, however, may exacerbate global warming as it acts as a greenhouse gas—although there is still some uncertainty about its potency and significance.[62] In addition, CFCs and HFCs are potent greenhouse gases and each contributes to climate change, so in that respect one harmful chemical has been replaced by another.

If the response to stratospheric ozone depletion by the international community was relatively swift and effective, progress towards a global agreement to cut greenhouse gas emissions has been slow in comparison. As early as 1896, the Swedish scientist Svante Arrhenius first described how burning coal to fuel the Industrial Revolution would result in a build-up of carbon dioxide (CO_2) with the potential to raise the average temperature of the planet. An idea echoed in 1938 by the British engineer and amateur climatologist Guy Stewart Callendar, who warned members of the Royal Meteorological Society in London that CO_2 emissions from fossil-fuelled human industry were changing the climate. Global warming, however, was a marginal issue at this time, and Callendar himself thought that a small rise in global temperature might even be beneficial for humankind by boosting food production in some regions of the northern hemisphere. It was not until the late 1970s, when the first World Climate Conference met in Geneva, that human-induced climate change began to emerge as a serious environmental problem. Organised by the World Meteorological Organisation (WMO) and the United Nations Environment Programme (UNEP), in 1979 more than 300 scientific experts from 50 countries came together to discuss the possibility that human activities were causing global warming. They reached a consensus that an increase in anthropogenic CO_2 emissions could cause significant long-term climate change, although delegates did not agree about how urgent the need was for preventive measures to be put in place. And overall the meeting attracted very little media or public attention.[63]

Record high temperatures in 1988 with, for example, heat waves and droughts reminiscent of the Dust Bowl era in the United States, saw public awareness of climate change begin to increase (although no single weather event can be

[61] Jacobson (2002) and United Nations Environment Programme (2009).

[62] Gao et al. (2010).

[63] Weart (2003) and Kessel (2006).

attributed solely to global warming). The Intergovernmental Panel on Climate Change (IPCC), a scientific body set up by WMO and UNEP in the same year to advise the world's governments on the topic, later reported that twentieth-century temperature increases were unusually rapid when compared with those for the previous two millennia. Over the last century, average global temperatures have risen by 0.3–0.6 °C, and they are projected to pass the critical threshold of 2 °C by 2100 (the 'tipping point' for 'dangerous' changes in climate).[64] Left unchecked, human-induced global warming will result in a whole range of 'natural disasters' that include: desertification, droughts, forest fires, species extinctions, sea-level rise, and a destructive change in weather patterns.

By living off the accumulated energy capital of the past (fossil fuels) instead of 'current income' (renewable energies), humankind released huge quantities of CO_2 that had been securely locked up underground, profoundly altering the global carbon cycle. Ice core data from Greenland and Antarctica show conclusively that atmospheric concentrations of the main 'greenhouse gas' CO_2 now far exceed preindustrial values, up from around 280 to more than 390 parts per million, with most of this increase coming after 1950 (well above the 350 parts per million mark that many scientists say is the highest 'safe' level for carbon dioxide in the atmosphere). Concentration levels exceeded the symbolic 400 parts per million threshold at several Arctic monitoring stations for the first time in 2012.[65] Other anthropogenic greenhouse gas emissions have also increased substantially since 1750, particularly methane from irrigated agriculture, livestock, decomposing garbage, gas pipelines and coal mining, as well as nitrous oxide from agricultural fertilisers, car exhausts and industrial smokestacks. In addition, CFCs and HFCs account for a small percentage of human-generated warming (see above). According to the World Resources Institute, in the year 2005 agriculture was responsible for almost 14 % of greenhouse gas emissions; transport (especially private automobile use) emitted just over 14 %; industry and industrial processes produced 19 %; changing land use (primarily deforestation) just over 12 %; while energy consumption for electricity and heating (particularly in residential and commercial buildings) generated almost 25 % of the total. Because most humans now live and work in cities, and the world's urban population will continue to grow in the future, they are the 'front lines' in the battle to halt dangerous climate change.[66]

At the global level, between 1850 and 2000 the United States, Britain, Germany, Japan and other industrialised countries accounted for about 75 % of greenhouse gas emissions. But industrialising countries like China and India—with their rapidly growing populations and booming economies—will drive carbon dioxide build-up in the atmosphere for decades to come. Effective international policies and

---

64 Weart (2003) and Intergovernmental Panel on Climate Change (2013).

65 Mosley (2010) and World Meteorological Organization (2013).

66 Penna (2010), World Greenhouse Gas Emissions (2005), Worldwatch Institute (2009) and UN-HABITAT (2008).

substantial investment in clean, renewable technologies are urgently needed to tackle climate change. However, the globalisation of economic life and consumer culture, with China and other developing countries manufacturing low-priced goods for export to the West (as previously mentioned, developed nations have essentially 'outsourced' much of their atmospheric pollution), and the great expense of designing and disseminating environmentally-sound technologies, makes negotiations to radically reduce emissions of greenhouse gases complex. Huge disparities in average carbon dioxide emissions per person between developed and developing countries—over 19 tons in the United States, around 5 tons in China, and less than 2 tons in India—also means that a fair and equitable agreement to combat climate change will be difficult to reach.[67] The matter is further complicated by the need to regulate the activities of large multinational corporations (some more powerful than small nation states). But the cost of failure, as the former World Bank economist Nicholas Stern has warned, could be catastrophic. As well as the human suffering caused by the increased incidence of severe floods, forest fires, droughts and food shortages, if global warming is left unchecked it could shrink the world's economy by up to 20 %. The impacts of climate change, though, will not be evenly distributed. The poorest people and countries, Stern stressed, will suffer most and earliest.[68]

To date, however, progress towards a binding global pact to cut greenhouse gas emissions has been slow. States and societies still prioritise economic growth over protecting the environment. A mechanism for discussion, the Framework Convention on Climate Change, was put in place following the 1992 United Nations Conference on Environment and Development—better known as the Earth Summit—held in Rio de Janeiro. But the resulting Kyoto Protocol, signed in Japan in 1997, which committed participating nations to a collective 5.2 % reduction in carbon dioxide emissions (against 1990 levels) by 2012, was flawed and ineffective, with no tough penalties for non-compliance. Nonetheless, the United States withdrew from the process in 2002, worried about damage to its economy. More than half of the countries that signed are unlikely to meet their modest reduction targets. Although both China and India were involved, they did not have to curtail their emissions. In 2013, the fifth IPCC assessment report—prepared by 259 researchers from 39 countries around the world—reported with 95 % certainty that human activities since the Industrial Revolution were responsible for climate change. If current 'Beyond Kyoto' negotiations do not produce a more robust international agreement to considerably reduce greenhouse gases, serious global warming may become irreversible.[69]

---

67 Mosley (2010), Penna (2010) and UN-HABITAT (2008).

68 Stern (2007).

69 Weart (2003), Giddens (2009) and Intergovernmental Panel on Climate Change (2013).

References

Acid Precipitation Monitoring Network in East Asia (EANET) (2013) First expert meeting. http://www.eanet.asia/event/expert/expert01.html. Accessed on 18 Oct 2013

Ackermann ME (2002) Cool comfort: America's romance with air-conditioning. Smithsonian Institution Press, Washington

Andersen A (1994) Die Rauchplage im deutschen Kaiserreich als Beispeil einer versuchten Umweltbewältigung. In: Jaritz G, Winiwarter V (eds) Umweltbewältigung: Die Historische Perspektive. Verlag Für Regionalgeschichte, Bielefeld, pp 99–129

Anderson MR (1995) The conquest of smoke: legislation and pollution in colonial Calcutta. In: David A, Ramachandra G (eds) Nature, culture, imperialism: essays on the environmental history of South Asia. Oxford University Press, New Delhi, pp 293–335

Ashby E, Anderson M (1981) The politics of clean air. Clarendon Press, Oxford

Brenner JF (1974) Nuisance law and the industrial revolution. J Legal Stud 3:403–433 (Quote on page 414)

Brimblecombe P (1988) The big smoke: a history of air pollution in London since medieval times. Routledge, London

Brimblecombe P (2008) Air pollution history. In: Sohki RS (ed) World Atlas of atmospheric pollution. Anthem Press, London, pp 7–18

Brüggemeier F-J (1994) A nature fit for industry: the environmental history of the Ruhr Basin, 1840–1990. Environ Hist Rev 18:35–54

Chakrabarti T (2007) Fire, fume and haze: environmental disorder in Southeast Asia and ASEAN response. In: Chakrabarti R (ed) Situating environmental history. Manohar, New Delhi, pp 359–385

Colbeck I (2007) Air pollution: history of actions and effectiveness of change. In: Pretty J et al (eds) The sage handbook of environment and society. Sage, London, pp 374–384

Davis D (2002) When smoke ran like water: tales of environmental deception and the battle against pollution. Perseus Press, Oxford

Dingle AE (1982) The monster nuisance of all: landowners, alkali manufacturers, and air pollution, 1828–64. Econ Hist Rev 25:529–548

Dorf RC (2001) Technology, humans, and society: toward a sustainable world. Academic Press, San Diego

Dupuis EM (ed) (2004) Smoke and mirrors: the politics and culture of air pollution. New York University Press, New York

Elsom DM (1992) Atmospheric pollution: a global problem. Blackwell, Oxford, p 309

Evelyn J (1976) Fumifugium; or the inconvenience of the Aer and Smoak of London dissipated. Rota, Exeter, 5 (First published in 1661)

Frenzel G (1985) The restoration of medieval stained glass. Sci Am 252:126–135

Gao W, Schmoldt DL, Slusser JR (2010) UV radiation in global climate change: measurements, modeling and effects on ecosystems. Springer-Verlag, Berlin

Garwood C (2004) Green crusaders or captives of industry? The British alkali inspectorate and the ethics of environmental decision making, 1864–95. Ann Sci 61:99–117

Giddens A (2009) The politics of climate change. Polity Press, Cambridge

Gugliotta A (2000) Class, gender and coal smoke: gender ideology and environmental injustice in Pittsburgh, 1868–1914. Environ Hist 5:165–193

Hawes R (1995) The control of alkali pollution in St. Helens. Environment History 1:159–171

Hong S, Candelone J-P, Patterson CC, Boutron CF (1994) Greenland ice evidence of hemispheric lead pollution two millennia ago by Greek and Roman civilizations. Science 265:1841–1843

Hong S, Candelone J-P, Patterson CC, Boutron CF (1996) History of ancient copper smelting pollution during Roman and medieval times recorded in Greenland Ice. Science 272:246–249

Hughes JD (1996) Pan's travails: environmental problems of the ancient Greeks and Romans. Johns Hopkins University Press, Baltimore

Intergovernmental Panel on Climate Change (2013) Fifth assessment report http://www.ipcc.ch/index.htm#.UmeXOfmTguc. Accessed 18 Oct 2013
Jacobson MZ (2002) Atmospheric pollution: history, science and regulation. Cambridge University Press, Cambridge
Jenner M (1995) The politics of London air: John Evelyn's *Fumifugium* and the restoration. Hist J 38:535–551
Kessel A (2006) Air, the environment and public health. Cambridge University Press, Cambridge
Lambright WH (2005) NASA and the environment: the case of ozone depletion. NASA, Washington D.C
Lezema JL (2004) The social and political construction of air pollution: air pollution policies for Mexico City, 1979–1996. In: Dupuis EM (ed) Smoke and mirrors: the politics and culture of air pollution. New York University Press, New York, pp 324–336
Luckin B (1990) Questions of power: electricity and environment in inter-war Britain. Manchester University Press, Manchester
Luckin B (2003) The heart and home of horror: the great London fogs of the late nineteenth century. Soc Hist 28:31–48
Lundgren L (1998) Acid rain on the agenda: a picture of a chain of events in Sweden, 1966–1968. Lund University Press, Lund
Mamane Y (1987) Air pollution control in israel during the first and second century. Atmos Environ 21:1861–1863
McCormick J (1997) Acid Earth: the politics of acid pollution. Earthscan, London
McLaren JPS (1983) Nuisance law and the industrial revolution—some lessons from social history. Oxf J Legal Stud 3:155–221
McMichael T (2001) Human frontiers, environments and disease: past patterns, uncertain futures. Cambridge University Press, Cambridge
McNeill JR (2000) Something new under the sun: an environmental history of the twentieth-century world. Penguin, Harmondsworth, p 89
Morag-Levine N (2003) Chasing the wind: regulating air pollution in the common law state. Princeton University Press, Princeton
Mosley S (2003) Fresh air and foul: the role of the open fireplace in ventilating the British home, 1837–1910. Plann Perspect 18:1–21
Mosley S (2007) The home fires: heat, health and atmospheric pollution in Britain, 1900–1945. In: Jackson M (ed) Health and the modern home. Routledge, New York, pp 196–223
Mosley S (2008) The chimney of the world: a history of smoke pollution in Victorian and Edwardian Manchester. Routledge, London.
Mosley S (2009) 'A network of trust': measuring and monitoring air pollution in British cities. Environment History 15:273–302
Mosley S (2010) The environment in world history. Routledge, New York
National Acid Precipitation Assessment Program (2005) National acid precipitation assessment program report to congress: an integrated assessment. NAPAP, Washington D.C.
Nye DE (1998) Consuming power: a social history of American energies. MIT Press, Cambridge
Obermeyer H (1933) Stop that smoke! Harper & Brothers, New York
Osborn M (2004) Uplands downwind: acidity and ecological change in the Southeast Lancashire Moorlands. In: Dupuis EM (ed) Smoke and mirrors: the politics and culture of air pollution. New York University Press, New York, pp 77–99
Papaioannou D, Sapounaki-Drakaki L (2001) Policies for cleaner air: the air pollution nuisance in Athens. In: Bernhardt C (ed) Environmental problems in European cities in the 19th and 20th centuries. Münster, Waxmann, pp 211–224
Penna AN (2010) The human footprint: a global environmental history. Wiley-Blackwell, Chichester
Rajan SC (2004) A fine balance: automobile pollution control strategies in California. In: Dupuis EM (ed) Smoke and mirrors: the politics and culture of air pollution. New York University Press, New York, pp 203–222
Ranlett J (1981) The smoke abatement exhibition of 1881. Hist Today 31:10–13

Read C (ed) (1994) How vehicle pollution affects our health. Ashden Trust, London
Robinson NA (2005) Air pollution control laws: common but differentiated responsibilities for managing the atmosphere. In: Bradbrook AJ et al (eds) The law of energy for sustainable development. Cambridge University Press, New York, pp 124–137
Rosen CM (1993) Differing perceptions of the value of pollution abatement across time and place: balancing doctrine in pollution nuisance law, 1840–1906. Law Hist Rev 11:303–381
Sallares R (1991) The ecology of the ancient Greek world. Cornell University Press, Ithaca
Schiedel W (2001) Death on the Nile: disease and the demography of Roman Egypt. Lieden, Brill
Schmandt J, Clarkson J, Roderick H (eds) (1988) Acid rain and friendly neighbours: the policy dispute between Canada and the United States. Duke University Press, Durham
Schwela D (1983) Vergleich der nassen Deposition von Luftverunreinigungen in den Jahren um 1870 mit heutigen Belastungswerten. Staub-Reinhalt Luft 43:135–139
Shaw BD (1996) Seasons of death: aspects of mortality in imperial Rome. J Rom Stud 86:100–138
Sheail J (1991) Power in trust: the environmental history of the Central Electricity Generating Board. Oxford University Press, Oxford
Sieferle RP (2001) The subterranean forest: energy systems and the industrial revolution. The White Horse Press, Cambridge
Smil V (1994) Energy in world history. Westview Press, Boulder
Smith RA (1859) On the air of towns. J Chem Soc 11:232
Sperling D, Gordon D (2010) Two billion cars: driving towards sustainability. Oxford University Press, New York
Stern N (2007) The economics of climate change: the Stern review. Cambridge University Press, Cambridge
Stradling D (1999) Smokestacks and progressives: environmentalists, engineers, and air quality in America, 1881–1951. Johns Hopkins University Press, Baltimore
Stradling D, Thorsheim P (1999) The smoke of great cities: British and American efforts to control air pollution, 1860–1914. Environ Hist 4:6–31
Tarr JA (1996) The search for the ultimate sink: urban pollution in historical perspective. Akron University Press, Akron
Tarr JA, Zimring C (1997) The struggle for smoke control in St. Louis. In: Hurley A (ed) Common fields: an environmental history of St. Louis. Missouri Historical Society Press, St. Louis, pp 199–220
Te Brake WH (1975) Air pollution and fuel crises in pre-industrial London, 1250–1650. Technol Cult 16:337–359
Thorsheim P (2002) The paradox of smokeless fuels: gas; coke and the environment in Britain, 1813–1949. Environment History 8:381–401
Thorsheim P (2004) Interpreting the London fog disaster of 1952. In: Dupuis EM (ed) Smoke and mirrors: the politics and culture of air pollution. New York University Press, New York, pp 154–169
Thorsheim P (2006) Inventing pollution: coal, smoke and culture in Britain since 1800. Ohio University Press, Athens
Uekötter F (1999) Divergent responses to identical problems: businessmen and the smoke nuisance in Germany and the United States. Bus Hist Rev 73:641–676
Uekötter F (2009) The age of smoke: environmental policy in Germany and the United States, 1880–1970. University of Pittsburgh Press, Pittsburgh
UN-HABITAT (2008) State of the world's cities 2008/2009: harmonious cities. Earthscan, London
United Nations Economic Commission for Europe (2004) Handbook for the 1979 convention on long-range transboundary pollution. United Nations, New York and Geneva
United Nations Environment Programme (1972) Declaration of the United Nations conference on the human environment, Stockholm. http://www.unep.org/Documents.Multilingual/Default.asp?documentid=97&articleid=1503. Accessed 18 Oct 2013
United Nations Environment Programme (2002) The Asian brown cloud: climate and other environmental impacts. UNEP, Pathumthani

United Nations Environment Programme (2009) Handbook for the Montreal protocol on substances that deplete the ozone layer. UNEP, Nairobi

United Nations Environment Programme and Organisation for Economic Cooperation and Development (1999) Older gasoline vehicles in developing countries and economies in transition: their importance and the policy options for addressing them. UNEP, Paris

Weart SR (2003) The discovery of global warming. Harvard University Press, Cambridge

Wirth JD (2000) Smelter smoke in North America: the politics of transborder pollution. University Press of Kansas, Lawrence

World Greenhouse Gas Emissions (2005). http://www.wri.org/resources/charts-graphs/world-greenhouse-gas-emissions-2005. Accessed 18 Oct 2013

World Health Organisation (2013) Air Pollution. http://www.who.int/ceh/risks/cehair/en/. Accessed 18 Oct 2013c

World Health Organisation, Health and Environment Linkages Initiative (HELI) (2013) http://www.who.int/heli/risks/urban/urbanenv/en/index.html. Accessed 18 Oct 2013

World Meteorological Organization (2013) Observed concentrations of CO_2 cross 400 parts per million threshold at several global atmospheric watch stations. http://www.wmo.int/pages/mediacentre/news/documents/400ppm.final.pdf. Accessed 18 Oct 2013

Worldwatch Institute (2009) State of the world 2009: into a warming world. Norton, New York

Chapter 6
Urban Development and Environment

Dieter Schott

6.1 Introduction

Current environmental policy aiming for sustainable development postulates that cities and urban dwellers reduce their ecological footprint,[1] their use of non-renewable resources, above all of carbon-dioxide emitting fuels in order to comply with global targets of reduction of Carbon-dioxide emission to mitigate climate change. Cities with a share of world population of just over 50 % are responsible for over 75 % of global resource use. The fact that many cities not only in the most developed countries of North America, Europe and East Asia but also the mega-cities of developing countries have a huge ecological footprint raises in historical perspective the question, how the environmental impact of urban civilization has been in pre-industrial cultures and how cities have been able to sustain themselves, to procure the necessary resources for their reproduction without threatening the ecological stability of their hinterlands. This chapter will provide an overview of the relationship between urban development and the environment with a specific focus on Europe since the High Middle ages (160).

6.2 Urban Metabolism

In order to analyze the relationships between cities and their environment in a systematic way and with a long-term perspective, the concept of 'social metabolism' linked with 'colonization of nature'[2] has proven to provide a useful

[1] Rees and Wackernagel (1997).
[2] Fischer-Kowalski et al. (1997).

D. Schott (✉)
History Department, Technische Universität Darmstadt, Darmstadt, Germany
e-mail: schott@pg.tu-darmstadt.de

M. Agnoletti and S. Neri Serneri (eds.), *The Basic Environmental History*,
Environmental History 4, DOI 10.1007/978-3-319-09180-8_6

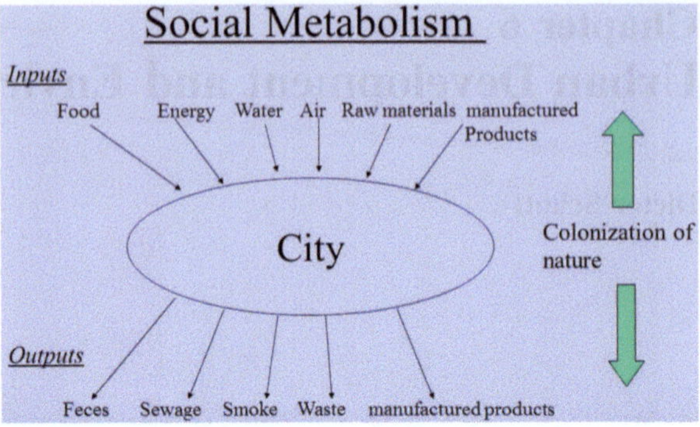

Fig. 6.1 Social metabolism of a city. Adapted from Schott (2011)

framework for a differentiated understanding of all kinds of environmental interventions (Fig. 6.1).

'Metabolism' of a society is defined as, 'the sum of all input and output between the biosphere/geosphere and society.' Colonizing interventions are defined 'as the sum of all purposive changes made in natural systems that aim to render nature more useful for society'.[3]

The concept looks at resources which are essential for the reproduction of a city on both the level of physical reproduction of the urban residents (including animals), i.e. their 'biological' metabolism as well as collective reproduction of the city as a social, economic and cultural system, i.e. the construction and maintenance of houses, collective buildings such as churches, streets, walls etc., the material production of goods for the needs of the urban residents themselves or for trade to import necessary resources from other places. On the *input* side, food stuff, raw materials, water, energy and air flow into the city to be consumed, processed and transformed there. On the *output* side through biological metabolism and through material production all kinds of products are set free which—since their purpose was to generate revenue outside the city, as they were no longer immediately useful or even potentially dangerous for the city and its residents—have to be discharged to the environment of the city. The focus of this concept lies on material flows and their transformation over time. The concept of 'colonization of nature' brings further dynamic temporal as well as spatial dimensions into this relationship: If cities and their population grow—as has been a repeated dominant phenomenon over the last ten centuries—they will need to reach beyond their immediate surroundings in order to fulfill their basic needs. They will tend to exercise either political dominance by extending the territory they control, or use market power to attract production surpluses from further distant regions. Thus cities mobilize in a variety of ways resources of an ever widening hinterland for their social metabolism. Doing so they frequently transform even remote ecological systems of these hinterlands, for instance by lowering the water level through large-scale water

[3] Winiwarter (2001).

extraction, by deforestation, by polluting rivers and dumping urban wastes on dumps and sinks at a distance from the city. In the 21st century this 'colonization of nature' has gone so far, that the whole globe has become 'hinterland' of large cities in the western world, which is for example testified by the range of countries of origin of products sold in urban super-markets.

6.3 Where Cities Are Located: Access to Resources

The basic structure of the European urban system developed between 1100–1300. This was the period when the large majority of cities still existing were first mentioned in sources as an 'urban settlement' with terms like 'urbs' or 'civitas'.[4] Of course the urban system did not result from any planning master-mind but evolved in response to pre-existing settlement structures, topographical conditions and new needs. Pre-existing were Roman cities in large parts of Southern and Western Europe where Roman occupation had left its lasting traces. Within Europe the degree of continuity of urban civilization beyond the collapse of the Roman Empire varies massively: In Spain, southern France and Italy, Roman population persisted to dominate even after the great migration, thus Roman cities frequently were kept populated, even if changed in form and appearance. North of the Alps within a zone demarcated by the Seine and the Rhine there is a strong imprint by Roman culture but on the other hand massive perturbations by successive waves of Germanic migration which resulted in a rather mixed picture of continuity and rupture. East of the Rhine and north of the Danube, finally, where the imprint of Roman culture was only short-lived and partial, almost no continuity can be noted. Christian church, particularly the Episcopal organization provided an important bridge of continuity between the late Roman period and the stabilization under the Carolingians. Frequently during times of political instability the bishop acted as a protecting and unifying agency for urban dwellers.[5] When medieval society started on a slow but steady growth path after 1000 we can observe a large number of settlements located at fortuitous places developing into 'cities'. Crucial location factors were transport and defense: since transport is an essential factor for almost all commodities other than water and air, locations which were situated on trade routes over land or—better still—on water (or both!) were preferred. Thus we can frequently find cities established at crossings of trade routes, particularly were trade routes crossed a (navigable) river.[6] Transport acted as an essential bottle-neck of medieval and early-modern economy: In contrast to Roman times roads were badly maintained and frequently also dangerous; robbery and way-laying were endemic in large parts of Europe. Moreover, road transport was extremely expensive and inefficient: if

[4] Hohenberg and Lees (1995) and Clark (2009).

[5] Ennen (1987).

[6] Hohenberg and Lees (1995, p. 31) and Schott (2011).

grain was transported more than 7 days over land by cart, the draught animals had eaten more in caloric value than they had transported.[7] Waterways, in contrast, were almost free of maintenance, downstream the current took care of propulsion and even upstream a horse could pull ten times as much weight on a barge than on a cart. Thus locations on navigable rivers, and the notion of navigability in medieval times included fairly small rivers, were highly favored, as the potential hinterland for the provision of a city with foodstuff, raw materials, wood and other bulky goods was greatly expanded by water transport.

'Defense' or rather 'defensibility' was essential for locating fortified houses of secular or clerical lords, which needed to be defended against attacks by enemies. Frequently such fortified seats formed a nucleus around which urban settlements developed, particularly where favorable trade locations and defensible sites combined. To give but one example: Würzburg in Southern Germany was founded in 741 by the missionary Boniface on flood-protected land near a river ford of the Main and below an older castle on the mountain. Here the cathedral as the seat of the new diocese formed the nucleus of the city which developed in the ninth century along a widened street market linking the cathedral with the river ford. Since the early 12th century a bridge facilitated river crossing and Würzburg succeeded in attracting significant long-distance trade onto it on account of that bridge. In other instances a fortified castle or royal palatinate secured such places as the Danube crossing at Ulm, from which grew a mint and a market settlement by the 11th century, or the crossing of the Main at Frankfurt.[8] Markets always were significant functions of new cities and frequently a city is first mentioned in sources when it was granted rights to hold markets by the king or the regional lord. Thus, in most cases, a favorable trade location giving easy access to resources, the fortified seat of a clerical or secular lord and a market which developed protected by this lord and was intended to cater for the needs of his court, came together to enable the setting-up of successful urban settlements in the high middle ages. In the 15th/16th century other types of city foundations arouse such as mining towns which were essentially determined by the access to mineral resources (Freiberg, Joachimsthal) or fortification resp. naval towns which performed a special military function for the emerging territorial states.[9]

6.4 Waters of the City

Water is one of the essential resources a city cannot do without. Thus for pre-modern times we can usually assume that no settlement was established without access to some water. But its limited availability could force urban authorities to try

[7] Sieferle (2008), Sieferle cites Ohler (1986, 141). Other authors give an even smaller spatial range (35–40 km) for grain transport over land cf. Irsigler (1991).

[8] Schmieder (2005).

[9] Rosseaux (2006). On the European level see Clark, European Cities, 112–114.

to manage this scarce resource in a variety of ways or to attempt to increase the supply by conducting additional water to the city from further away. The latter was the well-known solution to the water problem in Rome and the city of Rome could not have grown to such dimensions in the Imperial Period without the aqueducts overcoming the local mismatch of supply and demand.[10] However, this solution of transferring water over longer distances was not suitable in the light of the highly fragmented political structure of European medieval society. Cities frequently had to face being besieged by enemies; being cut off from their hinterland a dependency on aqueducts would have rendered cities highly vulnerable due to lack of water. Thus we can identify a preference for using locally controllable water resources. And rather than opt for centralized unitary systems, medieval cities looked for a multiplicity of resources and water supply systems to decrease their dependency and enhance their resilience.[11]

In the medieval city water served a multitude of purposes[12]: Besides the use of drinking by humans and animals, water was used in processes of making food and drink, of producing goods, particularly textiles etc. Water also constituted a potential source for mechanical energy, driving mills which were the main types of machines. Mills were not only used for grinding flour, of course a very important use to keep urban population well fed, but also for all kinds of industrial processes, particularly in textile production (fulling) and metal-processing (hammer-mills). The current of rivers was used to power mills, located on the shores or sometimes also—as ship mills—in the middle of the stream. In hilly cities with an abundance of smaller rivers and creeks these were frequently re-directed and divided up into smaller channels to enable as many users as possible to make use of the kinetic energy. And in trading cities on rather flat ground canals made water transport available for as large a portion of urban houses as feasible. The Flemish city of Bruges, one of the major nodes of long-distance trade in the late middle ages, had established a complicated system of water provision and transport serving a multitude of functions, such as providing water power for a variety of crafts as well as feeding a large number of public fountains on streets and squares.[13] A specific feature was the cloth hall which was built across a little canal so that the valuable cloth could be unloaded and handled under the arcades of that building protected from the weather. A critical issue for Bruges was the access to and from the sea: the Zwijn, the river linking Bruges with the sea, was increasingly silting up, which made it more and more difficult for sea-going vessels to reach Bruges or its out-port Sluis. In consequence the city, one of the leading European entrepots and gateway-cities in the 14th century suffered long-lasting decline.[14]

[10] Stahl (2008).

[11] Grewe (1991).

[12] Cf. Guillerme (1988) and Schott (2014), esp. Chap. 5.4.

[13] De Witte (2004).

[14] Girouard (1987) and Blockmans (1992).

South of the Alps frequently scarcity rather than abundance of water was the problem cities had to cope with. But scarcity of water did not necessarily prevent urbanization: we can find a highly complicated system of water provision and use in the city of Siena, one of Italy's foremost city republics with a population of over 50,000 by the late 13th century. Situated on a hill-top along the highly frequented via *Francigena*, the road linking Rome with French cities, Siena has no significant river and only the water which it receives from precipitation, mostly in winter. The geological build-up of the Siena hills causes rainwater to collect in the soil from where it surfaces as springs at certain places since the impermeable clay prevents its further immersion into the ground. Given the need to manage a scarce resource, Siena developed a highly regulated water policy. Among a specified hierarchy of uses drinking came first; the city tried to ensure compliance to rules by means of guards watching over public fountains and wells, secret informers and severe penalties. The water supply rested on a system of urban conducts of spring water which was conferred to suitable places where they were accessible first as drinking fountains for humans, then as horse troughs and facilities for bathing horses, then as a series of basins for industrial purposes such as tanning and dying wool. Finally, after leaving the city the water was used again to drive fulling mills. Thus the purest water was to be preserved for the most sensitive uses such as human consumption, whereas polluting uses were located further down the chain of uses. The water supply rested on three tiers: springs with running water, which was considered highest quality, wells of ground water and—least regarded—water from cisterns, storing earlier rain-falls for the dry season, which however still could be used for washing or industrial uses. The elaborate system of water networks but especially of water regulations reflects the ideology of 'good government' as it was illustrated by Ambrogio Lorenzetti's famous frescos in the Palazzo Pubblico of Siena, produced in the 1340s. And this 'good government', which rested on a remarkable degree of citizen participation, obviously included the 'beauty' of the city as a high value and reflected a high appreciation of public cleanliness despite difficult resource conditions.[15]

Frequently elite house-holds or special institutions organized their own water provision through small-scale water networks which transported spring water from just outside the city gravity-fed into the city. There it was conducted by pipes to private house-holds, monasteries, hospitals, breweries or public fountains. Such public fountains, often donated by local aristocrats, could be used by the general urban public to fetch their water and to secure an ample supply of water in case of fire. Public fountains were also symbolically charged places of jurisdiction, for making contracts and public announcements; cities frequently employed expert fountain masters who were responsible for the functioning and maintenance of the city's water system. Polluting or poisoning fountains and wells was a capital crime, punished by drowning.[16] In most cities the largest quantity of water, at least for

[15] Kucher (2005).

[16] Grewe, Wasserversorgung.

general household purposes except drinking was taken from private or public wells dug in the court-yards of buildings or at small squares to reach to the ground-water. Private wells had to be maintained by the owner. In the case of public wells, we frequently find some sort of collective responsibility: neighborhood residents who used the well, had to take care of it. Once a year the well was emptied of all the filth which had accumulated in it and after this collective work a well party was celebrated.[17] For Vienna, we have records of 10,000 private wells still in existence by 1900 although public water supply from the mountains had been in operation for several decades already by then.[18] With rapid population growth in the 18th and early 19th century this system of water provision proved less and less adequate: Fetching water from public fountains absorbed large parts of the working day of servants and house-wives. A new profession emerged, in larger cities already in the middle-ages, the water-carriers who collected drinking water from public fountains or from springs or rivers outside the city in barrels and sold it from carts to private house-holds ready to pay for water carried into their home. In Paris, when under the prefect Haussmann a modern pressurized water provision system was introduced in the 1850s, 20,000 water carriers lost their livelihood.[19]

6.5 Feeding the City

For early phases of urbanization and in some places up to the 19th century it would be mistaken to imagine cities as completely non-agrarian places. Sizeable portions of urban residents still were farmers or, if their main occupation was non-agricultural, they nevertheless cultivated gardens or small fields within or at the periphery of cities. Furthermore, animals abounded in medieval and early modern cities: it was normal and acceptable, that a baker held ten pigs.[20] Apart from pigs, which also acted as natural refuse cleaners, devouring any organic waste in streets and courts, goats and sheep, geese and chicken were held in cities and daily driven for feeding onto meadows and into woods close to the city. Cattle was less frequent, being restricted to urban farmers or dairy farms on the fringes, but horses were present where carting and transport was ubiquitous, such as around inns close to city gates. Cats were held to keep mice and rats at bay, although frequently this was a losing battle as the successive waves of plague in the late medieval and early modern era demonstrate. Thus cities were not devoid of natural habitats; in many cities and towns large private tracts were given to agriculture, viniculture or orchards. Public green, however, was virtually unknown; we only rarely find trees on streets and

[17] Schmid (1998) and Malamud and Sutter (2008).

[18] Koblizek (2005).

[19] On water carriers in Paris and London in 12th century (Keene 2001); on the replacement of water carriers in Paris of 19th century (Hall 1998).

[20] Schubert (2012).

squares. On maps from the 16th century we can still identify large tracts of vineyards and orchards within the walls of major cities such as Cologne and Augsburg.[21]

Nevertheless cities could not be fed only from the produce of their own territory. To feed a city of 50,000 inhabitants, already a very large city by standards of the urbanization north of the Alps (Cologne, the largest city of the Holy Roman Empire had 40,000 inhabitants by 1400), a hinterland of 45 km radius was needed.[22] But since within this hinterland other cities might compete for the surplus and harvest failures or transport problems might impede the provision, larger cities developed trade links further afield to ensure particularly the grain provision as the most sensitive issue. Thus the Dutch cities received one seventh of their grain from Gdansk and the Vistula Basin in the 17th century.[23] In the 16th century when population grew but grain harvests declined and fluctuated, partly due to the effects of the 'little ice age', municipal authorities started to set up granaries to stock grain for their population in order to be able to buffer harvest failures or transport problems.[24]

Urban food provision created specific spatial structures of agricultural usage and cultivation. Which land was used for which crop or cultivation of course partly depended on its soil properties and fertility but also the distance to the city as a centre of consumption and marketing and the time and expenses involved in transport thereto played a decisive role. In the 1820s the Prussian economist von Thünen sketched a diagram of concentric rings around a city as a consumption centre by which he intended to calculate, in which area which crops could be grown most profitably (Fig. 6.2).

Assuming a level and homogenous surface around the city with soil of even quality (diagram a), a first ring of cultivation closest to the city would be occupied by market gardeners supplying the urban market with fresh vegetables, fruit, hay, potatoes and beets. Their products were easily perishable and had to reach the urban market quickly to be saleable. In the case of hay, the large quantity at low value, i.e. high transport costs, was the decisive factor. The next zone would be occupied by dairy farms, since milk and butter were also, without artificial cooling systems, easily perishable. The third zone was frequently occupied by forests devoted to fire-wood, as again the high transport costs would prevent—if no water-transport was available—the import of fire-wood from longer distances. The fourth zone then is dominated by wheat-growing farmers. Given the model assumptions of Thünen, it would be difficult to find exact matches to his model in reality. But in the case of London, one of the fastest growing European capitals in the early modern period, from its topography close to diagram B, we can clearly identify zones of specialization, which are arranged at different distances from the capital: at distances of

---

[21] Kießling and Plaßmeyer (1999).

[22] Clark, European Cities, 140; the city of Cologne needed 1,800 km^2 for its provision with grain, a low figure due to the very fertile soils. Eiden and Irsigler (2000).

[23] Unger (1999).

[24] Dirlmeier (1988).

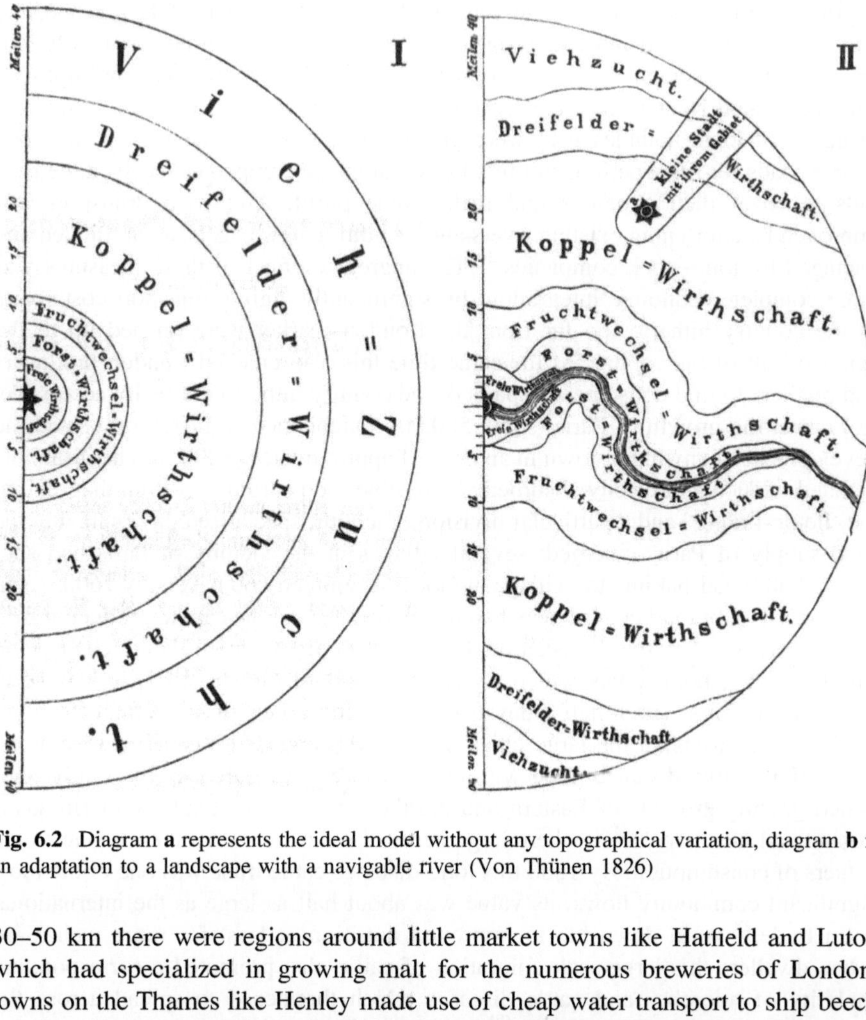

Fig. 6.2 Diagram **a** represents the ideal model without any topographical variation, diagram **b** is an adaptation to a landscape with a navigable river (Von Thünen 1826)

30–50 km there were regions around little market towns like Hatfield and Luton which had specialized in growing malt for the numerous breweries of London, towns on the Thames like Henley made use of cheap water transport to ship beech wood and grain to London, originating from the Chilterns. The southern Midlands, by the 17th century almost exclusively grazing land, reared livestock for the London meat market. Since cattle could provide its own transport, thousands of oxen were driven to London where they were fattened up again on the marshes downstream of London before being sold for butchering in London. But the market pull of a city like London with 575,000 inhabitants by 1700 radiated even further: in 1695 the cheese mongers of London asked Parliament for a bill to improve navigation of the river Derwent in Yorkshire, by which they hoped to reduce transport costs for cheese and butter bought there during transport to Hull on the East coast, from where it was taken to London by sea-going ships. Thus large cities like London exercised enormous influence over their hinterland and promoted regional specialization, as soon as the provision of local residents in those specializing regions with basic foodstuff could be taken care by imports from surplus regions at acceptable costs.[25] Improvement of transport thus was of the essence, and this set in earliest and most thorough in Britain, where the 'transport revolution' preceded and accelerated the industrial revolution. Road transport was improved and accelerated by means of the turnpike trusts, joint stock companies, frequently made up of local notables who were given concessions to manage and improve certain roads for fixed periods of time. In exchange the companies received the road tolls to cover their expenses and make some profit. And water transport was improved by shortening existing rivers and by cutting new canals, again funded and managed by joint-stock companies.[26] The aggregate effect of these measures was more complete economic integration: by significantly cutting transport costs parts of the country hitherto too far from the London market were opened up to the demand pull of the capital. At the same time this also enabled London producers and retailers to find consumers in parts of the country hitherto not really accessible. In France, the growth of Paris (some 500,000 inhabitants by 1700), prevented the development of any major town in an orbit of approximately 100 km. The insatiable demand of such a large city absorbed all possible food surpluses within that region, the 'Ile-de-France' and a particular division of labour, specialized on organizing the food supply of Paris, emerged: several villages in the vicinity of Paris had specialized on bread-baking, the village of Gonesse counted 160 bakers by 1600. As in London a ring of market gardeners had developed around the capital producing fruit and vegetables for the Parisian tables, peaches in Montreuil and cherries in Montmorency, both towns which today form part of Greater Paris, but were at 13 km from Paris within half a day from the fruit markets of the capital.[27]

For meat provision the large cities of Europe increasingly came to rely on an international long-distance trade with cattle, which was reared on the thinly populated grazing grounds of Eastern and Northern Europe, particularly in Hungary, Poland and Denmark, from where large herds of oxen were driven each year to the centers of consumption. By 1600 this long-distance cattle trade was one of the most significant commodity flows, its value was about half as large as the international grain trade.[28] As a downside to improved nutrition at least for more well-to-do urban dwellers this large-scale migration of cattle also promoted a corresponding mobility of microbes and cattle diseases which threatened to spread across the whole European stock of animals.[29]

---

[25] Reed (1996).

[26] Reed, London; Bagwell (1988).

[27] Jacquart (1996).

[28] Blanchard (1986).

[29] See measures by public authorities in North Germany to check the spread of cattle diseases while still allowing the trade with healthy animals from Denmark to the centres of consumption in South Germany: Hünninger (2011).

6.6 Wastes and Diseases: The City as a Population Sink

If we conceive the city as a place where metabolism takes place, we need to reflect what happens to the output of physical metabolism of human and animal residents in medieval and early modern cities. Humans and animals produced feces as result of their physical metabolism. These feces were usually dumped in cess-pools located in court-yards. As long as cities were relatively thinly populated, and there was enough agricultural land within and outside the city walls, on which these feces could be used as fertilizing dung, the problem of waste disposal could be contained. For economies of scarcity such as the European well into the 19th century, feces were not simply 'waste' but rather 'resources' since it was essential to regain and maintain soil fertility. But as cities grew towards the late middle ages and then—after a period of population decline following the Black Death in the 14th century—again from the late 15th century, spaces within the city were increasingly built up, houses were built higher. Thus drinking wells in court yards, which originally had been dug at a distance from cess-pools, moved closer to them and this could mean, that the ground water became infiltrated by the liquid parts of the contents of cess-pools contaminating this water supply. Germs from human and animal feces could thus re-enter the organism via the water supply but of course medicine then had no knowledge of 'germs' which were only identified by Pasteur and Koch from the 1860s onwards. With the disappearance of agricultural land inside the walls and the spatial growth of cities it became more difficult to use the contents of cess-pools as fertilizer. Thus wastes were stored in disused cess-pools or dumped into rivers. This was only one of many reasons why cities were rather unhealthy places, despite numerous regulations by municipal authorities on cleansing streets and public places. Apart from massive pandemias of plague, which haunted European cities from the mid fourteenth to the late 17th century in successive waves, a plethora of dangerous diseases particularly of the digestive tract (such as typhus or typhoid fever) was endemic to European cities, taking a heavy toll every year particularly among the infants. Life expectancy in large cities was significantly lower than in the country and the high residential density linked with rather undeveloped standards of personal hygiene enhanced mortality. Cities—particularly in the early modern period—can be seen as 'population sinks'; mortality was normally higher than natality, thus the city could not keep the level of population by its own fertility.[30]

For London it has been calculated that the city needed a net in-migration of 8,000 people per year by the late 17th century in order to stabilize its long-term growth rate. Each year a town of 8,000 people was absorbed by the capital, at a time when most other English towns had almost no growth. For Paris, its growth was particularly maintained by influx from the Ile-de-France where most towns and villages showed very little growth between 1600–1800.[31]

[30] Knittler (2000) and Knoll (2008).

[31] Boulton (2000).

Given the abundance of organic matter in cess-pools, but also the waste and feces on city streets cities stank badly. Of course this stench was not uniform all over the city; representative squares and streets were kept cleaner and were also frequently paved, but the fact of stench, linked with particular high incidence of diseases and fatalities in those places with particular bad smells, caused contemporaries to develop etiological theories linking stench with disease. The bad smell, called 'miasma' by contemporary experts, was seen as direct cause of diseases, akin to an understanding of poisoning. By the late 18th and early 19th century the fight against stench and for cleanliness acquired a new sense of urgency and significantly contributed to the broad sweep of urban sanitary reforms in the second half of the 19th century.[32]

6.7 Cities in the Wood Age

For the development of pre-industrial cities wood constituted a basic resource which could not be substituted except at unreasonable expenses. The forest constituted in multiple ways the economic foundation of medieval cities.[33] By the year 1000 large tracts of the European landscape north of the Alps were still covered by forests. These forests provided for cities and other human settlements a broad range of functions: they offered wood for domestic and industrial fuel, for construction purposes and as material for the production of tools, instruments, furniture and all artifacts of daily life. But forests were also "nutrition forests" for the people living close by. In them residents could collect berries, mushrooms and spices, set up bee hives to gain honey, the only sweetener available before the import of cane sugar from the Americas. Forests also supported cattle, pigs, sheep or fowl, provided habitats for game and although hunting was a seigniorial privilege, sources which report about poaching are abundant.[34]

Medieval cities, at least north of the Alps, were predominantly wooden cities: since wood was the cheapest building material, most easily available and most easily handled, the large majority of urban houses were of timber and building techniques with wood became increasingly sophisticated. The timber-frame house used significantly less wood than the original log house. Only selected buildings such as churches, monasteries and, by the late middle ages, the guild houses were built in masonry fashion. Roofs were covered in wooden tiles, straw or thatch. Since in all house-holds and work-shops open fires were almost constantly burning, such a 'wooden city' stood of course in very high risk of fire; city fires were among the most frequent causes of large scale disasters which struck cities.[35]

---

[32] Hamlin (1998).

[33] Schubert (1986).

[34] Lorenz (1993) and Küster (1998).

[35] Boockmann (1994), Zwierlein (2011) and Schott (2013).

In the light of this high risk of a conflagration city authorities from the late Middle Ages on were trying to oblige house owners to use brick-built fire places and to have their roofs covered by stone tiles or slates which however were considerably more expensive than wooden tiles. Considerations of economy and safety were at constant battle in the regulation of how cities were to be built, not only in respect to fire but also to other hazards.

Providing sufficient quantities of fuel and construction wood had been relatively easy in the high Middle Ages when cities were comparatively small in terms of population and forests fairly close by. But the long growth period between 1000 and 1300 meant that the agricultural area had massively expanded at the expense of forests. In the 14th century the area covered by forests in Central Europe had been smaller than in the 20th century. Pushing back the 'forest frontier' also had ecological consequences such as increasing erosion on sloping fields and a higher rate of flooding because the forest could not retain as much rain water as before. Forests in the vicinity of cities also had changed their character: due to grazing and repeated logging of trees in short intervals the forests were loosened up and frequently turned into a mixture of pasture and forest with a predominance of medium growth trees.[36]

The fire-wood needs of towns were quite substantial. A town of 10,000 inhabitants needed up to 50 cartloads of fuel-wood every day.[37] By the late 13th century municipal authorities started to realize the potential dangers of overusing the forests in the proximity of the city. Nuremberg, one of the most dynamic and prosperous German cities in that period, developed a forest protection policy for the Imperial forests which Nuremberg had managed to gain control over and tried to protect these forests against excessive use. Elements of such protective policies could be the banning of collieries or saw mills. Since the prime-growth forests of oaks and beeches, species which naturally dominate forests in that region, were almost totally depleted, the city of Nuremberg decided to replant the forests with faster growing pine trees. The technology of extracting pine seeds and replanting forests, developed for this protection policy, became a lucrative business venture; a company dealing in seed of trees was highly successfully after 1400 selling its products from Nuremberg all over Europe.[38]

Nuremberg was in a privileged position, having these Imperial forests in its immediate vicinity, but for many other cities there were no sufficient forests close by, over which they could exercise control. Providing enough fire-wood and timber for construction thus was one of the permanent challenges for urban authorities up to the period of industrialization.[39] The problem was less one of absolute shortage of wood but rather of affordable and efficient transport. For pre-industrial transport costs of overland haulage by cart were about 10 times as high as transport by sea

36 Schubert, Alltag, 43–45; Küster, Geschichte.

37 Clark, European Cities, 141.

38 Sporhan and von Stromer (1969).

39 Freytag and Piereth (2002).

ships.[40] Thus before the advent of the railway wood could only be transported by water over longer distances. A city like Vienna, the capital of the Habsburg Empire, rapidly growing in population since the siege by the Turks in 1683 (1680: 70,000, 1783: 209,121 inhabitants), received 80 % of its fire wood via the Danube, only 20 % were transported from the regional *Wienerwald*. But even for that area, fairly close to Vienna, water transport was preferred: the firewood was—cut in pieces of about 90 cm—washed down little creeks and rivers towards the Danube where it was caught in special rakes and transported on floats or ships towards Vienna. In many instances also special slides, tunnels or channels to facilitate wood transport were constructed. Also construction wood was transported this way: originating from upper Austria, Tyrol and Bavaria, the logs were tied together to larger floats of 400–700 m^3 wood which were driven down the Danube and eventually stored in suburbs of Vienna.[41]

In international comparison Vienna was in a fairly advantageous situation having rich forest resources in distances of only some 100 km. The Flemish and Dutch cities, as a contrast, had only very little, sometimes no forests in their vicinity, due partly to the particular ecology of these lowlands. Here we can observe a long-distance trade in construction timber already developing in the late middle ages. Cities like Bruges, Gent and Antwerp, major players in the European trade networks of that period, were provided with wood from the Black Forest, from Scandinavia and the Baltic, particularly from the Vistula. Gdansk became the gateway-city for the opening up of the resources of the Vistula basin for the benefit of the Flemish cities with their large need for construction wood and grain.[42] Thus we can already observe an international division of labor developing in staple commodities such as wood and grain between the highly urbanized North–West of Europe and the more agrarian and less densely populated Baltic and Eastern European regions. But this network depended on water-borne transport; as soon as carts had to be used, market integration of bulky goods tended to be local rather than regional.

Whereas Vienna had solved its wood problems by investing in transport systems, London, another rapidly expanding capital in the early modern period, took a different route: Already in the late Middle Ages coal was used as an alternative to fire-wood since it was cheaper as regional forests started to be depleted. After the Black Death which brought a massive population drop, coal disappeared from the London market only to reappear after 1500. Firewood saw a price rise which was double the general average in the period 1500–1700, a clear indicator for growing scarcity.[43] Coal was still competitive although it had to be transported from Newcastle upon Tyne in the north-east of England, some 400 km distance from London. There the coal could be mined on the steep hillsides of the river Tyne and

---

40 Sieferle, Transportgeschichte.

41 Johann (2005).

42 Galloway et al. (1996); Unger, Feeding.

43 Allen (2010).

easily transported to boats on the river by slides and early railways. Since almost the entire transport could be made by ship, coal, termed by contemporaries 'sea-coal' to distinguish it from (char-) coal, was still cheaper than fire-wood on the London market.[44] By mid-16th century 250 coal ships from Newcastle unloaded in London each year, by 1690 this had increased tenfold to 2,500.[45] Coal replaced fire-wood first for industrial processes with high energy requirements such as salt-making, smelting iron ore, brick-making, but eventually spread to other uses including domestic fireplaces. Its wider introduction for domestic heating required a technological innovation, the diffusion of a new style of house, the 'coal-burning house'. This house had a special masonry chimney, a fire-place (or several) which partly enclosed the coal-fire for better draft. This innovation occurred in London during the rapid population growth of the 16th and 17th century, which was accompanied by a massive rebuilding boom.[46] Introducing coal as the main fuel massively changed the quality of the environment: Since coal from Newcastle contained a higher percentage of sulphur, coal fumes produced sulphur dioxide and led to a significant degradation of urban climate. Thus air pollution was already a massive problem of 17th century London.[47] John Evelyn, a famous landscape architect and philosopher at the court of Charles II, published a critical pamphlet on the ubiquitous air pollution of London under the title "Fumifugium" in the 1660s.[48] Air pollution must have been so bad that the new king William of Orange, after the 'Glorious Revolution', did not take his residence in Westminster Palace but preferred the more rural and less polluted suburb Kensington.[49] London thus experienced the transition from wood to coal as basic energy resource, a major feature of the Industrial Revolution, occurring in continental Europe normally in the middle decades of the 19th century, already more than 150 years earlier! This eventually meant that a huge area otherwise required to grow fire-wood, was released and became available for other uses; economic growth could thus delink from the limited availability of renewable energy resources and make use of the energy treasure of fossil fuels which had accumulated over a very long period of geological history.[50]

6.8 Crisis of the Cities and Public Health

It was pointed out above in Sect. 6.6, that the material metabolism of urban dwellers tended to create a rather unhealthy urban milieu with high mortality rates and considerable individual insecurity about future perspectives. In the first half of the

---

44 Sieferle (2001) and Middlebrook (1950).

45 Boulton, London, S. 323.

46 Cf. Allen, Industrial Revolution, 90–96; Allen (2013).

47 Brimblecombe (2011).

48 Jenner (1995).

49 Braunfels (1979).

50 Sieferle, Forest.

nineteenth century this situation further aggravated due to the continuing urban growth which now was linked with the socially disrupting effects of early industrialization.[51] Whereas older forms of inter-class solidarity and reciprocal relations between patrons and clients were eroding, no new institutions or cultural norms had as yet taken their place. The excessive supply of landless labourers created mass pauperism as a situation where large parts of the population could not earn their living despite working hard. This social crisis of pauperism, affecting large parts of Europe, not just in the cities but most easily visible there, was joined by the acute pandemias of cholera, haunting Europe from the 1830s in several waves. Reacting to this crisis, but more specifically to the failure of a sweeping reform of the Poor Law a few years earlier to achieve its goals, Edwin Chadwick, the secretary of the Poor Law Board, in 1842 published a 'Report on the Sanitary Conditions of the Labouring Population of Great Britain', which was submitted to Parliament and claimed to document the health conditions of the poorer sections of British society. With this report Chadwick laid the foundation for the so-called 'Sanitary Movement', which promoted major infrastructural changes to cities in order to improve their health situation. Chadwick was a firm believer, as many of his contemporaries, in the miasma theory of disease causation. Infectious diseases, in his report frequently just termed 'fever', were caused by bad smells emanating from the ubiquitous dirt and filth in cities. To prevent diseases, this filth must be removed and Chadwicks recipe was to bring fresh, clean water in all house-holds, to install water-closets there and to remove feces and urine from residential quarters by an underground system of sewers. Cleaning up cities thus would reduce the incidence of infectious disease, lower mortality rates and enable more poor people to earn their livelihood by work.[52] This far-reaching and very costly programme was eventually, with modifications in details, implemented all over Europe and North America in the second half of the 19th century and helped, together with other factors such as improved nutrition, to significantly reduce the frequency of epidemics, to lower mortality and improve health standards.[53] It is remarkable, that a reform movement, based on a false scientific theory—with the discovery of bacteria as carriers of disease by Pasteur and Koch the miasma theory was effectively falsified—could muster such wide support and effect such far-reaching changes to the fabric of cities, but also to the culture of cleanliness and hygiene. Considerations of hygiene acquired top-priority in the last decades of the 19th century and the physical transformation of cities, the building of water provision systems, of sewage systems, the paving and cleansing of streets and squares, the introduction of regular public waste removal systems, the construction of public slaughter-houses and public baths were all governed by the goal to improve public hygiene.[54]

---

[51] Lees and Lees (2007).

[52] Hamlin (1998).

[53] Hardy (2005).

[54] Schott (2012).

What were the effects of this massive transformation on the environment, how did it change urban metabolism? For reasons of practicality and economy in most cities the sewage system was implemented as a "mixed" or "combined" system, where all fluid wastes and rain water coming down in a city were to be collected in a single system. Moving solid feces in the tubes needed large quantities of water as a carrier, so besides the water flushed down the water toilets, rain water was quite welcome to flush through the tubes and prevent solid fecal matter from getting stuck inside. Originally Chadwick had envisaged a recycling system: after being collected and transported out of the city, the waste waters would be dispersed on land outside the city and the nutrients contained in the waste water would fertilize the soil thus improving agricultural yields. This arrangement, highly plausible under the auspices of a threatening food shortage in the early 19th century, was to counter arguments advanced by agricultural chemists such as Justus von Liebig who warned against robbing the soil of its necessary dung by building sewers. However, in most cities such a system of sewage farms was not implemented, either because of lack of suitable land—only sandy soils qualified well for this practice—or because other cheaper solutions were at hand such as the dumping of waste waters into rivers and the sea. In London as well as in Hamburg, two cities where sanitary systems had been set up fairly early, the discharge into the river was preferred which led to considerable pollution downstream.[55] On the input side of urban metabolism, the modern sanitary systems brought a massive increase in per capita demands of water. Up to mid-19th century 15–20 l per inhabitant and day had normally been sufficient to fulfill the water needs of urban dwellers. With the water closet in place and an increasing range of water-consuming amenities filling houses and apartments, daily per capita consumption jumped to 150–200 l per day.[56] Cities, particularly larger cities thus could no longer rely on their local water resources but had to bring water from further afield, tap into water resources of distant and frequently rural regions. Paris under Haussmann built impressive aqueducts which transported water from tributaries of the Seine at a distance of over 200 km.[57] Thus, whereas cities became cleaner, epidemics such as cholera and pandemic diseases such as typhus were checked, The overall environmental effects of the 'cleaning up of cities' which undoubtedly took place in the later 19th and early 20th century, were to externalize pollution, to transfer much of the dirt to the periphery, to pollute the land of sewage farms, rivers and the sea with the filth of urban populations and to massively increase water extraction from far away, frequently with considerable ecological as well as social consequences. These consequences could be even more pronounced in arid regions of Europe or in those parts of other continents, where European migrants established modern urban societies and implemented these sanitation technologies. Thus in Southern California a very complicated system of water provisioning emerged to provide water

55 Breeze (1993) and Büschenfeld (1997).

56 Schott, Urbanisierung, Chap. 9.

57 Hall, Cities, 724.

for the rapidly growing metropolis Los Angeles since the early 20th century when William Mulholland constructed his famous Sierra Nevada Pipeline over 400 km, opened in 1913. Nowadays Los Angeles water comes over several 1,000 km from as far afield as Wyoming, the Colorado river, the San Francisco Bay and the Sierra Nevada. And the almost unquenchable thirst of the still growing metropolis with its lush and water-needy private green spaces has severe implications for regional agriculture where large tracts have problems to secure sufficient water for their intensive cultivation, for instance in Owen's Valley and several lakes have already dried up completely.[58] In colonial cities modern (European) sanitation technology was usually only implemented for restricted parts of the city, those quarters where the European colonizers (and sometimes their indigenous collaborators) lived. Whereas in the former Dutch colonial entrepot Batavia, today's Indonesian capital Jakarta, modern sanitation is still limited to the formerly European quarters where now the Indonesian government resides, the former district "Weltevreden", in Singapore the British public health policy of late nineteenth and early twentieth century, combined with a rigidly effective but authoritarian policy since independence has achieved to significantly improve public health as well as the state of the environment.[59]

6.9 Networking the City

Installing comprehensive systems of water provision and sewage collection underneath the cities was only part of a larger drive towards improving the state of cities by implementing modern network technologies.[60] Since the state of cities appeared critical by mid-19th century not only in terms of health but also in terms of overly condensed traffic and problematic residential densities, urban reformers looked for solutions to relieve the congested city centers and to improve the quality of urban living. They found a range of technologies which improved the lighting and thus the perceived safety in cities, which facilitated urban transport, which offered alternative means of industrial power besides the steam engine. Starting with gas light in early 19th century England, a series of technical networks was set up in cities which offered a range of services to urban dwellers, shops, industrial companies or administrative services.[61] Frequently these services were first developed by private entrepreneurs or share-holder companies, who entered contractual arrangements with the cities, granting entrepreneurs and companies rights to use urban land for their tubes and pipes. When these contracts came up for renewal—frequently after periods of 25–30 years—municipal administrations

[58] MacDonald (2007), Orsi (2004) and Gottlieb (2006).

[59] Toyka-Seid (2008), Spreithofer and Heintel (1997) and Schott (1998).

[60] Hughes (1983), Tarr and Dupuy (1988), König (1990), Melosi (2000) and Schott (1999).

[61] Paquier and Williot (2005), Schivelbusch (1983) and Goodall (2005).

decided in many cases to take these services in their own hands and municipalized gasworks, water works or tramway companies. The motive for this pan-European movement of 'municipal socialism' was the dissatisfaction with the level of services received so far and the recognition on the side of municipalities, that the municipal ownership of such utilities might improve the financial situation of the city and might put it into a position to better apply the formative potential of these infrastructures in order to promote and direct urban development.[62] This particularly applies to traffic infrastructures, where decisions over which lines to build and what fares to charge were crucial to shape patterns of future development. Thus cities in Europe and North America were turned into mega-machines, they were perforated and structured by multiple systems of technical infrastructures which—in their totality—transformed daily life of urban dwellers completely by 1900, as compared to 1850.[63] This transformation can be seen as part and parcel of the 'civilizing Process' (Elias); public urinals on city squares offered urban residents the opportunity to relieve themselves in what came to be considered 'decent' fashion, while at the same time urban codes and bylaws penalized indecent behavior in the public, such as urinating in shady corners, which had been quite common at mid-century.[64] The electric tramway introduced a new speed and a heightened sense of urgency and acceleration into urban traffic which had been dominated by the pace of pedestrians and horses up to 1890. People now expected their co-urbanites to adapt their public behavior to this new speed and new mechanic rhythm, a letter to the editor of a gazette in Mannheim (Germany) demands his fellow-citizen to develop 'metropolitan discipline' in order not to lose the time-gains offered by the new means of transport.[65] Setting up the networked city also brought massive changes to the structure and functioning of urban administration. Until the 1870s municipal administrative bodies had been fairly small and they had restricted themselves to maintaining order and safety. With this multitude of networks and services urban administration became a much larger administrative body with comprehensive responsibilities for the welfare of urban residents; therefore it also professionalized and bureaucratized. The position of 'Oberbürgermeister' (Mayor) of a large German city became quite an attractive career perspective for middle-class professionals with a background in Law or Economics.[66] In a transitional phase, dating roughly from the 1890s to the 1930s, setting up the networked city gave considerable powers to municipal administrations to direct urban development and influence the daily life of urban residents. With the growth and formation of statewide electricity and gas networks, however, the municipal level was progressively marginalized in the shaping and directing of power networks, apart from keeping, what frequently was the case, the local distribution networks in their

62 Krabbe (1990) and Kühl (2002); Schott, Vernetzung.

63 Hård and Misa (2008) and Otter (2008).

64 Payer (2000).

65 Schott, Vernetzung, 448; Schmucki (2012).

66 Reulecke (1985) and Hofmann (2012). See for Europe Roth and Beachy (2007).

hands. The experience of both World Wars and the huge power requirements of war industry as well as the difficulty to plan and fund the massive investments required by large power plants and extensive distribution networks had demonstrated in the eyes of the decision-making forces of the time, that regulatory power had to be removed from the cities and placed either at the level of nationalized power utilities, as it happened in the UK, France and Italy after World War II, or that utilities had to be tightly regulated at the level of national government, leaving ownership with large regional utilities, frequently with a mixture of public and private shareholders, as was the case in Germany after 1935.[67] Only in the last two decades, the reluctance or resistance of large power companies towards 'green' energy policies and the deregulatory drive of the European Union have contributed to a process of re-municipalizing power systems which is supposed to make energy policies more susceptible to democratic pressure for more ecologically oriented energy policies.[68]

6.10 Planning the City

In the last decades of the 19th century another response to the 'urban crisis' of mid-century developed, town planning. Whereas networking the city was primarily concerned with removing obnoxious matter out of the city as quickly as possible and providing urban residents with energy, water and transport services which would facilitate their daily life and make it less unhealthy, town planning was concerned with spatial ordering of the city.[69] Urban life had become so problematic and unhealthy partly due to increased population densities in central districts of the cities, partly also due to the effects of industrialization. Factories had been set up at convenient places, where the decisive locational factors (water, transport accessibility, labour) were given and had henceforth structured their surroundings during their growth. Due to long working-hours and the absence of affordable public transport, workers normally had to live close to factories; they and their families were thus permanently exposed to the unhealthy emissions of smoke, noise and fluid discharges, emanating from factories. This led to a clear pattern of environmental injustice, damaging the health of workers and their families living close to factories most intensely, a fact which can be deduced from mortality figures.[70] One central motive for early town-planning thus was to disentangle and separate, as far as possible, industrial sites from places of residence. Since the environmental pollution from factories was then considered a fact which had to be accepted, the only solution was seen in physical separation. Another problem which had to be

---

[67] Schott (2008).

[68] Bauer et al. (2012).

[69] Sutcliffe (1981) and Ward (1994).

[70] Platt (2005) and Mosley (2001). For a wider discussion of 'environmental justice' see Massard-Guilbaud and Rodger (2011).

solved by town planning was to provide a clear and stable framework for private investors, wanting to build houses, factories etc. in expanding cities. Already the redevelopment of Paris under Napoleon III and his prefect Eugène Haussmann had given an exemplar to other cities, how to design modern city streets and square. But Paris had been an exception in regard to the extraordinary powers of confiscation and redistribution which had been at Haussmann's disposal in the early years.[71] In other countries, strong guarantees for the 'sanctity' of property made such large-scale urban renewal next to impossible. Thus town-planning as an academic discipline developed more around the issue, how to divide up new land at the periphery of cities. Very influential not just for Germany but later on for the general body of planning theories became Reinhart Baumeister's handbook of 1876 on city extension. Basically, it turned the very important Prussian law from 1875 on "alignment lines", demarcations between private land, which could be developed, and public land which was reserved for streets and square or other public uses, into guide lines for the practice of the engineer who has to set up an efficient and practicable street plan. Hygienic considerations on sufficient ventilation of streets and the use of natural gradients to construct well-working sewers were high on the agenda of this kind of planning. More sophisticated and comprehensive was then in 1890 the handbook published by Joseph Stübben, a German civil engineer who had carried out the planning for the extension of Cologne, which took down its massive fortification walls only in the 1880s.[72] At the same time we can already observe a growing critique of the kind of orthogonal and uninspiring urban landscape which was generated along these planning principles. Camillo Sitte, an Austrian architect and trades teacher, published a massive critique of these planning principles in his book "Urbanism according to artistic principles" (1889), in which he invited his readers to understand urban squares and streets as pieces of public art. His plea for crooked streets, for respect of natural topography and historical boundaries, for irregularity and asymmetry in the design of monuments and public squares was quickly taken up by many planners and resonated in many European countries.[73] The Garden city concept, developed by Ebenezer Howard and realized in Letchworth, close to London, from 1904 onwards, brought a new vision of suburban living and of the simple and 'authentic' cottage house into planning debates. Many social reformers saw in garden cities or in leafy settlements away from the bustle and smoke of the city a powerful instrument to solve the social question, not only by improving the health of workers, but also by turning them away from revolutionary ideas of overthrowing the capitalist system and integrating them into society.[74] 'Housing reform' thus became in the early 20th century, to a degree, what Public Health had been in the second half of 19th century, a panacea for all major social problems. By 1910 the major strands of town planning came together on an

---

[71] Hall (1998) and Jordan (1995).

[72] Sutcliffe, Towards; Albers (1997) and Ladd (1990).

[73] Wilhelm and Jessen-Klingenberg (2006).

[74] Ward (2011) and Hall (2009).

international level and formed what came to be a very influential international reformist discourse. In the UK town planning became a statutory practice by municipalities through the "Housing, Town Planning etc. Act" of 1909. The degree, to which cities supported and supplemented their planning policies through strategic land purchase policies and public transport policies was very different in Europe.[75] An almost universal tendency, however, was to devote more attention and considerably more public funds to housing after World War I. The revolutionary mood of large sections of the European working-class after 1917/18 needed to be placated, thus the rationale of that policy, by generous welfare policies, among which the improvement of housing provision was a major factor. Besides, better housing promised to yield better, more healthy workers, soldiers and mothers, so this investment also promised returns in terms of economic prosperity and national strength. Thus we can identify areas of social housing estates in new architectural forms and partly also with new communal services (laundry, library, meeting places etc.) at the then periphery of cities being developed in the 1920s and 30s, and linked to city centers and factory districts by electric tramways or urban light-rail systems. We now know, that with rare exceptions (Vienna) this social housing policy did not help to solve the housing problems of the poorest sections of populations since they simply—given their unstable income situation and frequent unemployment—could not afford to live there. But the social housing estates of the interwar period frequently helped to establish rather stable social milieus of qualified workers and lower middle class employees. Even today, many of the estates are still highly popular and sought-after residential quarters, whereas social housing estates of a later period, frequently in large high-rise complexes, have had a more mixed history.[76]

For the second half of the 20th century a planning vision which emerged of the international planning discourse mentioned before, became fundamental: With the "Charte of Athens", worked out at a conference of CIAM (Congres Internationaux d'Architecture Moderne) in 1933, an organization of decidedly modern architects and planners, and trimmed into a manifesto by their major spokesman, Le Corbusier, in 1943, a very clear and pronounced program of urban planning was published.[77] Le Corbusier and his colleagues envisaged the city of the future in analogy to a machine: Primary principle was optimal functioning which could—as the manifesto assumes—be accomplished by taking the different functions of the city, working, living, recreation, transport, apart and reassembling them at different locations of the city in order to prevent interference between them. They should then be linked by efficient public transport and wide streets giving room for private motor cars. Such a guiding vision necessitated a comprehensive redevelopment of existing urban structures, a consensus shared by most planners and architects in North America and Europe till the 1970s. This paradigm dominated urban planning

---

[75] Saunier (2007) and Schott (2009).

[76] Meller (2001), Von Saldern (1995), Kähler (1995) and Haumann and Wagner-Kyora (2013).

[77] Koch (1984).

globally and—with some variations—also in socialist countries. Although the outcome of World War II with massive destruction in many European cities seemed to provide an opportunity for such comprehensive redevelopment, the lack of funds and the weakness of public authorities prevented the realization of such radical plans in many cities and favored a more conservative mode of reconstruction, making use of existing foundations and infrastructures as far as possible.[78] With the long period of economic prosperity in the West European countries from the 1950s to the 1970s, the unexpected pressure from motorized traffic enforced a second wave of redevelopment and destructions in the 1950s–1970s, aimed to adapt the urban fabric for the requirements of motor traffic. The consensus on the need for comprehensive redevelopment, however, was challenged and then quickly broken in in the mid-1970s, In many cities young people started to occupy empty houses, earmarked for demolition and defend them against attempts by the police to evacuate the buildings. This protest movement, wanting to defend affordable housing against speculation, was back-grounded by a major cultural shift in values. In a climate of crumbling technological optimism after the Oil-Crisis and in the midst of a long-drawn period of mass unemployment, old houses and old things suddenly gained new popularity. This new values went along with new needs and new styles of living. Young people discovered for themselves modes of communal living outside classical family structures, for which the older style of mass housing was far more suitable than modern social housing blocks. The change of paradigm became very visible in a new style of urban renewal which was first practiced on a larger scale at IBA-Alt[79] in Berlin (1979–1987).

Since the Rio conference of 1992 which formulated "sustainable development" as the general goal of world community, "sustainability" has been taken on as a new target of urban planning which means that all planning measures have to be reflected according to their effects on resources, particularly energy consumption. General principles of urban planning which had internationally dominated planning thinking for large parts of the 20th century such as de-concentration of population and separation of function have been critically reevaluated over the last 20 years. The emphasis is now being placed on bringing people back into the cities in order to avoid long and unsustainable journeys to work and to be able to implement collective solutions of heating, communication and transport which work much better in a higher-density urban environment than in peri-urban sprawl. Also the idea of separating urban functions has given way to an at least partial reintegration of work and living, which is also due to the fact that many industrial or other processes are much less emitting in terms of pollution and noise and can thus be reintegrated into residential neighborhoods with only little mutual interference. This change of paradigm also corresponds to cultural shifts in larger parts of population over the last decades. Many people, particularly the growing number of senior citizens, now

[78] Von Beyme (1987).

[79] Translates into 'Internationale Bau-Ausstellung Altbau', international building exhibition, for old buildings.

place more value on proximity of services and a high variety of contacts and impressions present in an urban setting rather than extensive private green space and lack of intellectual and cultural stimuli more typical for suburban settings.

Over the last 20 years many European and North American cities have significantly improved their environmental qualities in terms of air, water and other pollution standards, due to more sophisticated technology, de-industrialization and the transfer of heavily polluting production processes to the 'global South'. On the other hand, the large megalopolises of the Third World still pose major environmental problems, partly because those infrastructural technologies which have helped to stabilize the urban environment in European and North American cities, have only been very partially implemented, partly also because the technological solutions developed in Europe in moderate climates with sufficient natural precipitation cannot be implemented in other world regions in the same ways without running the danger of unsustainable depletion of natural resources. It thus remains a major and hitherto unresolved challenge to enhance the quality of urban living for urban residents in the developing world while observing principles of sustainable development.

References

Albers G (1997) Zur Entwicklung der Stadtplanung in Europa. Vieweg, Braunschweig/Wiesbaden

Allen RC (2010) The British industrial revolution in global perspective, 4th print. Cambridge University Press, Cambridge, pp 80–105 (part chapter 4 'The cheap energy economy')

Allen RC (2013) Energy transitions in history. The shift to coal. In: Richard WU (ed) Energy transition in history. Global cases of continuity and change. (= RCC Perspectives), Munich, pp 11–15

Bagwell PS (1988) The transport revolution from 1770. Routledge, London

Bauer H, Büchner C, Hajasch L (2012) Rekommunalisierung öffentlicher Daseinsvorsorge. KWI-Schriften 6, Potsdam. http://opus.kobv.de/ubp/volltexte/2012/5806/

Blanchard I (1986) At the continental European cattle trades, 1400–1600. Econ Hist Rev XXXIX:427–460

Blockmans W (1992) Brügge als europäisches Handelszentrum. In: Vermeersch V (ed) Brügge und Europa. Mercatorfonds, Antwerpen, pp 41–56

Boockmann H (1994) Die Stadt im späten Mittelalter, 3rd edn. Beck, München

Boulton J (2000) London 1540–1700. In: Clark P (ed) The Cambridge urban history of Britain, vol II: 1540–1840. Cambridge University Press, Cambridge, pp 315–346, 317

Braunfels W (1979) Abendländische Stadtbaukunst. Herrschaftsform und Baugestalt. DuMont, Koeln, p 291 (3rd print)

Breeze LE (1993) The British experience with river pollution, 1865–1876. Peter Lang, New York

Brimblecombe P (2011) The big smoke. A history of air pollution in London since medieval times. Methuen, London (3rd print)

Büschenfeld J (1997) Flüsse und Kloaken. Umweltfragen im Zeitalter der Industrialisierung (1870–1918). Klett-Cotta, Stuttgart

Clark P (2009) European cities and towns: 400–2000. Oxford University Press, Oxford

de Witte H (2004) Some notes on the infrastructure of Brugge, with the emphasis on the water supply. In: Gläser M (ed) Lübecker Kolloquium zur Stadtarchäologie im Hanseraum IV: Die Infrastruktur. Schmidt-Römhild, Lübeck, pp 107–115

Dirlmeier U (1988) Lebensmittel—und Versorgungspolitik mittelalterlicher Städte als demographisch relevanter Faktor. Saeculum 39:149–153

Eiden H, Irsigler F (2000) Environs and hinterland: cologne and nuremberg in the later middle ages, In: Galloway J (ed) Trade, urban hinterlands and market integration c.1300–1600. Centre for Metropolitan History, London, pp 43–57

Ennen E (1987) Die europäische Stadt des Mittelalters, 4th edn. Vandenhoeck & Ruprecht, Göttingen

Fischer-Kowalski M et al (1997) Gesellschaftlicher Stoffwechsel und Kolonisierung von Natur. Ein Versuch in Sozialer Ökologie. Gordon & Breach Fakultas, Amsterdam

Freytag N, Piereth W (2002) Städtische Holzversorgung im 18. und 19. Jahrhundert. Dimensionen und Perspektiven eines Forschungsfeldes. In: Siemann W, Freytag N, Piereth W (eds) Städtische Holzversorgung. Machtpolitik, Armenfürsorge und Umweltkonflikte in Bayern und Österreich (1750–1850). Beck, München, pp 1–8

Galloway JA, Keene D, Murphy M (1996) Fuelling the city: production and distribution of firewood and fuel in London's region, 1290–1400. Econ Hist Rev XLIX:447–472

Girouard M (1987) Die Stadt. Menschen, Häuser, Plätze. Eine Kulturgeschichte. Campus, Frankfurt/Main, pp 85–100

Goodall F (2005) Gas in London: a divided city. In: Paquier S, Williot J-P (eds) L'Industrie du Gaz en Europe aux XIXe et XXe Siècles : L'innovation Entre Marchés Privés et Collectivités Publiques. PIE Peter Lang, Bruxelles, pp 121–138

Gottlieb R (2006) The present as history. In: Deverell W, Hise G (eds) Land of sunshine. An environmental history of metropolitan Los Angeles. University of Pittsburgh Press, Pittsburgh, pp 288–293

Grewe K (1991) Wasserversorgung und-entsorgung im Mittelalter. In: Frontinus-Gesellschaft e V (ed) Die Wasserversorgung im Mittelalter. von Zabern, Mainz, pp 11–88, 31

Guillerme A (1988) The age of water. The urban environment in the North of France, AD 3000–1800. Texas A&M University Press, College Station

Hall P (1998). Cities in civilization. Culture, innovation and urban order (chapter 24 The city of perpetual public works. Paris 1850–1870). Weidenfeld & Nicolson, London, pp 706–745

Hall P (2009) Cities of tomorrow. an intellectual history of urban planning and design in the twentieth century, 3rd edn. Blackwell, Oxford

Hamlin C (1998) Public health and social justice in the age of Chadwick. Britain, 1800–1854. Cambridge University Press, Cambridge

Hård M, Misa T (ed) (2008) Urban machinery: inside modern European cities. MIT-Press, Cambridge

Hardy AI (2005) Ärzte, Ingenieure und städtische Gesundheit. Medizinische Theorien in der Hygienebewegung des 19. Jahrhunderts. Campus, Frankfurt

Haumann S, Wagner-Kyora G (2013) Westeuropäische Großsiedlungen—Sozialkritik und Raumerfahrung. Informationen zur modernen Stadtgeschichte 1:6–12

Hofmann W (2012) Oberbürgermeister als politische Elite im Wilhelminischen Reich und in der Weimarer Republik. In: Idem Bürgerschaftliche Repräsentanz und kommunale Daseinsvorsorge. Studien zur neueren Stadtgeschichte. Steiner, Stuttgart, pp 121–138

Hohenberg P, Lees LH (1995) The making of urban Europe 1000–1994. Harvard University Press, Cambridge

Hughes TP (1983) Networks of power. Electrification in western society 1880–1930. John Hopkins University Press, Baltimore/London

Hünninger D (2011) Die Viehseuche von 1744-52. Deutungen und Herrschaftspraxis in Krisenzeiten. Wachholtz, Neumünster, pp 135–150

Irsigler F (1991) Bündelung von Energie in der mittelalterlichen Stadt. Einige Modellannahmen. Saeculum 42:308–318, 311

Jacquart J (1996) Paris. First metropolis of the early modern period. In: Clark P, Lepetit B (eds) Capital cities and their hinterlands in early modern Europe. Ashgate, Aldershot, pp 105–118

Jenner M (1995) The politics of London air. John Evelyn's fumifugium and the restoration. Hist J 38:535–551

Johann E (2005) Das Holz-Zeitalter. Die städtische Holzversorgung vom 17. bis zum 19. Jahrhundert. In: Brunner K, Schneider P (eds) Umwelt Stadt. Geschichte des Natur—und Lebensraumes Wien. Boehlau, Wien, pp 170–179

Jordan DP (1995) Transforming Paris. The life and labors of Baron Haussmann. Free Press, New York

Kähler G (ed) (1995) Geschichte des Wohnens. 4. 1918–1945. Reform, Reaktion, Zerstörung. DVA, Stuttgart

Keene D (2001) Issues of water in medieval London to c. 1300. Urban Hist 28:161–179

Kießling R, Plaßmeyer P (1999) Augsburg. In: Behringer W, Roeck B (eds) Das Bild der Stadt in der Neuzeit. 1400–1800. Beck, München, pp 131–137

Knittler H (2000) Die europäische Stadt in der frühen Neuzeit. Verlag für Geschichte und Politik, Wien/München

Knoll M (2008) "Dicke Luft und lachende Fluren." Überlegungen zur Umweltgeschichte der europäischen Stadt. Themenportal Europäische Geschichte. www.europa.clio-online.de/2008/Article=318. Accessed 08 Dec 2013

Koblizek R (2005) Lauwarm und trübe. Trinkwasser in Wien vor 1850. In: Brunner K, Schneider P (eds) Umwelt Stadt. Geschichte des Natur—und Lebensraumes Wien. Böhlau, Wien, pp 188–193

Koch T (1984) Le Corbusiers „Charta von Athen". Texte und Dokumente. Vieweg, Braunschweig

König W (1990) Massenproduktion und Technikkonsum. Entwicklungslinien und Triebkräfte der Technik zwischen 1880 und 1914. In: Idem, Weber W (eds) Netzwerke. Stahl und Strom 1840–1914. Propyläen, Berlin, pp 263–552

Krabbe WR (1990) Städtische Wirtschaftsbetriebe im Zeichen des ‚Munizipalsozialismus'. Die Anfänge der Gas—und Elektrizitätswerke im 19. und frühen 20. Jahrhundert. In: Blotevogel HH (ed) Kommunale Leistungsverwaltung und Stadtentwicklung vom Vormärz bis zur Weimarer Republik. Böhlau, Köln/Wien, pp 117–135

Kucher M (2005) The use of water and its regulation in medieval Siena. J Urban Hist 31:504–536

Kühl U (ed) (2002) Der Munizipalsozialismus in Europa. Oldenbourg, München

Küster H (1998) Geschichte des Waldes. Von der Urzeit bis zur Gegenwart. Beck, München

Ladd B (1990) Urban planning and civic order in Germany, 1860–1914. Harvard University Press, Cambridge

Lees A, Lees LH (2007) Cities and the making of modern Europe, 1750–1914. Cambridge University Press, Cambridge

Lorenz S (1993) Wald und Stadt im Mittelalter. Aspekte einer historischen Ökologie. In: Kirchgässner B, Schultis JB (eds) Wald, Garten und Park. Vom Funktionswandel der Natur für die Stad. Thorbecke, Sigmaringen, pp 25–34

MacDonald GM (2007) Severe and sustained drought in Southern California and the west: present conditions and insights from the past on causes and impacts. Quatern Int 173–174:87–100. doi:10.1016/j.quaint.2007.03.012

Malamud S, Sutter P (2008) Existenziell, repräsentativ, konfliktbeladen. Öffentliche Brunnen im spätmittelalterlichen Zürich. In: Rippmann D, Schmid W, Simon-Muscheid K (eds) *„... zum allgemeinen statt nutzen". Brunnen in der europäischen Stadtgeschichte*. Kliomedia, Trier, pp 89–106

Massard-Guilbaud G, Rodger R (2011) Reconsidering justice in past cities: when environmental and social dimensions meet. In: Massard-Guilbaud G, Rodger R (eds) Environmental and social justice in the city. Historical perspectives. White Horse Press, Cambridge, pp 1–40

Meller H (2001) European cities 1890–1930s. History, culture and the built environment. Wiley, Chichester

Melosi MV (2000) The sanitary city. Urban infrastructure in America from colonial times to the present. John Hopkins University Press, Baltimore

Middlebrook S (1950). Newcastle upon Tyne. Its growth and achievement. S.R. Publishers, East Ardsley

Mosley S (2001) The chimney of the world. A history of smoke pollution in Victorian and Edwardian Manchester. White Horse Press, Cambridge

Orsi J (2004) Hazardous metropolis. Flooding and urban ecology in Los Angeles. University of California Press, Berkeley
Otter C (2008) The Victorian eye. A political history of light and vision in Britain, 1800–1910. University of Chicago Press, Chicago
Paquier S, Williot J-P (eds) (2005) L'Industrie du Gaz en Europe aux XIXe et Xxe Siècles. L'innovation entre marchés privés et collectivités publiques. Peter Lang, Bruxelles
Payer P (2000) Unentbehrliche Requisiten der Großstadt. Eine Kulturgeschichte der öffentlichen Bedürfnisanstalten von Wien. Löcker, Wien
Platt H (2005) Shock cities: the environmental transformation and reform of Manchester and Chicago. Chicago University Press, Chicago
Reed M (1996) London and its hinterland 1600–1800: the view from the provinces. In: Clark P, Lepetit B (eds) Capital cities and their hinterlands in early modern Europe. Ashgate, Aldershot, pp 51–83
Rees W, Wackernagel M (1997) Unser ökologischer Fußabdruck. Wie der Mensch Einfluß auf die Umwelt nimmt. Birkhäuser, Basel
Reulecke J (1985) Geschichte der Urbanisierung in Deutschland. Suhrkamp, Frankfurt, pp 118–131
Rosseaux U (2006) Städte in der Frühen Neuzeit. Wissenschaftliche Buchgesellschaft, Darmstadt, pp 35–43
Roth R, Beachy R (eds) (2007) Who ran the cities? City elites and urban power structures in Europe and North America, 1750–1940. Ashgate, Aldershot
Saunier P-Y (2007) Transatlantic connections and circulations in the 20th century. The urban variable. Informationen zur modernen Stadtgeschichte 1:11–23
Schivelbusch W (1983) Lichtblicke. Zur Geschichte der künstlichen Helligkeit im 19. Jahrhundert. Hanser, München/Wien
Schmid W (1998) Brunnen und Gemeinschaften im Mittelalter. Historische Zeitschrift 267:561–586
Schmieder F (2005) Die mittelalterliche Stadt. Wissenschaftliche Buchgesellschaft, Darmstadt, pp 21–26
Schmucki B (2012) The machine in the city: public appropriation of the tramway in Britain and Germany, 1870–1915. J Urban Hist 38:1060–1093
Schott D (1998) Zur Genese der Kolonialstadt in Südostasien: Batavia und Singapur. Trialog 56:13–19
Schott D (1999) Die Vernetzung der Stadt. Kommunale Energiepolitik, öffentlicher Nahverkehr und die "Produktion" der modernen Stadt. Darmstadt–Mannheim–Mainz 1880–1918. WBG, Darmstadt
Schott D (2008) Empowering cities: gas and electricity in the European urban environment. In: Hård M, Misa TJ (eds) Urban machinery. Inside modern European cities. MIT-Press, Cambridge, pp 165–186
Schott D (2009) Die Stadt als Thema und Medium europäischer Kommunikation: Stadtplanung als Resultat europäischer Lernprozesse. In: Roth R (ed) Städte im europäischen Raum. Verkehr, Kommunikation und Urbanität im 19. und 20. Jahrhundert. Steiner, Stuttgart, pp 205–225
Schott D (2011) Städte und ihre Ressourcen in der Geschichte: Blicke über und aus Europa. In: Hoppe A (ed) Raum und Zeit der Städte. Städtische Eigenlogik und jüdische Kultur seit der Antike. Campus, Frankfurt, pp 95–116, 97–102
Schott D (2012) The 'Handbuch der Hygiene': a manual of proto-environmental science in Germany of 1900? In: Berridge V, Gorsky M (eds) Environment, health and history. Palgrave Macmillan, Basingstoke, pp 69–93
Schott D (2013) Katastrophen, Krisen und städtische Resilienz. Blicke in die Stadtgeschichte. Informationen zur Raumentwicklung H.4:297–309
Schott D (2014) Die Urbanisierung Europas 1000–2000. Eine umweltgeschichtliche Einführung. Böhlau, Köln
Schubert E (1986) Der Wald: wirtschaftliche Grundlage der spätmittelalterlichen Stadt. In: Herrmann B (ed) Mensch und Umwelt im Mittelalter. DVA, Stuttgart, pp 252–269

Schubert E (2012) Alltag im Mittelalter, Natürliches Lebensumfeld und menschliches Miteinander, 2nd edn. Wissenschaftliche Buchgesellschaft, Darmstadt, p 99

Sieferle RP (2001) The subterranean forest. Energy systems and the industrial revolution. White Horse Press, Cambridge

Sieferle R (2008) Transport und wirtschaftliche Entwicklung. In. Idem. Transportgeschichte. 1–38:7

Sporhan L, von Stromer W (1969) Die Nadelholzsaat in den Nürnberger Reichswäldern zwischen 1469 und 1600. Zeitschrift für Agrargeschichte und Agrarsoziologie 17:79–99

Spreithofer G, Heintel M (1997) Jakarta: Der „Big Apple" Südostasiens. In: Feldbauer P et al (eds) Mega-Cities. Die Metropolen des Südens zwischen Globalisierung und Fragmentierung. Brandes und Apsel, Frankfurt, pp 151–176

Stahl M (2008) Die antike Stadt und ihre Infrastruktur. Die Wasserversorgung. In: Schott D, Toyka-Seid M (eds) Die europäische Stadt und ihre Umwelt. Wissenschaftliche Buchgesellschaft, Darmstadt, pp 27–45

Sutcliffe A (1981) Towards the planned city. Germany, Britain, the United States and France 1780–1914. Blackwell, Oxford

Tarr J, Dupuy G (eds) (1988) Technology and the rise of the networked city in Europe and America. Temple University Press, Philadelphia

Toyka-Seid M (2008) Die europäische Stadt in der imperialen Peripherie. In: Schott D, Toyka-Seid M (eds) Die europäische Stadt und ihre Umwelt. Wissenschaftliche Buchgesellschaft, Darmstadt, pp 145–168

Unger RW (1999) Feeding low countries towns: the grain trade in the fifteenth century. Revue belge de philologie et d'histoire 77(2):29–358

Von Beyme K (1987) Der Wiederaufbau. Architektur und Städtebaupolitik in beiden deutschen Staaten. Piper, München/Zürich

Von Saldern A (1995) Häuserleben. Zur Geschichte städtischen Arbeiterwohnens vom Kaiserreich bis heute. Dietz, Bonn

Von Thünen JH (1826) Der isolierte Staat in Beziehung auf Staat und Nationalökonomie, oder Untersuchungen über den Einfluß, den die Getreidepreise, der Reichthum des Bodens und die Abgaben auf den Ackerbau ausüben ([Hamburg 1826], reprint from the 3rd edn, Schumacher H (ed), [Berlin 1875], Darmstadt 1966, Part I, 390 and 391)

Ward S (1994) Planning and urban change. Paul Chapman Publishing, London

Ward SV (ed) (2011) The garden city. Past, present and future. Spon, London

Wilhelm K, Jessen-Klingenberg D (eds) (2006) Formationen der Stadt. Camillo Sitte weitergelesen. Bauverlag, Gütersloh

Winiwarter V (2001) Where did all the waters go? The introduction of sewage systems in urban settlement. In: Bernhardt C (ed) Environmental problems in European cities in the 19th and 20th century. Waxman, Münster, pp 105–119, 107

Zwierlein C (2011) Der gezähmte Prometheus. Feuer und Sicherheit zwischen Früher Neuzeit und Moderne. Vandenhoek & Ruprecht, Göttingen

Chapter 7
History of Waste Management and the Social and Cultural Representations of Waste

Sabine Barles

Abstract The history of waste mirrors that of the societies that produced it, and their relationship with the environment and the resources they mobilized. Until the industrial revolution, the management of urban excreta was predominantly linked with urban salubrity, from the Roman *cloaca maxima* to the Parisian *motta papellardorum*. The quantity of waste produced remained small and the methods for collection and discharge often unsatisfactory, which led to frequent denunciations of urban dirtiness. Neo-Hippocratic medicine, which considered the tainted environment and air to be the principal causes of urban excess mortality, prompted the implementation of new policies and management techniques in Europe to clean up the cities. In addition, the value of most urban excreta intended either for agriculture or industry increased. Thus, from about the 1770s to the 1860s, salubrity and excreta recovery went hand in hand. From the 1870s onward, the fertilizer revolution, the rapid development of coal and, later, that of the petroleum industry and the search for more convenient and plentiful materials, undermined the recycling industry. Although some cities at first tried to fight the devaluation of urban by-products, they gave up during the interwar years. What was once a source of profit became a cost to society, and, until the 1960s, the aim of waste management was to reduce this cost. The environment became the receptacle for waste. The 1960s and 1970s were marked by an environmental crisis, a growing concern for the limits of the planet and a criticism of the industrial city. In this context, waste was regarded as the symbol of the aberrations of a consumer society. The production of waste continued to grow and the sanitary accidents as a result left a deep impression. Waste policies were implemented with mixed results. Developing countries also began to suffer from this curse of developed countries.

S. Barles (✉)
UMR Géographie-Cités, Université Paris I Panthéon-Sorbonne, Paris, France
e-mail: sabine.barles@univ-paris1.fr

M. Agnoletti and S. Neri Serneri (eds.), *The Basic Environmental History*,
Environmental History 4, DOI 10.1007/978-3-319-09180-8_7

7.1 Introduction

The histories of waste, and of the words that have been used and continue to be used to describe it, are inseparable from one another. Indeed, a quick survey shows that three different types of vocabulary have emerged to describe what we now call waste. In the first category, terms are associated with the themes of loss and uselessness: *déchet* in French from the verb *choir* (to fall), *refuse* and also *garbage* in English (which primarily refers to animal offal), *rifiuti* in Italian, *residuo* in Spanish, *Abfall* in German. In the second category, terms emphasize the dirty or repulsive nature of these particular materials: *immondice* in French, *immondizia* in Italian, from the Latin *mundus* which means clean; *ordure* in French from the Latin *horridus*, meaning horrible. Finally, terms in the third category describe the materials that make up the waste: *boues* in French, *spazzatura* in Italian, *Müll* and *Schmutz* in German, *rubbish* in English derived from *rubble*.[1]

The word *waste* belongs in the first category. From the old French *vastum*, which means empty or desolate, it was first used to depict a desolate, ruined or neglected region. Later, the term was used to describe a wasteful expenditure (and, in this sense, it had the same meaning as *déchet* in French). It finally acquired its current meaning in the 15th century. The fact that the original meaning of *waste* has a spatial dimension in that it described a place, similarly to *spazzatura* from the verb *spazzare* (to make room, remove clutter), is likely not neutral. It is also undoubtedly the case with the rich vocabulary, which has only been touched on lightly here, used to describe various wastes. Indeed, the issue of waste has long been closely linked to (even confused with) both the issue of salubrity and sanitizing of urban space and the management of urban urine and excrement.

After a quick overview of the period between Antiquity and the eve of the Industrial Revolution, we will focus on the period between 1770 and 1860, during which the value of excreta, particularly urban excreta, thanks to its agricultural and industrial importance, increased. Next, the birth of waste in the form of abandoned junk and materials from the 1870s to the 1960s will be presented. Finally, we will show that since the 1960s and 1970s, the environmental crisis has translated into a waste crisis for which only imperfect solutions have been found. The story of waste is an international one, however, here we focus on the history of waste in Europe and North America[2] (with the exception of the last chapter) and do not address its history in the former East Bloc.[3]

[1] See also Harpet (1998).

[2] Even if the literature about the history of waste has developed since the end of the 20th century, it remains relatively scarce compared to other urban environmental history topics. See for instance, for the last two centuries: Melosi (2005), Strasser (1999), Barles (2005) and Giuntini (2006).

[3] See for instance: Gille (2007).

7.2 From Antiquity to the Eve of the Industrial Revolution

Streets, and more generally open spaces in cities, have often been used as receptacles for urban waste: human and animal urine and excrement, other organic materials from domestic or artisan activities, rubble from demolitions, various mineral debris, etc., such that the composition of these soils provide an account of a city's history. The impregnation of waste into the soil was particularly significant because streets and squares were not always surfaced and could absorb much rainwater or because urbanized areas were built on low, even marshy, ground.

The need to clean up polluted urban space was at the root of the famous Roman *cloaca maxima*, built under Tarquin the Proud (7th–6th century BC) to drain the Velabrum and the lowlands located between the Capitoline Hill and Palatine Hill. First through an uncovered canal network, then with a subsurface sewage system, it collected urban refuse and materials from latrines and drained them into the Tiber. Subsidiary lines, such as ditches originating from houses, led to the *cloaca maxima* and contributed to the cleaning up of Rome. Thus perched on these subterraneous passages, Rome was described as a "hanging city" (*urbs pensilis*) by Pliny the Elder.[4] The maintenance and cleaning of sewers, a job given to convicted criminals, was the basis for the *cloacarium* tax. Many cities at that time were equipped with similar community facilities.

The use of these underground pipes to drain and clean urban areas declined in varying degrees during the Middle-Ages in Europe and was replaced by surface runoff for rainwater and drainage waters. Urban brooks (still identifiable today in France by the name *Merdereau* or *Merderet*) and moats acted as sewers. Many cities diverted, canalized and created networks of drainage systems in order to allow for the development of their artisan activities (at the time this water played a mechanical role, later, according to its composition, it took on a chemical and biological role). Because these canals contributed to drainage, they were simply considered sewers; however, they had a much more significant role in that they founded urban prosperity.[5]

Furthermore, the status of human excremental materials varied in space and time. Some cities retained a combined sewerage system used since Roman time; many cities adopted, during the Middle Ages or the Renaissance, pit privies, which were at first simple holes and later underground reservoirs placed under dry latrines. The growing use of these cesspools led to the development of a new profession: the cesspool emptier (although in some cities local growers did this job). Moreover, the necessity for salubrity led many cities to prohibit the disposal of human waste into sewers and rivers—this was the case in Paris where, since the 13th century, the Great Sewer ("Grand Égout"), a former backwater of the Seine River, drained the Right Bank. However, these bans, as well as possible sweeping and cleaning obligations, were often ignored by urbanites.[6]

---

4 In his *Natural History*, XXXVI.

5 Guillerme (1983, 1988).

6 Boudriot (1990).

In these cities, where only a few streets were paved, where the slope of streets was not regulated, where both human and animal populations were extremely dense and where cart and other tipcart traffic contributed to the formation of a putrid mud, a significant elevation of the ground led to ground floors, even second floors, of houses to become buried. This partly unintentional elevation of the ground level was the result of an accumulation of urban waste and rubbish at the surface. It occurred at varying rates and often accelerated following demolitions in times of conflict. It also tended to increase as a result of artificial embankment construction which transformed marshy areas into developable land and where construction material often was itself a type of waste: excavated material from moats, demolition rubble and urban mud.

During the Middle Ages, dumpsites, formed from bulky refuse and the drainage of pit privies or through the deployment of street cleaning services, could be found in some cities.[7] These dumps, originally established at the city gates, then later surrounded by the growing city and replaced by sites outside of the new urban limits, often grew into real hills. This is the case in Paris where these mounds have been completely integrated into the urban landscape; because they are raised above the general ground level, they can accommodate more efficient windmills (Fig. 7.1). The labyrinth of the Jardin des Plantes is another example of a historical dumpsite that is still visible today. Other mounds have disappeared: this is the case for the *motta papellardorum*, located on the western point of the Île de la Cité and for the *Monceau Saint-Gervais*, located behind the city hall (Hôtel de Ville).[8]

Generally speaking, salubrity levels dropped in European cities from the 15th to 18th century.[9] It was during the 18th century that two movements were set in motion that eventually resulted in a reassessment of the management of urban excreta.

7.3 1770s–1860s

7.3.1 Neo-Hippocratism and Hygienics

During the 18th century in Europe, medical thinking was characterized by a growing interest in Hippocrates' theories. In particular, his treatise "Airs, Waters, and Places",[10] in which he emphasized the primary role of the environment in

---

[7] Chevallier (1849).

[8] Belgrand (1887).

[9] For more precisions about Middle Ages and Renaissance, see (among others): Leguay (1999); Assainissement et salubrité publique en Europe méridionale (fin du Moyen Âge, époque moderne). 2001. *Siècles—Cahiers du centre d'histoire "Espaces et cultures"* 14; Magnusson (2006) and Jorgensen (2008).

[10] Coray (1800).

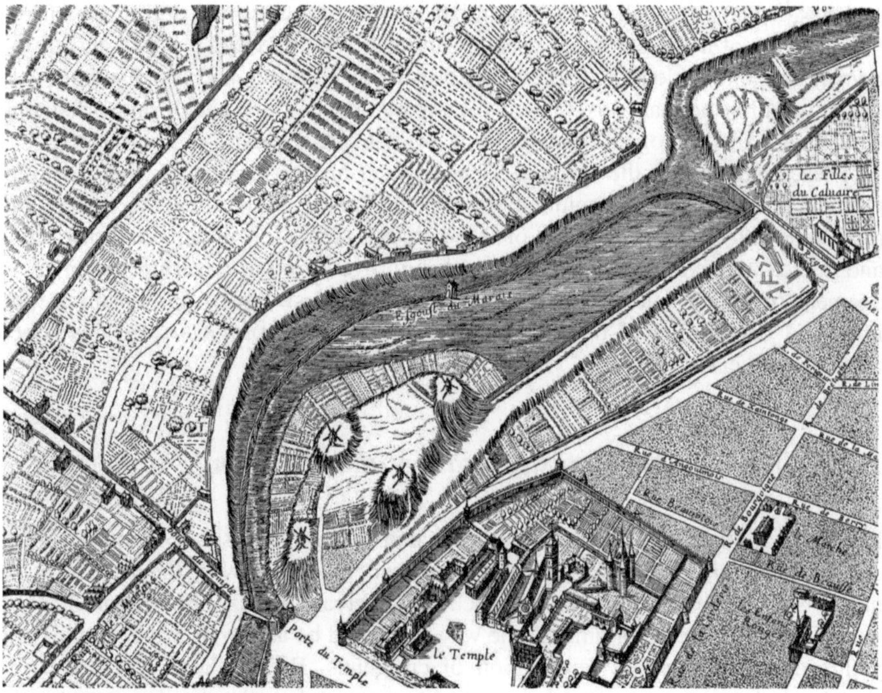

Fig. 7.1 Paris, map of Jacques Gomboust (partial), 1653

health, was frequently referenced.[11] Many doctors followed his recommendations and looked to the environment to explain morbidity and mortality. Many of them considered air, and its intimate and frequent contact with the body, a transmission medium for miasma or sulfurous pollution emitted by the soil whose fumes were often considered morbid, even deadly.[12] Their analysis was confirmed by the high frequency of intermittent fevers in wetlands where a generalized weakening of the body shortened the life of their residents.[13]

Eighteenth century doctors paid particular attention to cities whose statistics—which at that time were being developed extensively—revealed excess mortality: not only was the natural balance negative (more deaths than births), but the life expectancy of their residents was lower than in the neighboring countryside.[14] It became the natural tendency for doctors to view these environments through this neo-Hippocratic lens and to consider them as extreme and artificial types of

[11] Corbin (1986, 1988), Barles (1999) and Lécuyer (1986).

[12] Boissier de Sauvages (1754) and Méphitis (1765).

[13] Lancisi (1717).

[14] Poussou (1992) and Kunitz (1993).

marshes.[15] Thus, the belief was that urban excess mortality was due to the cumulative effects of a contaminated ground saturated with putrefying waste and of the human and animal density. Indeed, discoveries by Priestley and Lavoisier showed that respiration permanently tainted the air by consuming oxygen and producing CO_2 (then called phlogisticated air) in a process characteristic of combustion.[16]

These analyses led the medical establishment and, in large part, the scientific, political and intellectual communities to establish new requirements to correct and rectify the deleterious conditions of these cities.[17] They recommended airing cities and decreasing their putrefaction through improved ground covering, better management of human and urban excreta, universalized sweeping and cleaning, and improved distribution of these services. These types of projects were developed throughout the 19th century.

7.3.2 Urbanization, Industrialization and Recycling

Both industrialization and urbanization separately raised the issue of food resources and raw materials required for industrial use. Demographic growth, the increase in the number of urbanites, required a concurrent increase in agricultural production. According to future agronomists, one way this could be achieved was by improving yields through improved fertilization of croplands. By the late 18th century, a shortage in farm manure prevailed, leading to a search for other fertilizing materials. Indeed, the numerous studies on putrefaction during the 18th century and the identification of the great material cycles by early chemistry showed that death was critical to life and that human and animal excreta, as well as food residues, could be used as fertilizers.[18] As these wastes were most concentrated in cities, it was there that mud from streets, drainage of pit privies, beef blood, old shoes, indeed any organic waste was collected. Throughout Europe and North America, scientists and intellectuals stressed the need for cities to return their food as fertilizer to the countryside. Recovering those "materials which the cities owe to the earth"[19] was the only way to ensure both salubrity (through an efficient collection of organic materials scattered throughout cities) and food production. Throughout the 19th century, the chemists Jean-Baptiste Dumas, Jean-Baptiste Boussingault, Justus von Liebig, Alexander Müller (among others), the lawyer and social reformer Edwin Chadwick, engineers like Adolphe-Auguste Mille, all promoted human and urban fertilizers.[20] Later, even public figures addressed the issue, such as Victor Hugo in *Les Misérables*.

---

[15] Baumes (1789).

[16] Priestley (1774) and Lavoisier (1782).

[17] Fortier (1975).

[18] Wines (1985) and Tarr (1996).

[19] Dumas (1866–1867).

[20] von Liebig (1862), Müller (1860), Chadwick (1842) and Paulet (1853). See also: Mårald (2002), Goddard (1996) and Hamlin (2007).

Furthermore, an important part of emerging industry was reliant on using raw materials that could be supplied only by cities. This is the case, for example, with vegetable rags used for papermaking for several centuries, but became much more needed once the papermaking machine was developed.[21] In the 19th century, rags became a strategic industrial issue (1.5 kg of rags were needed to produce 1 kg of paper), such that France banned their export from 1771, followed by Belgium, Holland, Spain, Portugal and a few other countries during the first half of the 19th century.[22] Great Britain and North America fought over the international markets, their local resources insufficient to meet the growing industrial demand, forcing them to look for rags in countries that did not produce or produced little paper: in 1850, North America imported 50,000 tons of rags, more than 60,000 tons in 1875.[23] Rag collection was therefore an urban activity: an urbanite produced on average more rags than a rural resident which, in addition to the typically more concentrated population of cities, made the collection of urban rags more profitable.[24] Thanks to urban rag collection, production doubled during the first half of the 19th century (Table 7.1).

Similarly, there was a growing industrial use of animal bones (Fig. 7.2), which were also concentrated in urban areas since slaughtering took place in the city (in butcheries then later in specialized slaughterhouses). Bones were increasingly needed for their classical use—the manufacture of objects, grease, glue—as well as for new market opportunities: from the 1820s, phosphorus was used to make matches ignited by friction; animal charcoal to refine sugar whose consumption was growing—from 1 kg/capita/year[25] in 1788 to nearly 5 kg/capita/year in 1856 in France, three times more in England[26];—gelatin (identical to glue except for its use) for food preparation and later for photographic negatives; and later, superphosphates for agricultural fertilization (first in England and Germany then in France). Other butchery by-products found market opportunities in the manufacture of candles and later of stearic candles, Prussian blue, glue, ropes, combs, etc.

Urban by-products emerged from these new industrial products and led to other market opportunities: used paper gave rise to cardboard industries, tin cans to metal toys, town gas (obtained from the distillation of coal) to tar which was used in the manufacture of numerous chemical compounds and, soon after, for surfacing sidewalks and later streets. Many other examples of the city as a source of raw materials for industry and for agricultural fertilizers exist.[27]

---

21 André (1996), Hills (1988) and Strasser (1999), op. cit. p. 80 sq.

22 Turgan (1860–1885).

23 Strasser, op. cit. p. 85.

24 Esquiros (1861) and Barberet (1866–1887).

25 Chaptal (1819).

26 Payen (1859).

27 For more details about urban raw materials, see: Barles (2005). op. cit. About industrial use of urban by-products, see: Guillerme (2007). See also: Simmonds (1862).

Table 7.1 World production of paper and cardboard in 1850

Country	Production (tons)
England	62,960
Scotland	14,300
Eire	3,310
France	41,680
Zollverein	37,200
Austria	22,320
Denmark	1,680
Sweden	1,530
Belgium	6,132
The Netherlands	4,200
Spain	5,310
Italy	7,992
Switzerland	13,000
Turkey	180

Source Picard (1891)

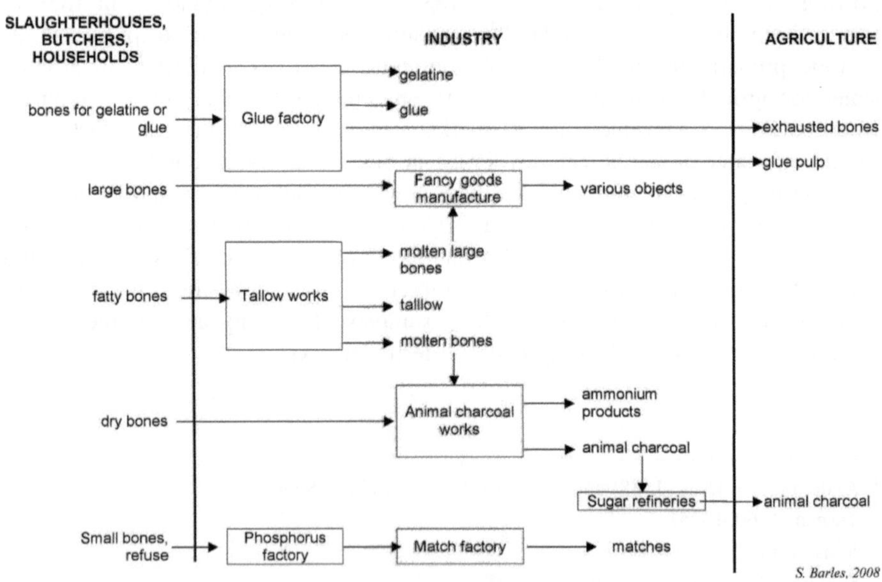

Fig. 7.2 Life cycle of animal bones, second third of the 19th century

To multiply their profits, manufacturers sought out market opportunities in products with no value, i.e. waste from their primary operation. Animal charcoal dirtied during the clarification of sugars could not be revived endlessly: it was considered a waste up until 1810 when it began to be used as a fertilizer (containing a high concentration of phosphorus). In such a way, used animal charcoal soon became more expensive than unused animal charcoal. Cotton waste from textile mills—particularly in Manchester—began to be used, like rags, for papermaking. As late as 1882, the hygienist Henri Napias summarized the goal: "In industry, there must not be any actual scrap, and everything must be used either for industry itself or for agriculture [translation]".[28]

7.3.3 An Uneven Situation

Hygienic requirements on the one hand and agricultural and industrial demand for urban excreta on the other, had unequal effects on the management of waste. In big cities, and particularly in the capitals, hygienic policies were quickly established (which does not imply that small and mid-size cities did not demonstrate innovation, as illustrated in the English borough of Croydon[29]). In large part, these policies led to the recalibration of arterial systems, street paving, water distribution, collection of wastewaters in gutters or sewers, and reorganization of the collection services of urban excreta.[30]

A distinction can be made among these cities between those who chose to immediately adopt the combined sewerage system and those who resisted. In fact, home water distribution was established in London as early as the 18th century and created a corollary need for water drainage. The solution quickly became to discharge the used water into sewers, then in the Thames. The growing use of water-closets with flushing systems prompted the removal of pit privies (a gradual and relative removal such that the systems were often used together). As a consequence, drainage became an issue of wastewater and the need for recycling led, in many cases, to the use of wastewater for agricultural irrigation purposes in sewage fields (as in Milan or Chambéry). In other cities, including Paris, water was distributed to homes much later: it was first used to clean streets, and the system of pit privies remained prevalent. Here, innovations in the management of sewage removal from cesspools and the production of human fertilizers were numerous and assisted in making sewage disposal a very lucrative economic sector. The "Flemish method",[31] the fertilization of land by spreading fresh sewage, was used in Northern France and Holland. Widely used in China, this method was adopted throughout Europe during

---

[28] Napias (1882).

[29] Goddard and Sheail (2001).

[30] See for instance: Melosi (2000), Vögele (1998) and Frioux (2013).

[31] Moll and Mille (1857).

the 19th century to such a degree that some regions chartered "sewage trains" to increase their radius of use around cities. In other areas, a fertilizer obtained from drying night soil for a few years was produced—Paris' dry night soil (*poudrette*) was sold in a 200 km radius around the city. Later, sulfate of ammonia was manufactured, a compound obtained by the distillation of the liquid part of night soil in a process developed in Paris during the 1830s and commercialized in England.[32] This fertilizer was later made from by-products of coal distillation and became one of the major suppliers of nitrogen at the turn of the 19th and 20th centuries. Animal-seaweed (a mixture of brown algae and night soil) produced in Marseilles, was transported by sea and commercialized as far as Italy. Large waste removal companies were thus doubly profitable: they profited not only from emptying cesspools, but also from commercializing fertilizers—which explains why they were opposed to future combined sewerage system projects. Furthermore, cities often collected taxes on waste removal and these monies greatly contributed to their annual revenues.[33]

The collection of refuse and mud from streets also noticeably evolved. In many cities, the sweeping, usually required of residents, was gradually replaced (where it was not already the case) by the collection of tax and the organization of a public service that could be ceded to private enterprises. Refuse heaps, made either by urbanites or by employees of the sanitation department, were searched by rag-and-bone men whose profession reached its apex between 1830 and 1870. Hygienists considered scavenging a necessary evil: when practiced on the street it was harmful to salubrity, but as a supplier to industry it was one of the factors of economic growth. During this period, the revenues of rag-and-bone men generally increased and wholesalers of rags and other salvaged materials acquired at times considerable wealth. Material not salvaged by the rag-and-bone men—mud—was picked up by public or ceded waste removal services. Intended for agriculture, mud was often sold by waste removal companies or services to farmers, like in Edinburgh, Lille, Manchester, Paris—whereas it was given freely a century before—such that in several cities (including Paris) cleaning costs actually decreased in the first two-thirds of the 19th century despite substantial increases in cleaning requirements and the growing population.[34]

This improvement—admittedly relative in terms of the sanitary issue—to the management of waste was not equal among cities. Two factors were at stake in the growth of recycling measures in big cities. On the one hand, recycling was cost effective only if the supply was sufficient, such that the population producing the materials had to be large. On the other hand, these materials were most often intended for agriculture use (and in particular for market garden production) and local industries, which depended on a large market and thus also needed a large population base. Elsewhere, the modalities in the management of urban excreta

---

[32] Commission des engrais (1865–1866).

[33] Barles (2005). op. cit.

[34] Du Mesnil (1884).

were hardly different than those that characterized the 18th century. As a consequence, growth in the recycling sector can be considered an indicator of the spread of the first wave of industrialization.

7.4 1870s–1960s

7.4.1 Industry and Agriculture Turn Their Backs on the City

From the 1870s onward, the doctrine that the recycling of by-products is a condition of industrialization, food production and salubrity, was undermined by the mobilization of new resources and raw materials.

Urban waste, responsible for the growth of many industrial branches, was increasingly considered the limiting factor. This was the case for rags, whose supply was insufficient to meet the industrial demand from the 1860s onward and whose price continued to rise as a consequence. Paper manufacturers had little control over wholesalers of rags and looked for vegetable-based substitutes to escape these constraints. At first, straw and alfa were used and later, as a result of work by the German papermaker Henri Vœlter, wood.[35] Initially, both the new types of pulp and rag pulp were used; later, at the end of the 19th century, rag pulp was gradually replaced entirely. Use of the new pulp led to a tenfold increase in world production between 1850 and 1890.[36] At the same time, market opportunities for animal bones also declined. Sugar refiners considered animal charcoal too expensive and started to use activated charcoal (from plant sources) and mechanical refining. Plastic materials replaced bones for the manufacture of some objects: celluloid, first developed in 1869 from cellulose and camphor to replace the ivory in billiard balls, and later Bakelite, fabricated in 1907 from phenol and formol (formaldehyde), both invented in the United States.[37] From Bakelite on, all plastic materials were made from by-products of the coal and oil industries. As early as the 1860s, these gave rise to the dye industry and led to the demise of the Prussian blue industry.[38] Electrical lighting was in use by the late 19th century in cities, thus competing with stearic candles; vegetable glue and later synthetic glue replaced animal glue; fish gelatin took the place of bone gelatin. With refrigeration, large slaughterhouses moved away from the cities and relocated closer to where animals were being raised. This move led to the emergence of the powerful meat and bone meal industry that utilized useless butchery by-products. The durability of this

[35] Figuier (1873). See also: André, op. cit.; Hills, op. cit.

[36] *Exposition universelle internationale de 1900 à Paris. Rapports du jury international. Introduction générale. Tome II. 3e partie: Sciences. 4e partie: Industrie*. 1903. Paris: Imprimerie nationale, p. 366.

[37] Friedel (1983).

[38] Bensaude-Vincent and Stengers (2001). (Trad. Bensaude-Vincent and Stengers 1996).

industry remained unquestioned until the 1990s following the Bovine Spongiform Encephalopathy crisis.[39]

Issues surrounding fertilizers also evolved.[40] Beginning in the 1850s, based on the well established role of phosphorus in plant growth (recognized as early as the 1840s), there was a frantic world-wide search for fossil phosphates to supply a need not filled by bones. Animal bones were no longer needed for phosphorus fertilization following the discovery of large deposits in North Africa and the expansion of mines in the United States at the beginning of the 20th century.[41] Furthermore, during the last 3 decades of the 19th century, sulfate of ammonia became essential for nitrogen fertilization; it was increasingly extracted from ammoniated water during the manufacture of town gas and later from large industrial coking plants.[42] At the same time, imports of sodium nitrate from South America into Europe and the United States soared. The growing concern that this latter supply would soon be depleted[43] turned the search towards the largest known reservoir of nitrogen: air (it contains 80 % nitrogen). The Haber–Bosch process,[44] developed on the eve of the First World War, appeared to provide an infinite amount of nitrogen for agriculture and the war (it was also used in the manufacture of explosives). With a growing trend to trust mineral fertilizers over organic ones, these new nitrogen sources competed with urban fertilizers and more generally with those from recycled materials. In fact, this competition became fiercer as urban growth loosened the link between city and agriculture.

7.4.2 New Methods of Recovery

In cities that had based their waste management on recycling, these changes put into question the entire economy and management of urban excreta that had been, until then, a source of tax revenue or, at the very least, of municipal savings. The crisis was particularly severe as the issues of recycling and of salubrity, which had for a long time seemed convergent, conflicted with increasing frequency. Hygienists severely criticized the act of dumping refuse on streets and, in a few cities from the 1860s, it became necessary to use boxes or refuse bags (Lyons in 1855, Paris in 1883, Saint-Petersburg).[45] This made scavenging more difficult and less lucrative as mixed materials took on an altered quality.[46] The gradual disappearance of animals

39 See for instance: van Zwanenburg and Millstone (2005).

40 Wines, op. cit.

41 Matignon (1931).

42 Ibid.

43 See for instance: Crookes (1917).

44 Smil (2001).

45 Jugie (1993).

46 Fontaine (1903).

in the urban space, encouraged by hygienists, and made possible by the mechanization of transportation, the use of tar or asphalt-based surfacing designed for the new automotive traffic, and the improvement to garbage collection, reduced some of the fertilizer value of urban mud. At the same time, growing household consumption resulted in increased amounts of waste being produced by urbanites as well as the emergence of new types of waste. References to packaging, for example, can be found as early as the beginning of the 20th century. Nevertheless, not all cities abandoned the concept of recycling; on the contrary, many more cities that did not yet recycle began to. In both cases, these initiatives aimed to factor in the new industrial and agricultural constraints.

The first strategy was to look for land farther away and with greater yields for agricultural market opportunities that had been previously available from market gardening at the city gates. Marseilles, France's second largest city with half a million inhabitants at the end of the 19th century, was renowned for its insalubrity. It was one of the last cities in France to suffer from a cholera epidemic and was unable to dispose of its mud and garbage within its immediate outskirts; rather it was piled up in two dumps, which worsened the city's situation. A first project set out to discharge refuse into the sea—a solution that was adopted by many coastal cities throughout the world. Meanwhile, the Crau plain, located 60 or so kilometers northwest of the city, was the focus of a significant agricultural development project. Its progress, however, was slowed by the scarcity and high cost of fertilizers. In the 1880s, a private company undertook the construction of a railway line to dispose Marseille's mud in this plain. The trains returned to the city loaded with stones collected from the fields, and these stones were used in public works. The city made minimal financial contributions to the project because the company's primary revenues came from selling the mud and because it received contributions from the manufacturers of Marseilles who wished to improve the city's standards of hygiene in order to boost economic development.[47] Around the same time, Paris disposed of its mud in the north of France where a thriving beet cultivation required low cost fertilizers—the profitability of the operation was possible only because railway companies were obligated by the State to charge very low tariffs on this type of transport. Similarly, the adoption of a combined sewerage system in many cities led to the establishment of sewage fields becoming located increasingly further away from urban areas in order to benefit from sufficient surfaces and affordable land prices. In the 1920s, the sewage fields of Paris were located at several tens of kilometers away from the city and the engineers even contemplated fertilizing the region of *Champagne sèche*, approximately 120 km away.[48] Another solution was to manufacture more reliable and effective fertilizers to compete with mineral and industrial fertilizers. Zymothermic fermentation, where optimal conditions of

[47] *Enlèvement journalier par chemin de fer et conduite dans la plaine de la Crau des immondices, balayures et vidanges de la ville de Marseille. Observations sur le projet Montricher*. 1886. Marseille.

[48] Védry (1992).

Fig. 7.3 Waste crushing plant near Paris, Pantin, early 20th century

ventilation and humidity are used to accelerate decomposition in dungheap pits, was tested in Florence in 1912 and later in many cities. Using this method, fertilizer could be obtained in 40 days or 9 cycles per year, later as quickly as 30 days and 12 cycles per year, compared to 4 or 5 months for mud kept in heaps (2–3 cycles per year).[49]

Another avenue, investigated both in France and the United States, was to industrialize scavenging. The traditional scavenging site of rag-and-bone men, streets (as in Paris) or dumps (as in New York), were moved into a specialized establishment (Fig. 7.3): a salubrious sorting plant where sorting operations could be carried out in a more rational and efficient way. From the 1880s on, there was an increasing number of these plants that aimed to offer manufacturers and farmers products that had been carefully sorted and processed and were therefore competitive. In 1923, a sorting plant built in Nice (in the South of France), a city that had once discarded its waste into the sea,[50] produced fertilizers and separately salvaged papers, cartons, rags (sorted by type), corks, bones, various scrap metals, tin scrap, other metals, bread and other food remains. Processes were further improved with the use of magnets for the recovery of ferrous metal and belt conveyors for the transport of waste past the workers. Each product had its specialized collector and rag-and-bone men went from being self-employed (freedom was one

[49] Joulot (1946), see: Giuntini, op. cit., pp. 65–67, 70–71.

[50] Courmont et al. (1932).

of the arguments made by rag-and-bone men to practice their profession in the 19th century) to salaried workers.[51] In North America, swine-feeding and other recycling programs developed, specially during World War I.[52]

Finally, as early as the 1870s, incineration was tested in England.[53] The first *British destructor* was used in Nottingham in 1874, another in Manchester 2 years later. The huge advantages of these *destructors* were their capacity to reduce the volume and weight of waste to smoke, and the fact that they could be established in urban areas, even highly populated urban areas, thanks to their (supposed) great salubrity. Indeed, it brought processing closer to the site of collection and decreased the cost of service compared to long-distance solutions that were often more costly than initially anticipated and that gave rise to complaints from residents of the roads that were being used by dumpers, and later, transport trucks. At first, incinerators adopted in Great Britain and the United States were not used in France because of the generally held opinion that they resulted in the loss of many useful materials, converted into smoke. Nevertheless, energy recovery tests proved to be fruitful and revealed that incineration provided a new opportunity for the recovery of waste (because it is autocombustible). The most striking example is from Liverpool where the incineration of 53 % of household waste, 174,090 tons in 1907, produced 9.2 million kWh that was used to power tramways. Ash (33 % of the incinerated tonnage) was used for the manufacture of mortar, concrete structures and concrete slabs for sidewalks.[54] From that time on, many sorting plants were equipped with incinerators and only materials with high value were salvaged (in particular metals), the rest was burned. Moreover, central and urban district heating offered a new market venture for incinerators. In fact, this type of heating was developed in North America as early as the 1870s and later in Germany in 1900. In Canada and the United States, 300 cities were equipped with incinerators in 1932; 20 cities in Germany by 1927.[55] Central heating, which according to a French advertisement of the 1920s "allows comfort even if a woman works and is not home to maintain the stove [translation]",[56] was a sign of modernity.

7.4.3 The Abandonment of Waste

Despite the hope urban administrators had placed in these innovations, there was a growing disinterest in recovery. There were several reasons for this. By the late 19th century, the usefulness of urban excreta became impaired, and as discussed

[51] Joulot, op. cit., p. 82.
[52] Melosi (2005), op. cit., p. 154, 163.
[53] Melosi (1988).
[54] Paris (1909).
[55] Gallo (1994).
[56] Quoted by Gallo, op. cit.

earlier in this text, its use decreased rapidly during the inter-war period. Moreover, new recovery processes became obsolete, particularly because waste collection had led to the debasement of materials. In fact, there was an unprecedented increase in the volume of garbage, much more rapid than its increase in weight. This consistently lower density was due to the proliferation of packaging, newspapers and, in some cities, to the decline in the quantity of ash released, which was very variable as it was dependent on which specific local heating processes were used at the time. Vehicles traditionally used for garbage collection were often motorized, and no longer appropriate as they filled up quickly resulting in an increase in service required and higher collection costs. Packer trucks (with compressing device) gained popularity in Europe from the 1930s onward[57] and provided the solution as they compressed garbage; however, they also altered it in the process.

At the same time and partially as a result of the situations described above, many experiments ended in financial failure. In 1923, entrepreneurs of Nice's recovery plant, mentioned above, requested the modest sum of 20,000 francs per year and a free supply of household rubbish and mud from the local government in order to proceed with what they thought to be a highly profitable recovery project. In 1926, when the plants were commissioned, the city was required to increase its funding by 50 times, to one million francs per year. In 1930, all the plants were replaced by a mass-burning plant.[58] By and large, the operating costs of the recovery plants proved to be much higher than the revenues from the sale of their products. The same was true for sewage farms and led to their eventual abandonment in Europe and North America (they continued to be used in Mexico however). As for energy recovery, the costs outweighed the returns: to use an example from Europe, the 1953 study by the Organization for European Economic Co-operation revealed that a cost of 1,150–1,715 FF ton^{-1} (amortization, interest on the investment, operation and maintenance costs) had to be factored in for revenues of 270 FF ton^{-1} up to, infrequently, 1,500 FF ton^{-1}.[59]

The prevailing view of administrators was that urban excreta was no longer profitable, and henceforth an "unavoidable burden"[60] for society. This evolution is reflected in the use of the French word "déchet" for household garbage from the 1930s. The idea of profit from waste appears to have been abandoned even earlier in Great Britain and the United States. Consequently, techniques associated with garbage collection were no longer developed with usefulness in mind, but rather with the aim for disposal or storage at a lower cost. Four techniques were primarily used: incineration without utilization for energy purposes (although some cities continued to use this process) with the sole objective being to reduce volume (as described earlier); the garbage grinder; drainage direct to sea, and to a lesser degree, drainage direct to river (mainly concerning domestic sewage); and disposal on land.

---

[57] Organisation européenne de coopération économique (1953).

[58] Joulot, op. cit., p. 82–83.

[59] Organisation européenne de coopération économique, op. cit., p. 101.

[60] Joulot, op. cit., p. 175.

The garbage grinder was developed to link the disposal of solid waste (that can be ground up) with the disposal of wastewater and thus to realize the ancient dream of networking garbage collection and waste transport. It is an electric device customized for a kitchen sink that allows for the drainage of shredded refuse to the sewers. It was the subject of many experiments in Great Britain and the United States but, in the end, experienced only limited development. The technique was comparatively inconvenient, concerned only one part of refuse (kitchen waste), required an increase in the capacity of wastewater treatment plants (where these existed) and resulted in increased water consumption (round 7 l $inhab^{-1}$ day^{-1}).[61]

New York is an example of a city that since 1872 practiced drainage direct to sea: barges were used to carry refuse 25 miles from shore (prior to this, a part of the refuse was discharged into the East River) up to 1934.[62] The process was defended for many years. Not only would it allow for the development of marine plankton, it appeared to be *the* solution to the new problem of bulky waste. Indeed, consumption of new objects resulted in the production of new waste, including electrical appliances and end-of-life motor vehicles. The sudden emergence of these monstrous objects appeared not to have been anticipated by policies which promoted their manufacture and sale to consumers. This problem, already recognized in the 1930s, became very worrisome after the Second World War.

Sanitary landfilling originates, like incineration, from England where it emerged in 1912 (the terms is used from the 1930s onwards). This method consisted of placing successive layers of waste, 1.5–2.0 m thick, separated by inert matter. Another layer was added only when the temperature of the previous one had stabilized (fermentation produces heat). The process was advantageous because of its low cost and its only constraint was its need for large spaces located close enough to towns to avoid excessive transportation costs. This explains why landfills are more frequent in vast countries like North America than in Europe where space is limited. Nevertheless, the number of landfills multiplied as never before during the inter-war period and they were gradually considered the best solution for garbage storage (60 % of English garbage was placed in landfills in 1950[63]). This was the case even in France where landfills were at first considered to be insalubrious and wasteful. Moreover, proponents of landfills argued that garbage dumping contributed to the urban development of waste grounds and uncultivated land. However, most of the time, a landfill was considered a no man's land. The Entressen site (South of France), used as a landfill up to 2010 for garbage from Marseilles, is the best example of this since it resulted from the discontinuation of the Plaine de Crau recovery project. It was at the beginning of the 21st century the largest landfill in Europe.

Thus there was a general trend towards an overproduction of waste (Table 7.2) and a devaluing of these materials. This trend was all the more remarkable because

[61] Summer (1968).

[62] But New York City continued to dump sewage sludge at sea. Melosi (2005), op. cit, pp. 181–182.

[63] Organisation européenne de coopération économique, op. cit., p. 144.

Table 7.2 Waste removed in some cities, *circa* 1930 (g inhab^{-1} day^{-1})

London and suburbs	830
Zurich	630
Paris	770
Berlin	Summer: 370
	Winter: 584

agricultural and industrial wastes added to the mountain of waste already created by households. In this area however, Germany and the Netherlands stood out. In Germany, until 1945 the political system under *autarchy* favored recycling;[64] yet landfills can still be found as parts of "the tectonic landscape [translation]",[65] such as the 60 m high mound of refuse that stands near Leipzig. In the Netherlands, a comprehensive program was started in the 1930s for the agricultural recovery of household garbage. In the 1960s, 200,000 tons per year of compost was produced from the garbage of two million people (20 % of the entire population). This greatly contributed to the development of horticulture and particularly, after 1955, to flower bulb cultivation.[66]

The Second World War also blurred the big picture. On the one hand, the shortages that characterized the war and the years following the conflict led to a reduction of garbage production and a renewed interest in certain recovery processes; however this was considered a short-term situation and, considering its causes, undesirable. On the other hand, the destruction caused by bombings posed a real problem to urban administrators once the conflict ended. This was particularly the case in Berlin where two-thirds of the buildings were destroyed in 1945. The near absence of transportation made the disposal of the rubble difficult, and it had to be piled up locally. Thus about 30 hills of rubble up to 100 m tall erected towards the end of the 1940s, mostly by female workers (the *Trümmerfrauen)*.

7.5 1970s–Today

7.5.1 The Environmental Crisis

During the environmental crisis, which originated in the 1960s and 1970s, scientists, intellectuals, artists, journalists and citizens in most developed countries (although to varying degrees in each country) finally denounced the pernicious effects of industrialization, consumption and even development. As early as 1948, Fairfield Osborn published his internationally acclaimed book, *Our Plundered*

64 The same occured in Italy some years before. Giuntini, op. cit., p. 70.

65 Joulot, op. cit, p. 53.

66 Houter and Stolp (1968).

Planet.[67] The number of written works of this nature multiplied from the 1960s, including Rachel Carson's *Silent Spring*,[68] published in the United States in 1962 and rapidly translated in many countries. These works did not specifically criticize solid waste, but opened the way for environmentalism and an awareness of the *Limits of the Earth*, the title of another book by Fairfield Osborn published in 1953.[69] A little later, the UNESCO Intergovernmental Conference of Experts on the Scientific Basis for Rational Use and Conservation of the Resources of the Biosphere, which became known as the Biosphere Conference, was held in Paris in 1968,[70] reopening and expanding these arguments. Both unity and uniqueness, and the idea of a *Spaceship Earth,* were stressed. As well, the observation that "the rationalization on a planetary scale of the utilization of the biosphere's resources is critical if one desires to ensure satisfactory living conditions for future generations [translation]" was emphasized at the conference. The 1972 publication of the American book, *The Limits to Growth*,[71] which also received international acclaim, confirmed the worries expressed by many stakeholders.

Works such as, *The City in History* by Lewis Mumford, in which he denounced the "Paleolithic Paradise: Coketown" and the "The Myth of Megalopolis",[72] and *The Death and Life of Great American Cities* by Jane Jacobs,[73] both published in 1961, added a violent criticism of industrial cities to these general considerations. Engineers like Abel Wolman and his famous article "The Metabolism of Cities"[74] published in Scientific American in 1965, and ecologists such as Eugene Odum in the United States or Paul Duvigneaud in Europe agreed with historians and urban planners. They all emphasized how the management of cities and their absurd metabolism had created an impasse. They argued that because these heterotrophic systems not only imported their food and most of the resources they needed, but also, through their use, transformed these materials into solid, liquid or gaseous waste, they would damage both urban and natural environments. In particular, Duvigneaud went further by linking the social crisis to the environmental crisis.[75] The former, according to him, was a consequence of the latter. It is possible to discern here the move from a hygienic approach—whose defenders considered the removal of waste the answer to the problem of urban salubrity—to the environmentalist approach that, while often remaining anthropocentric, revealed the limits of the first approach in view of the inseparability of human societies from the biosphere that supports them.

---

67 Osborn (1948).

68 Carson (1962).

69 Osborn (1953).

70 UNESCO (1969).

71 Meadows et al. (1972).

72 Mumford (1961).

73 Jacobs (1961).

74 Wolman (1965).

75 Duvigneaud (1980).

In this general setting, solid waste became the symbol of a consumer society and of its faults which were, in particular, accounted for by its ubiquitous presence in urban, peri-urban and rural landscapes. Art and literature seized garbage as an expression of the futility of a comfortable existence and as one of the contributing factors in the formation of grotesque landscapes and new environments (examples include *Les Météores* by Michel Tournier, published in 1975, and much more recently, Tristan Egolf's *Lord of the Barnyard: Killing the Fatted Calf and Arming the Aware in the Corn Belt*, published in 1998, the Brazilian film *Ilha das Flores*, produced in 1989, and the recent film for mass audiences *Wall E*, released in 2008). Social Science invented rudology and garbology, the study of waste, to reveal the societies that produce it.

7.5.2 Garbage Crisis, Garbage Policy

Despite the lull, even decline during the Second World War, there was a significant increase in the amount of waste generated throughout the 20th century (Fig. 7.4): in the United States, municipal waste production went from 1.2 kg $inhab^{-1}$ day^{-1} in 1960 to 2.1 kg $inhab^{-1}$ day^{-1} in 1990 where it seemed to reach a ceiling; in France, it went from less than 1 kg $inhab^{-1}$ day^{-1} in 1970 to 1.4 kg $inhab^{-1}$ day^{-1} in 2000.[76] The same year, it reached 1.6 kg $inhab^{-1}$ day^{-1} in the Europe of the 15 and later stabilized at roughly 1.5 kg $inhab^{-1}$ day^{-1}—although there was a sharp disparity between countries. In addition to municipal waste, there was waste from agriculture, industry, construction and public works. The quantity of these wastes, however, was largely unknown as it was not closely monitored—by comparison, in 2004, 16.4 kg $inhab^{-1}$ day^{-1} of waste was produced by economic activities (including households) in the Europe of the 27 of which municipal waste made up 1.4 kg $inhab^{-1}$ day^{-1}.[77]

As such, as early as the 1970s a waste crisis emerged and the public stakeholders were often powerless in the face of the piles of materials there were to manage. The crisis was also due to the increase in the number of accidents stemming from the toxicity of waste: the abnormally high cancer rate in the 1970s at Love Canal, a neighborhood in Niagara Falls (United States) where 21,000 tons of toxic waste had been dumped from 1942 to 1952[78] (see Oates 2004 novel *The Falls* which recounts the story of this tragedy); the 1983 discovery in France of 41 barrels of chemical waste which contained dioxin from Seveso (Italy) following an industrial disaster in 1976; the pathologies developed by inhabitants of houses built on ancient municipal landfills or the contamination of produce from vegetable gardens by polluted

---

[76] Melosi (2005), *op. cit.*, p. 206.

[77] According to Eurostat. http://epp.eurostat.ec.europa.eu/portal/page/portal/environment/data, accessed 3 January 2014.

[78] Levine (1982).

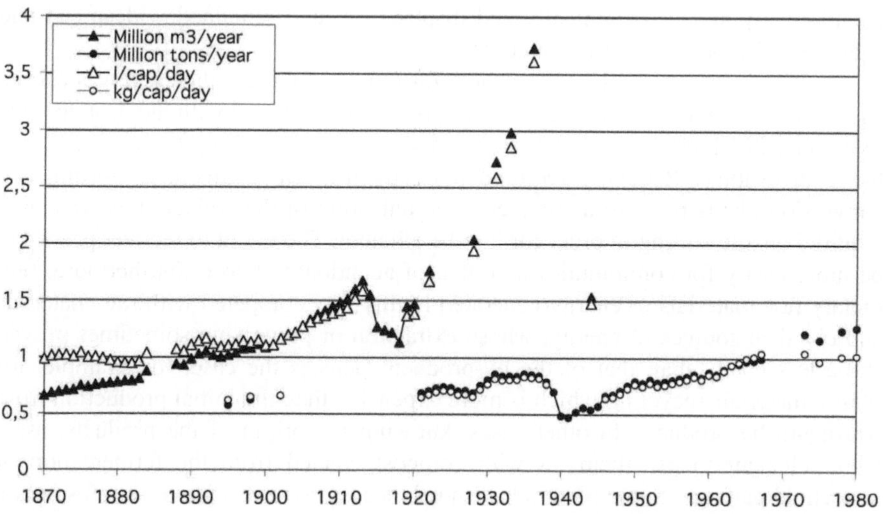

Fig. 7.4 Production of household waste, Paris, 1870–1980. *Adapted from* Barles (2005)

soils[79]… These were all factors that contributed to challenging the so-called ideal solution for disposal on land, and more generally, the management (and in some cases the absence of management) of solid waste, the *third pollution* (after water and air pollution) as described by William Small in 1970.[80]

The first large political bills concerning waste (in particular urban waste) were brought in during the 1960s and 1970s: the *Solid Waste Disposal Act* in 1965 and the *Resource Conservation and Recovery Act* in 1976 in the United States, the *European directive* of July 1975, to name only a few. These texts all emphasized the necessity to reduce the production of waste at its source—for example, by reducing waste from packaging—and recovering the waste collected—through recycling, utilization for energy purposes or biological conversion. They also recommended the implementation of a 'polluter pays' principle to the management and pricing of waste, which was universally adopted by the OECD in 1972.[81] According to the principle of internalizing the negative externalities, it was a question of placing a monetary value on the residual damage to the environment—damage remaining after collection and treatment of waste—and to charge this cost to the emitter of the waste (which could be an individual or an industry for example) in addition to the cost of the collection service and treatment. Nevertheless, the situation remained worrisome and evolved insubstantially until the 1990s: the quantity of waste did not decrease, or only a little, recycling stayed limited both

[79] Sowers (1968).

[80] Small (1971).

[81] Guiding principles concerning the international economic aspects of environmental policies. Recommendation adopted by the OECD Council on 26th May 1972.

quantitatively and geographically and disposal on land remained widespread and even gained ground in some countries.[82]

Additional cost was one of the major constraints of these policies. It was caused by the cost of implementing the recovery processes combined with the near absence of markets in which to sell these value-added products—contrary to the case during the 19th century. Recovery required not only the establishment of sorting and conversion plants but also a complete reorganization of the collection service as it required on-site sorting in order for it to be efficient. The intent to recover proved to be more costly for communities than that of abandoning waste. Furthermore, secondary raw materials or energy generated in this way competed with raw materials and classical sources of energy, whose extraction or processing sometimes proved to be less costly than that of the by-products. This is the case, for example, for plastic materials recycling which is more expensive than the initial production from petroleum by-products. In other cases, knowing the origin of the products, users were reluctant to use them, as with compost created from the fermentation of household garbage. Some sectors had more success however: glass, as refuse glass (cullet) has always been used to make new glass, and paper, thanks to the development of effective recycling processes and in view of a limited supply of wood. Nevertheless, the price of old papers has sometimes been negative in the last decades. The arbitrage of one recovery method over another was difficult for communities: utilization for energy purposes was all the more valuable because of the elevated energy value of garbage. Pre-sorting garbage for recycling, for example paper and plastics, proved to be ineffective for incineration of the residue as it had a reduced energy value.

Other difficulties were added to these. The process of separate collection required the involvement of citizens as they had to perform the initial sort of the recycled materials (whether collection was done curb-side or by drop-off at a depot). In fact, the technologization of society—especially if technology is considered a medium between man and his environment—undeniably has had the effect (if not the aim) of separating individuals from the chores of everyday life and to eliminate all contact with excreta, now handled by technical devices of varying degrees of sophistication and outsourced cleaning services. In some countries, the participation of urbanites was successfully (re-)achieved: this was the case in Germany, probably thanks to its past recycling policies (see above, Sect. 7.4.3). In some other countries however, households considered sorting a bother. At another level, new plants were now required because of the removal of landfills, and the implementation of recovery and the expansion of urban areas were frequently in opposition to the refusal of residents. Often attributed to the NIMBY syndrome (*not in my backyard*), this attitude also related to the distance between urbanites and their excreta (as touched on here) and the way in which these projects had been historically implemented: without dialogue or participation from the concerned populations.

[82] OECD Workshop on Waste Prevention: Toward Performance Indicators (2002).

At the source, waste reduction programs had very little success because it required the sphere of production to be rethought. While the removal of urban waste generally fell under the responsibilities of the local government, production belonged to the private sector (contrary to the case of liquid sanitation insofar as water distribution is generally a public utility; incidentally, all things being equal, in this case it is easier to coordinate a management policy that links upstream distribution and downstream recovery and treatment). Although, the private sector could certainly be subject to rules, incentives, taxes, etc., as it also remained under the control of these governing bodies and the State. Furthermore, applying the 'polluter pays' principle did not always prove to be a deterrent as the polluter's penalty was low in comparison to its overall budget or to the cost of reducing the quantity of emitted waste at the source.

These reasons explain why, despite their ambitions, waste management policies implemented in developed countries for the last 40 years have not been particularly successful.

7.5.3 A Global Problem

At the same time, the issue of waste, which had been mainly limited to developed countries until the 1960s, spread globally. Against the background of chronic poverty, inefficient management structures and weak regulations, the proliferation of garbage in developing countries made waste a sanitary, social, environmental and economic issue.[83]

Indeed, collection and storage conditions contributed to urban and generalized insalubrity. The services were generally undersized (not enough vehicles, storage areas insufficiently large) and unsuitable, such that collection was frequently only partial. As a consequence, streets and gutters became landfills. The non-appropriation of excremental materials and the absence of monitoring of industrial activities increased the sanitary risk of waste. The situation was made worse by equipment dating from the colonial period, often unsuitable within the local context. At the same time, the so-called informal sector played a major role in on-site handling, collection, sorting, recycling, and recovery. Millions were engaged in these activities and made a major contribution—even though it was imperfect—to the management of waste, often in deplorable sanitary conditions. The scavenging activities typical of European and North American cities during the 19th century moved geographically; the conditions of work also deteriorating.

For city councilors, the activities of the informal sector, and in particular recycling, have long been considered a sign of underdevelopment and poverty—which they are. Many cities thus tried to eradicate these activities in favor of disposal on land for all materials or, occasionally under pressure from international moneylenders and

[83] See for instance: Ngnikam and Tanawa (2006) and UN-Habitat (2010).

Fig. 7.5 Installation for the pre-sorting of household waste in a Shanghai district. *Photo* S. Barles, January 2008

large public services companies, for incineration. The cure was often worse than the disease: job losses for salvage dealers, lack of funds for the upkeep of the vehicle fleet and facilities, decline in service. In the last several years it has emerged that it would be best to recognize and integrate the activities of the informal sector in the management process, but this idea has not yet translated into any results. The situation in emerging countries is also of concern, where the adaptation of urban services does not match the speed of processing (Fig. 7.5).

7.6 Conclusion

The history of waste mirrors that of the societies that produced it, and their relationship with the environment and the resources they mobilized in the process. The very notion of waste has fluctuated before taking its contemporary meaning—a material that its owner intends to discard. What was at one time considered waste may no longer be considered so now, and what is waste today may not have been deemed so in the past. One hundred and fifty years ago, oyster shells were a fertilizer of repute, now they often end up in a dustbin, landfill or incinerator.

This history of waste also reflects the metabolism of societies and the linearization of material flows: industrialization translated into an increasing use of resources, first processed by industries and then consumed by urbanites. Each of these steps produced an array of waste materials, until their final disposal. However, during the first industrial age, the absence of synthetic fertilizers and a lack of knowledge of fossil fertilizers associated with certain industrial processes, such as the manufacture of paper, candles, dyes, etc., contributed to limiting waste production. At the time, the metabolism of societies was partially closed, in view of the agricultural and industrial importance of most urban, even industrial, by-products. As soon as industry and agriculture had the means to do without these by-products, they stopped using them. Thus it was only during the second industrial revolution that the opening of the bio-geochemical cycles became widespread and that waste had to be accepted as consubstantial with the rise in the standard of living and development. The environmental crisis of 1960–1970, and later, the depletion of the ozone layer, climatic change and the emergence of the concept of sustainable development led to a renewed consideration of waste. Waste became the symbol of the pernicious effects of a consumer society. Despite this new view of waste, the management of it has not considerably improved, especially since the amount of waste continues to grow in the South.

The issue of waste is now part of a larger discourse, that of non-renewable resources and the uniqueness of our planet. It is no longer enough to recycle or to recover excreta simply to limit the quantity of final waste. What matters now is to close the loops and, through recycling and recovery, to limit the extraction of resources at the source. Such a project, driven by industrial ecology and territorial ecology, will succeed only if the levels of consumption fall. It will require a profound reform in society and its way of viewing its waste.

References

André L (1996) Machines à papier. Innovations et transformations dans l'industrie papetière en France, 1798–1860. Éd. de l'EHESS, Paris

Barberet J (1866–1887) Le travail en France. Monographies professionnelles, vol 4. Paris, pp 59–104. (Chap. "Chiffonniers")

Barles S (1999) La ville délétère: Médecins et ingénieurs dans l'espace urbain (XVIIIe-XIXe siècles). Champ Vallon, Seyssel

Barles S (2005) L'invention des déchets urbains, France, 1790–1970. Champ Vallon, Seyssel

Baumes J-B (1789) Mémoire (…) sur la question (…): déterminer, par l'observation, quelles sont les maladies qui résultent des émanations des eaux stagnantes, et des pays marécageux (…), et quels sont les moyens d'y remédier. Nîmes

Belgrand E (1887) Les travaux souterrains de Paris, vol 5. Les égouts, les vidanges. Dunod, Paris, p 14

Bensaude-Vincent B Stengers I (2001) Histoire de la chimie, 2nd edn. La Découverte, Paris, pp 231–240 (1st edn 1992)

Boissier de Sauvages F (1754) Dissertation où l'on recherche comment l'air, suivant ses différentes qualités, agit sur le corps humain (…), Bordeaux

Boudriot P-D (1990) Les égouts de Paris aux XVIIe et XVIIIe siècles: les humeurs de la ville préindustrielle. Histoire, économie et société 9(2):197–211
Carson R (1962) Silent spring. Houghton Mifflin, Boston
Chadwick E (1842) Report on the sanitary conditions of the labouring population of Great Britain. London
Chaptal J-A (1819) De l'industrie françoise. Paris 2:180
Chevallier A (1849) Notice historique sur le nettoiement de la ville de Paris. Annales d'hygiène publique et de médecine légale 42:264–321
Commission des engrais (1865–1866) Enquête sur les engrais industriels, vol 1. Paris, p 43
Coray A (1800) Traité d'Hippocrate des aires des eaux et des lieux; traduction nouvelle, vol 1, 2. Imprimerie Baudelot et Eberhart, Paris
Corbin A (1986) Le miasme et la jonquille: l'odorat et l'imaginaire social. XVIIIe-XIXe siècles, 2nd edn. Flammarion, Paris (1st edn 1982)
Corbin A (1988) The foul and the fragrant: odor and the French social imagination. Harvard University Press, Cambridge
Courmont J, Lesieur C, Rochaix A-J (1932) Précis d'hygiène, 4th edn. Paris, p. 350
Crookes W (1917) The wheat problem. Longmans, Green and Co. London, New York
Du Mesnil O (1884) Nettoiement de la voie publique. Enlèvement et utilisation des ordures ménagères. Paris
Dumas J-B (1866–1867) Rapport conclusif. In: Enquête sur les engrais industriels, vol. 2, xxxi. Commission des Engrais, Paris
Duvigneaud P (1980) La synthèse écologique. Populations, communautés, écosystèmes, biosphère, noosphére. Douin, Paris
Esquiros A (1861) L'Angleterre et la vie anglaise. XII. L'industrie du papier. Les boutiques de chiffons, les fabriques du Kent et la poste de Londres. Revue des deux mondes 2:145–187
Figuier L (1873) Les merveilles de l'industrie, vol 2. Paris, p 270
Fontaine A (1903) L'industrie du chiffon à Paris. Paris
Fortier B (1975) La politique de l'espace parisien à la fin de l'ancien régime. Corda, Paris
Friedel R (1983) Pioneer plastic. The making and selling of celluloid. University of Wisconsin Press, London
Frioux S (2013) Les batailles de l'hygiène. Ville et environnement de Pasteur aux Trente Glorieuses. Presses Universitaires de France, Paris
Gallo E (1994) Le chauffage urbain du quartier des gratte-ciel de Villeurbanne, des années 1930 à nos jours. In: Proceedings of the conference "Citadins, techniques et espaces urbains du XVIIIe siècle à nos jours". Paris
Gille Z (2007) From the cult of waste to the trash heap: the politics of waste in socialist and postsocialist Hungary. Indianapolis, Bloomington
Giuntini A (2006) Cinquant'anni puliti puliti. I rifiuti a Firenze dall'Ottocento alla Sociétà Quadrifoglio. Franco Angeli, Milano
Goddard N (1996) "A mine of wealth"? The Victorians and agricultural value of sewage. J Hist Geogr 22(3):274–290
Goddard N, Sheail J (2001) Victorian sanitary reform: where were the innovators ? In: Environmental Problems in European Cities in the 19th and 20th Century/Umweltprobleme in europäischen Städten des 19 und 20. Jahrhunderts, ed. Christoph Bernhardt. Waxmann Verlag, Münster/New York, pp 87–103
Guillerme A (1983) Les temps de l'eau. La cité, l'eau et les techniques. Nord de la France, fin IIIe-début XIXe siècles. Champ Vallon, Seyssel
Guillerme A (1988) The age of water: the urban environment in the North of France. Texas A&M University Press, College Station
Guillerme A (2007) La naissance de l'industrie à Paris, entre sueurs et vapeurs, 1780–1830. Champ Vallon, Seyssel
Hamlin C (2007) The city as chemical system? The chemist as urban environmental professional in France and Britain, 1780–1880. J Urban Hist 33(5):702–728

Harpet C (1998) Du déchet. Philosophie des immondices. Corps, ville, industrie. L'Harmattan, Paris
Hills RL (1988) Papermaking in Britain, 1488–1988: a short history. Athlone Press, London
Houter PJ, Stolp DW (1968) Vente aux Pays-Bas du compost d'ordures municipales. In: Association internationale du nettoiement public. IXe congrès international. Proceedings of Paris congress. 26–30 June 1967, Paris, pp 129–150
Jacobs J (1961) The death and life of great American cities. Random House, New York
Jorgensen D (2008) Private need, public order: urban sanitation in Late Medieval England and Scandinavia. PhD dissertation, University of Virginia
Joulot A (1946) Les ordures ménagères: Composition, collecte, évacuation, traitement. Paris
Jugie J-H (1993) Poubelle-Paris (1883–1896): La collecte des ordures ménagères à la fin du XIXe siècle. Larousse, Paris
Kunitz SJ (1993) Deseases and the European mortality decline, 1700–1900. In: Kiple KF (ed) The Cambridge world history of human disease. Cambridge University Press, Cambridge, pp 287–293
Lancisi GM (1717) De Noxiis paludum effluviis (…). Rome, 1717
Lavoisier AL (1782–1783) Mémoire sur les altérations qui arrivent à l'air dans plusieurs circonstances où se trouvent les hommes réunis en société. Mémoires de la société royale de médecine 569–582
Lécuyer B-P (1986) L'hygiène en France avant Pasteur, 1750–1850. In: Claire S-B (ed) Pasteur et la révolution pastorienne. Payot, Paris, pp 67–142
Leguay J-P (1999) La pollution au moyen age: dans le royaume de France et dans les grands fiefs. Editions Gisserot, Paris
Levine AG (1982) Love Canal: science, politics, and people. Lexington Books, Lexington
Magnusson R (2006) Water and wastes in Medieval London. In: Terje T, Eva J (eds) Water control and river biographies. A history of water, vol 1. I. B. Tauris, London, pp 299–313
Mårald E (2002) Everything circulates: agricultural chemistry and re-cycling theories in the second half of the nineteenth century. Environ Hist 8:65–84
Matignon C (1931) Les progrès réalisés par l'industrie chimique dans le dernier demi-siècle. Chimie et industrie 26(2):255–272
Meadows D, Randers J, Meadows D (1972) Limits to growth. Universe Books, New York
Melosi (1988) Technology diffusion and refuse disposal: the case of the British destructor. In: Tarr JA, Dupuy G (eds) Technology and the rise of the networked city in Europe and America. Temple University Press, Philadelphie, pp 207–226
Melosi (2000) The sanitary city. Urban infrastructure in America from colonial times to the present. The Johns Hopkins University Press, Baltimore/Londres
Melosi MV (2005) Garbage in the cities: refuse, reform and the environment, 2nd edn. University of Pittsburgh Press, Pittsburgh
Méphitis (1765) In: Encyclopédie ou dictionnaire raisonné des lettres, des sciences et des arts. Neuchâtel, t X., p 357
Moll L, Mille A-A (1857) Application des vidanges à la culture, rapport présenté à M. le Préfet de la Seine. Paris, p 17
Müller A. (1860) Gödselboken eller grunderna för gödselämnenas behandling i städer och på landet [The manure book, or principles for the handling of nutrients in the cities and the countryside]. Stockholm
Mumford L (1961) The city in history. Its origins, its transformations, and its prospects. Harcourt, Brace & World, New York
Napias H (1882) Manuel d'hygiène industrielle. Paris, p 196
Ngnikam E, Tanawa É (2006) Les villes d'Afrique face à leurs déchets. Presses de l'UTBM, Belfort
Oates JC (2004) The falls. Harper Collins, New York
OECD Workshop on Waste Prevention: Toward Performance Indicators (2002) Proceedings of Paris workshop, 8–10 Oct 2001. OECD, Paris
Organisation européenne de coopération économique (1953) Collecte et évacuation des ordures ménagères. Nettoiement des voies publiques, Paris

Osborn F (1948) Our plundered planet. Faber and Faber, London
Osborn (1953) Limits of the Earth. Little Brown and Co, Boston
Paris C (1909) Rapport au nom des délégations de la 6e commission du Conseil municipal et de la Commission départementale des eaux et de l'assainissement chargées d'étudier l'épuration biologique des eaux d'égout et les fours d'incinération en Angleterre et en Écosse (août 1908). Paris
Paulet M (1853) L'engrais humain. Histoire des applications de ce produit à l'agriculture, aux arts industriels avec description des plus anciens procédés de vidanges et des nouvelles réformes, dans l'intérêt de l'hygiène. Paris
Payen A (1859) Précis de chimie industrielle à l'usage 1° des écoles d'arts et manufactures et des écoles d'arts et métiers, 2° des écoles préparatoires aux professions industrielles, 3° des fabricants et des agriculteurs, vol 2, 4th edn. Paris, p 176
Picard A (ed) (1891) Exposition universelle internationale de 1889 à Paris. Rapports du jury international. Groupe II, 2e partie. Matériel et procédés des arts libéraux. Classes 9 à 16. Paris, Imprimerie nationale, p 69
Poussou J-P (1992) La croissance des villes au XIXe siècle: France, Royaume-Uni, États-Unis et pays germaniques, 2nd edn. C.D.U. et SEDES, Paris
Priestley J (1774–1777) Experiments and observations on different kinds of air, vol 3. W. Bowyer and J. Nichols, Londres
Simmonds PL (1862) Wasted and undeveloped products, or hints for enterprises in neglected fields. London
Small WE (1971) Third pollution: the national problem of solid waste disposal. Praeger, New York
Smil V (2001) Enriching the Earth. Fritz Haber, Carl Bosch, and the Transformation of world food production. MIT Press, Cambridge
Sowers GF (1968) Foundation problems in sanitary land fills. J Sanit Eng Div-Proc Am Soc Civil Eng 94(SA1):103–116
Strasser S (1999) Waste and want: a social history of trash. Metropolitan Books, New York
Summer J (1968) Le dépôt et la collecte des ordures. Méthodes, pratique, développements techniques et tendances. Étude sur le plan international. In: Association internationale du nettoiement public. IXe congrès international. Proceedings of Paris congress. 26–30 June 1967, Paris, p 12
Tarr JA (1996) From city to farm, urban wastes and the American farmer. In: The search for the ultimate sink: urban pollution in historical perspective, Tarr. The University of Akron Press, Akron, pp 293–308 (1st published 1975, Agricultural history, 49(4):598–612)
Trad. Bensaude-Vincent B, Stengers I (1996). A history of chemistry. Harvard University Press, Cambridge
Turgan J (1860–1885) Les grandes usines, vol 1. Paris, p 158
UNESCO (1969). Use and Conservation of the Resources of the Biosphere. In: Proceedings of the intergovernmental conference of experts on the scientific basis for rational use and conservation of the resources of the biosphere. 4–13 September 1968, UNESCO, Paris
UN-Habitat (2010) Solid waste management in the World's Cities 2010. Earthscan, London
van Zwanenburg P, Millstone E (2005) Bovine spongiform encephalopathy: risk, science and governance. Oxford University Press, Oxford
Védry B (1992) Contribution à l'histoire des procédés d'épuration biologique des eaux résiduaires. Master thesis, CNAM
Vögele J (1998) Urban mortality change in England and Germany, 1870–1913. Liverpool University Press, Liverpool, p 150–189 (chap. 11 "Improving the environment: the impact of sanitary reforms on people's health")
von Liebig J (1862) Die Chemie in ihrer Anwendung auf Agricultur und Physiologie, vol 2. Braunschweig
Wines RA (1985) Fertilizer in America. From Waste Recycling to Resource Exploitation. Temple University Press, Philadelphie (Chap. 1: "The recycling system")
Wolman A (1965) The metabolism of cities. Sci Am 213(3):179–190

Chapter 8
Technological Hazards, Disasters and Accidents

Gianni Silei

Abstract Although some technological risks can be traced back to the ancient times, it was between the nineteenth and the beginning of the twentieth century that technical advancement and the process of industrialization posed the question of the management of the technologies and of their possible disastrous consequences. During these years there was an important change in approaching these issues: from the inevitability of disasters to the adoption of policies of prevention and risk management. This important change had as a consequence an increasing role of public institutions (national governments, agencies and authorities) in the control, prevention and emergency management of technological disasters. According to this new approach, scientists, the experts and the technicians that were required to "predict" using their special knowledge technological disasters, became central figures. The first post-war period represents an important turning point because this new and modern attitude towards technological hazards reached its full maturity. The spreading of new technologies also facilitated by the process of industrialization and the emergence of the era of mass consumptions, influenced a new discipline that, from different approaches, tried to address and resolve the various aspects of technological threats. Born in the postwar period, the disastrology and in general policies to ensure safety, found a systematic application after the Second World War. The increasing complexity of certain technologies used in industry, in the production of energy, in the transport sector and especially the potentially catastrophic consequences of technological accidents, imposed an additional effort in the field of regulation, prevention and management of emergencies. In some cases, such as the atomic energy for civilian use, an increasing role was played by national and international agencies that were created during this period. Since the 1970s but especially in the following decade, several major accidents (Three Mile Island, Seveso, Bhopal, Chernobyl, Fukushima, the environmental disasters caused by oil tankers) put forward the need for a standardization of rules and a greater international co-operation. The globalization of technological hazards at the time of the so-called "risk society" has fostered a more interdisciplinary approach to the issues of technological disasters.

G. Silei (✉)
Dipartimento di Scienze politiche e internazionali, Università di Siena, Siena, Italy
e-mail: gianni.silei@unisi.it

M. Agnoletti and S. Neri Serneri (eds.), *The Basic Environmental History*,
Environmental History 4, DOI 10.1007/978-3-319-09180-8_8

Moreover, the increased number of new hazardous substances and materials and the opportunities for human error inherent their use has determined an escalation of technological accidents. All this factors and the more and more unstable boundaries between natural disasters and man-made disasters has necessarily imposed growing efforts for harmonization policies at a national and an international level to ensure collective security, public health and environmental protection.

8.1 Introduction

There is not a universal definition of technological hazards and accidents. Even though some studies emphasize the complexity of the issue of individual responsibility in technological disasters, literature has commonly accepted a distinction between natural hazards (acts of god) and man-made (acts of man) hazards. According to some classifications natural hazards are threats determined by uncontrollable events, while man-made hazards are threats determined by artificial (technological) factors. Natural hazards can be defined as "those elements of the physical environment, harmful to man and caused by forces extraneous to him".[1] The term "natural hazards" refers to all atmospheric, hydrologic, geologic (especially seismic and volcanic), and wildfire phenomena, while the term "man-made hazards" refers to "artificial" phenomena caused by human action, inaction, negligence or error. These phenomena are also defined as technological hazards when determined by a technology (i.e. industrial, engineering, transportation) and as sociological hazards when they have a direct human motivation (i.e. crime, riots, conflicts).

As threat and potential danger, hazards are strictly connected to concepts of risk, disaster and catastrophe. The term risk (from the ancient Italian *risicare*) indicates the possibility of suffering a harmful event or loss or danger. While a risk involves uncertainty, a disaster (from the Italian *disastro*, literally "unfavorable to one's stars") is an unexpected natural or man-made event with harmful but temporary consequences. Disasters can be defined as the result of an extreme event that significantly disrupts the workings of a community. A disaster is "a tragic situations over which persons, groups, or communities have no control-situations which are imposed by an outside force too great to resist". This kind of events may have as a consequence deaths, material destructions and severe economic damages but can also determine situations of collective stress in a community and bring to the test the level of vulnerability of a society.[2] Some interpretations consider a disaster a consequence of peculiar social conditions, some others, consider man-made disasters mainly as socio-technical problems, as the product of a failure of foresight and a combination of technical, social and even institutional and administrative

[1] Smith and Petley (1990).

[2] Fritz (1961), Quarantelli (1966) and Dynes (1970).

factors. The Normal Accident Theory argues that the combination of high complexity and tight coupling must lead to failures. According to this theory, that has been integrated, empirically tested and verified, technological accidents are "inevitable and happen all of the time; serious ones are inevitable but infrequent; catastrophes are inevitable but extremely rare".[3]

Since the mid-1980s, starting with an approach opposite to that of the Normal Accident Theory, some researchers developed the High Reliability Theory, which says that is possible to create highly reliable systems capable of ensuring almost absolute security levels.

The science that deals with the study and prevention of disasters is called disastrology. Born at the beginning of the twentieth century but developed especially from the second post-war period, this discipline relies on the contribution of different specialists: physicists, geologists, geographers, planners, engineers, sociologists, psychologists, historians. In the 1980s it was developed a sort of new branch of this discipline, the kindunology (from the Greek *kindunos* that means "hazard"). This science is focused on the study of methods and means to know, understand, assess, classify and represent different aspects of hazards and disasters.

A catastrophe (from the Latin *catàstrofa* and the ancient Greek *katastrophē*, "to overturn") is a large harmful event with great and irreversible consequences.[4] According to some classifications the principal catastrophic risks can be divided into four homogeneous classes. The first class catastrophic risks consist of natural catastrophes (such as pandemics and asteroids) that are not directly determined by technology or human labor. The second class consists of scientific risks as "laboratory or other scientific accidents involving particle accelerators, nanotechnology [...] and artificial intelligence". Instead of the first class risks these catastrophic risks are directly caused by technology. The third class consists of unintentional man-made catastrophes that determine phenomena such as "exhaustion of natural resources", "global warming" or "loss of biodiversity". Finally, the fourth class of catastrophic risks consists of intentional or "deliberately, perpetrated" catastrophes such as "nuclear winter, bioweaponry, cyberterrorism and digital means of surveillance and encryption". Even though they are determined by the use of technology, these are warfare risks that can be considered intentional acts of violence and not accidental.

According to the International Society for Environmental Protection classification, hazards are physical or chemical agents capable of causing harm to persons, property, animals, plants or other natural resources. Technological accidents are the potential consequence of one of that events and are caused by technical, social, organizational or operational failures ranging from minor accidents (i.e. single toxic agents) to major accident (industrial, chemical or nuclear accidents). Some other observers consider technological accidents in a more strict sense as "accidental failures of design or management relating to large-scale structures, transport systems or industrial processes that may cause the loss of life, injury, property or

[3] Perrow (1984).

[4] Walter (2008).

environmental damage on a community scale". When these events have long-run effects they are considered chronic technological disaster.

Some studies identify seven major classes of technological hazards ordered on a three-fold scale of severity. According to this classification, the most severe technological hazards are the multiple extreme hazards (i.e. nuclear war, recombinant DNA, pesticides). In the second level of the scale there are the extreme hazards, respectively caused by intentional biocides (chain saws, antibiotics, vaccines), persistent teratogens (i.e. uranium mining, rubber manufacture), rare catastrophes (i.e. LNG explosions, commercial aviation crashes), common killers (i.e. auto crashes, coal-mining diseases such as black lung), diffuse global threats (i.e. fossil fuel and CO_2 release, ozone depletion). In the third and lower level there are the so-called simple technological hazards.

In late 1990s, trying to provide "technical and organizational tools for the prevention, mitigation and the relief of disasters an International Working group appointed by United Nations drafted an indicative list with different type of actions which can constitute technological hazards:

- Release of chemicals to the atmosphere by explosion, fire
- Release of chemicals into water (groundwater, rivers etc.) by tank rupture, pipeline rupture, chemical dissolved in water (fire)
- Oil spills in marine environment
- Satellite crash (radionuclides)
- Radioactive sources in metallurgical processes
- Other sources of releases of radionuclides to the environment
- Contamination by waste management activities
- Soil contamination
- Accidents with groundwater contamination (road, rail)
- Groundwater contamination by waste dumps (slowly moving contamination)
- Aircraft accidents
- Releases and contaminations as consequence of military actions (e.g. depleted uranium) or destruction of facilities
- Releases as consequence of the industrial use of biological material (e.g. viruses, bacteria, fungi).[5]

8.2 From the Nineteenth to the Early Twentieth Century

Natural hazards and disasters are phenomena with which human societies has always been accustomed to live since antiquity.[6] But even man-made threats, technological hazards and disasters cannot be considered a prerogative of modern

[5] Krejsa (1997).

[6] Kates (1971) and Nash (1976).

societies. Therefore, there is no doubt that the rise of industrial society, the modernization process and the spread of technology determined, after the First and the Second Industrial Revolution, a dramatic increase of harmful events. Hazard management in developed societies has consequently shifted from risks associated with natural harmful events to those arising from technological development and application.

Since early nineteenth century, but especially after the second half of the century, industrial accidents, maritime disasters, railway and public transportation wrecks became unavoidable aspects of most advanced societies. According to some studies major disasters occurred in this period can be classified into three main categories that can also be applied to some contemporary technological accidents. These main categories are respectively: large scale engineered structures (public buildings, bridges, dams), industry (manufacturing, storage and transport of hazardous materials, power production) and public transports (sea, rail, air).

Fire can be considered one of the most relevant agents in large scale structures disasters. From the Great Fire of London of 1666 (13,200 houses burned down) to the Great Chicago Fire of 1871 (18,000 houses burned down, about 300 victims) and from the Vienna Theatre Fire of 1881 (850 dead) to the Iroquois Theatre incident in Chicago (571 dead) these kind of disasters were extremely common in late nineteenth century and early twentieth century industrial societies. According to some of the first "disaster specialists", at the turn of the century the death toll resulting from theatre fires in the nineteenth century England was nearly thousand people.[7]

Between other major large scale structures disasters occurred in these years there are the collapse of Tay Bridge in Scotland in 1879 (75 dead) and the failure of South Fork Dam in Pennsylvania in 1889 (more than 2,000 victims). For the engineering elite, these calamities represented a sort of shock that eventually led to a more precise codification of building regulations. Another important consequence of these events was in the approach. "Scientific speech and the rhetoric of risk" supplanted "the didactic language of the pulpit": "instead of waiting for bridges to collapse or people to be burned alive in opera houses, structural engineers and social psychologists were employed to predict the effectiveness of design and the psychology of the crowds in danger".

Steam-boiler explosions, fires and other industrial accidents determined new kinds of threats and damages and putted at risk not only safety of the workers but also life and properties of communities close to factories and industrial plants. Even disasters in minefields were particularly frequent during the nineteenth century: in United Kingdom, particularly in Wales coalfields, there were recorded several accidents such as the gas explosions at the Albion Colliery in 1894 (almost 300 deaths) and at the Universal Colliery in 1913 (439 victims). In the United States the number of documented mine accidents with five or more deaths through 1876–1921 was 497. Many accidents occurred also in main European minefields especially in

[7] Gerhard (1899).

Belgium, Germany, Poland and Russia. One of the most deadly mine accident of the early twentieth century was the Courrières mine disaster occurred in 1906 (more than 1,000 victims, some of them children) near the French city of Lens.[8]

Besides the accidents directly caused by technical malfunctions or human negligence in using machinery or during production or extraction processes, should also be considered those disasters and emergencies caused by environmental events indirectly influenced by human actions. During the nineteenth century the poor land management and the urbanization process of major industrial cities in some cases multiplied the disruptive effects of floods and landslides causing damages and casualties. The severe impact of the great urban and industrial agglomerations on river basins and the lack of modern hygiene and health legislations were other factors that determined some environmental emergencies such as the Great Stink of London occurred during the summer of 1858, when the river Thames became a sort of huge sewer. This episode posed the question of the urban pollution and of the management and channeling of drinking water and wastewater to insure safety of the population and prevent the repeated outbreaks of cholera.

Directly related to the process of industrialization can also be considered phenomena such as air pollution episodes which occurred and had so much popular echo during nineteenth century. The pollutants emitted from the chimneys as a result of the different stages of the production process, mixed with the fumes produced by the coal for civilian uses, repeatedly caused huge emergencies that in some cases had dramatic consequences on public health. These events were registered especially in some great cities of the United Kingdom, in some areas of central and Western Europe, but also in more industrialized areas of the United States (Chicago, Pittsburgh, St. Louis, Cincinnati). One of the most serious episodes of nineteenth century was recorded in London between 1879 and 1880. Despite legislation on emission of smoke (introduced since the 1960s, following the studies of Robert Angus Smith on the effects of acid rains),[9] a heavy cloak of fog mixed with smoke remained for months on the city. The visibility was nearly zero: people that went out of the house to walk, was forced, to not get lost, to proceed along the walls of buildings. This phenomenon had also serious consequences on public health. According to some sources the London smoke of 1879–1880 increased the mortality rate of 220 %. The peculiar "London Pea Soup Fog" described in some novels by Charles Dickens or painted by Claude Monet, by 1905 was called with a new term: smog (smoke plus fog).

Besides the industrial hazards, the public transport hazards and disasters were probably the most relevant threats to public safety during nineteenth and early twentieth century. The development of modern mass transport systems influenced a relevant debate on the question of the safety of the passengers of the traditional means of transport (e.g. ships) and of the new ones: from train lines of urban

8 Karmis (2001).

9 Smith (1872).

transport, to the automobiles and other motor vehicles for transporting people and goods.

Even excluding war-time disasters and only considering those accidents occurred between late nineteenth century and the beginning of the 1900s, the list of maritime disasters is impressive: from the incident that involved the *Princess Alice*, a Thames river paddle steamer which sank after a collision in 1878 (about 700 victims) and the French passenger steamer *La Bourgogne* that was sunk after a collision on July 1898 (about 550 victims) to the Danish steamship *Norge*, sank near Rockall Island in 1904 (more than 600 victims). But the real annus horribilis for maritime disasters was 1912. In that year, in fact, occurred not only probably the most famous naval incident in the history of maritime civil transportation, the sinking of the *Titanic* (1,517 deaths), but also the sinking of the Spanish steamship *Príncipe de Asturias* (about 500 victims) and the disaster of the Japanese ship *Kirchemaru* (1,000 deaths).[10]

The list of rail disasters is equally long. Apart from early accidents that involved the early steam trains (in many cases because of the explosion of boilers), other disasters were caused by clashes, derailment of trains or the collapse of bridges. To this list must be added those disasters occurred on subway lines (e.g. the Paris Metro disaster of 1903, 84 victims).[11] From 1833 to 1918, at least 8,803 deaths are attributed to railroad crashes, about 35.7 % of total amount of accidents registered from 1833 to 1975.

All these disasters contributed to place the question of the adoption of safety standards to prevent further accidents and insure the safety of persons. The protection of workers and new legislative measures against industrial risks were also strongly demanded by trade union organizations and left-wing parties. This process involved both private and public subjects and generated a debate on technical and insurance matters that had a great influence on national governments and promoted the adoption of laws on prevention and safety, thus accentuating the role of the State in these areas.

The discussions on security and prevention and emergency management sanitation related to the production process were relevant aspects of the debate around the so called "unhealthy industries" that led to the first public health legislations: from the British Public Health Act of 1875 to the public health provisions contained in the legislation adopted by the Italian government of Francesco Crispi in 1888, which was in turn inspired by French legislation.

Another important indirect consequence of the industrial accidents was the development of the occupational medicine, that had a growing importance in prevention of occupational diseases starting to investigate on the relations between some diseases and certain manufacturing processes. For example, the link between the exposure to asbestos dust and some serious lung affections was emphasized and

[10] Schlager (1995).

[11] Schlager (1994).

confirmed in observations of doctors and experts in occupational diseases since early twentieth century up to *the Merewether and Price Report* of 1930.[12]

In most cases, however, until the second post-war years the health and the safety of the workers during the processes of production, were still considered some marginal issues. This was primarily because of the relationship between the workplace and certain diseases even when had dramatic connotations—e.g. the thousands workers that died from ancylostomiasis during the work of the St. Gotthard tunnel in 1888—was considered as inevitable. With the result that safety and health of the workers on the job was monetized or simply considered as a technical matter. This approach was partly due to the difficulties (in some cases to the impossibility) of making in the public domain the documentation of many of the environmental disasters which occurred in that period.

In general, the emergence of increasingly sophisticated techniques of "civil protection", generally applied in case of natural disasters, was further facilitated by technological and scientific progress, but also from a revision of knowledge and approach to professionalism and skills that, since mid nineteenth century, brought to the affirmation of the concept of expertise and of the figure of the expert, a professional with special knowledge and skill, specifically prepared and formed to solve technical questions, prevent and manage disasters.

8.3 The First Post-war Years

The beginning of the studies upon technological disasters is conventionally set in the first postwar period, after the so-called Great Halifax Explosion that occurred on December, 6th 1917.

On that occasion, the Canadian port of Halifax, a strategic stopover for ships engaged in the supply of the troops deployed in the Great War, was destroyed by what, at least until the launch of the atomic bomb on Hiroshima, would have been the most violent explosion ever caused by man. The explosion was the result of the collision between a Belgian military cargo and a French ship carrying thousands of tons of explosives, acids and benzene highly flammable.

Focusing on the disaster at the port of Halifax (the death toll was about 2,000 deaths and 9,000 wounded), Samuel Henry Prince, researcher at the Columbia University, published a study that is considered the first contribution of a new discipline focused on the study of disasters and their social implications.[13] Even this study contains some general claims about social consequences of a disaster that modern researchers on disasters have challenged, this study can be considered an important turning point. Until then, the studies that dealt with disasters, including technological ones, almost exclusively reconstructed the events, often following the

---

[12] Merewether and Price (1930).

[13] Prince (1920).

rituals and the "rhetoric of disasters" that traditionally characterize the chronicles of these events. This literature, mainly anecdotal, which boasted a huge production and often had a large number of readers, simply explained the technological disasters in terms of fate, incorrect use or technical malfunction rather than attempt to scientifically analyze the causes that led to the disastrous event.

Born as eminently sociological discipline, the "disastrology", for the multiple valences of its object of study, progressively developed a variety of approaches: from those that would have deepened analyze the impact of disasters, whether natural, technological or environmental or those that who would have studied the dynamics of a disaster (to better identify the most suitable answers) to those that would instead focus on the study of individual and collective reactions. During the first post-war years began to be studied also the psychological and behavioral aspects related to disasters, including technological ones. In this respect, these researches were largely influenced by the first studies upon social behavior, psychology of the crowds and the reactions of the masses in the face of exceptional situations and of danger, issued during the two last decades of nineteenth century. Focused on the collective behavior, the studies issued during these years were characterized by extremely detailed references on various disasters, and tried to explain individual and collective reactions to critical events occurred on environmental, social and situational fields.[14]

In this phase, the growing presence of industrial settlements in densely populated areas had as a consequence many serious accidents. The Great Molasses Flood of Boston in 1919, for example, was caused by the disastrous effects of the explosion of a tank containing molasses in a suburb of the city (21 killed and 150 injured). This incident, as the Triangle factory fire of New York in 1911 (146 young workers' death) contributed to the adoption of new laws on safcty at work. Another serious accident was the explosion of ammonium sulfate and ammonium nitrate fertilizer occurred at the BASF plant in Oppau, a suburb of Ludwigshafen, Germany, in 1921 (about 600 killed and 2,000 injured). Equally tragic (130 deaths) were the consequences of the Cleveland East Ohio Gas Explosion occurred on October 1944 in a district of the city.

During this period, besides to the traditional sources of industrial accidents occurred the more often incidents related to the consequences for the public of erroneous procedures in production processes. In addition to the recurring cases of poisoning from industrial food adulterated or contaminated, such as the scandal that at the turn of the century involved the Chicago meatpacking industry, also occurred the first episodes of mass poisoning caused by medicines. One of the most serious cases happened in 1937 in the United States and was caused by the Elixir sulfanilamide, a compound containing diethylene glycol, a highly toxic emulsifier. It was the worst incident of mass poisoning of medicine of the twentieth century, and claimed the lives of over 100 people. This case began in evidence all the limitations of the previous legislation on medicines, and led in the USA to the adoption of a

[14] LaPiere (1938).

modern legislation: the Federal Food, Drug, and Cosmetic Act. The risks to public health directly and indirectly resulting from industrial production began to be actually perceived by the collective consciousness as a social problem, and gradually involved the public authorities in assuring public and health security. This was a relevant turning point that had important consequences after the Second World War.

During this time, the technological development and the coming of mass society determined the emergence of new hazards and technological disasters. Particularly important were the new threats related to transports. The large diffusion of cars and other private motor vehicles caused, especially in the United States a dramatic increase of accidents. Moreover, the rail disasters began to be characterized not only, as in the past, by the clash between trains but also by explosions caused by derailments of freight trains used to transport flammable or highly hazardous chemicals.

The beginning of the civil aviation had also as a negative consequence some incidents. The most serious disaster that deeply impressed public opinion between the two world wars was certainly the explosion of the German airship Hindenburg, burned during the docking maneuver at Lakehurst (New Jersey) on May, 6th 1937 (36 victims).

Regarding to naval disasters, the sinking of the *Titanic* led to the adoption of more stringent safety standards, internationally recognized, on passenger ships: so even though the naval disasters still occurred the number of victims was reduced significantly. The Port of Halifax tragedy produces the same consequences for merchant ship carrying hazardous materials. This improvement in safety levels, however, did not prevent the recurrence of severe accidents.

8.4 The Second Post-war Years

In August 1945, the bombing of Japanese cities of Hiroshima and Nagasaki dramatically showed to the world the terrible, impressive and destructive power of the energy released by the process of atomic fission. For over a decade, from the program for the building of the atomic bomb (the Manhattan Project, launched secretly in 1942) until the early 1950s, the nuclear technology was exclusively a prerogative of military authorities of the United States and then, after 1949, of the Soviet Union and few other countries. During these years nuclear powers tended to discourage any potential application for civilian purposes of the atomic energy, fearing that this might lead to proliferation of nuclear weapons.

The first experiments to create electricity through nuclear reactors were held in 1951. In 1953, the President of the United States, Eisenhower, announced the intention of the USA government to promote international collaboration to exploit atomic energy for civilian purposes. A similar statement was made the following year by Lewis Strauss, that was the chairman of the Atomic Energy Commission. These declarations officially inaugurated the programs for the realization of the first electronuclear facilities. In June 1954, the Obninsk nuclear plant was the first site to product energy using this new technology.

The collaboration between scientists and technicians to exchange knowledge and experiences on this new form of energy was enshrined at the first United Nations conference on nuclear technology, which met in Geneva in 1955. Since mid 1950s first commercial nuclear plants became operational: the first, in 1956, was that of Calder Hall at Windscale, England, followed the next year by the Shippingport nuclear plant in Pennsylvania.

At this stage, in order to better coordinate research programs, adopt uniform security standards and promote the peaceful use of atomic energy, several important international organizations were created. In 1955, the General Assembly of the United Nations unanimously approved a resolution which established the United Nations Scientific Committee on the Effects of Atomic Radiation (UNSCEAR). Composed of an international pool of scientists, the Committee was entitled to collect and evaluate information on the ionizing radiation. In March 1957, the Treaties of Rome, that also created the European Economic Community (EEC), established the European Atomic Energy Community (EURATOM). In 1960, to ensure to the signatory countries the regular supply of material for the energy power production, this organization promoted the creation of a special Supply Agency. Eventually, in 1967, EURATOM was incorporated within the EEC. A few months after the creation of EURATOM, in July 1957, was also set up the International Atomic Energy (IAEA). In early 1958, the Council of the Organization for European Economic Co-operation (OEEC), few years later renamed Organization for Economic Co-operation and Development (OECD), also created the European Nuclear Energy Agency (ENEA), whose initial task was to promote the use of Atomic Energy for civilian use in the pioneering and experimental phase.[15]

The security of the exploitation of nuclear energy was obviously one of the core issues of the major international agencies. This particular attention to possible catastrophic implications of atomic energy exploitation derived from the character of the still experimental technologies, and by the high radioactivity of the waste materials produced from the process of nuclear fission, which began to be stored, depending on their risk and time decay, in special shielded deposits.

Despite the high security procedures already adopted, in this phase some significant incidents still occurred. Some, like the Castle Bravo test done in the Bikini atoll (Marshall Islands) in 1954, were the result of the release of radioactive material occurred as a result of military experiments. Other incidents involved directly nuclear facilities. Among the most serious can be mentioned the fire and the release of radioactive material into the surrounding environment that occurred in 1957 in the British nuclear central of Windscale and the steam explosion and meltdown of the Stationary Low—Power Reactor Number One (SL-1) occurred in 1961 in an experimental reactor of the United States Army.

[15] Sagan (1993).

The release of radioactive substances was a problem that emerged since the early 1960s, in civilian but especially in military facilities: items of possible contamination emerged in the case of US plants of Hanford, in Washington State, and especially in the Kyshtym radiation release occurred in 1957 in the nuclear facilities situated in Majak, near Chelyabinsk, in Soviet Union, considered the second most serious nuclear accident in history. In both cases, however, for reasons of strategic nature, investigations on possible effects on the environment and health of civilians were allowed only many years later.

The extraordinary economic and industrial development that occurred after the end of World War II led to a dramatic increase in industrial production and in internal and international trades, but also significantly increased the episodes of technological accidents. The impressive rise of the carriage of dangerous goods by rail, for instance, determined many incidents, some of them particularly serious. The list of train accidents caused by the explosion of hazardous substances is tragically long: among the most serious occurred during the second post-war years in the United States there are: the Meldrim disaster of 1959, when the derailment of a train load of butane caused 23 victims and the freight train fire in Rosedale, California, of 1960 (14 deaths). Similar accidents also occurred in Europe. One of the most impressive was the Langenweddingen rail crash of 1967, when near Magdeburg, at that time in the East Germany, a local train and a truck that transported petrol collided in a level crossing's barriers (94 victims).

In these years, the development of civil aviation multiplied the number of flights consequently rising the accidents. The air disasters occurred during this period were generally determined by technical or structural failure, human error or clashes (occurred during takeoff or landing) or adverse weather conditions. In response to these disasters, in order to prevent further accidents the public and private airlines companies and national and international authorities decided to increase safety standards and controls on civil aviation.

The growth in commercial traffic by sea of dangerous or potentially harmful substances caused relevant accident such as that one that occurred, in April 1947, in Texas City. The port was devastated by a terrible explosion caused by a fire on French ship *Grandcamp*, carrying a cargo of ammonium nitrate (576 victims). This kind of accidents also had a dramatic impact on the environment. Particularly serious were those that involved ships carrying crude oil. Among the most serious disasters were the sinking of the tanker *Torrey Canyon*, that carried more than 120,000 tons of oil, which occurred in 1967 off the coast of Cornwall and the collision in 1970 that involved the *Othello* tanker off the coast of Sweden. In both cases, the oil spilled from tankers caused serious environmental damages and contaminated the coastlines. As for the incidents of passenger ships, even in presence of far more serious naval disasters, the naval incident that dramatically impressed the collective memory was the collision between the cruise ship *Andrea Doria* and the Swedish steamer *Stockholm* occurred in 1956 off Nantucket (51 victims). This was because of the sinking of the luxury passenger ship *Andrea Doria* reminded the disaster of the *Titanic*.

Besides to these threats, the modernization process determined the emergence of some new typology of technological disasters such as those occurred during the development of rocket technology. This new form of propulsion, which was experimented in the early 1930s, definitely characterized the second post-war years and the space race. The first accident occurred during the pioneering development of these particular technologies can be considered the one that cost the live of Max Valier, a South-Tyrolean pilot who was killed by the explosion of a rocket in 1930. The most serious incidents took place, however, with the first space flights during the Cold War years and the space race between United States and the Soviet Union. For this reason, a quantitative assessment of the accidents occurred during this period must take into account the possibility that some events can be kept secret for reasons of national security. Excluding those happened during test flights on aircraft and those that involved ground staff, the most serious accidents occurred during the testing of space vehicles or training in special places. Among them those which involved the Soviet cosmonaut Valentin Bondarenko (burned in a special oxygen chamber in 1961) and the members of the mission Apollo 1 (Virgil Grissom, Edward White and Roger Chaffee) died in 1967 when a fire destroyed the spacecraft during a training exercise. Far more numerous, but with no victims, were instead the accidents that occurred during the missions.

Apart for these factors, the increase of technological disasters during the second post-war years was determined by the effects on public health of industrial production. This increase could undoubtedly be considered a sort of negative aspect of mass production and consumerism but it was also the result of the discovery (or the confirmation) that some pathologies were directly linked to some industrial production processes. This was particularly evident for some chronic occupational diseases. For example, by the gradual emergence of serious syndromes in many of those who were previously employed in the manufacture of asbestos, a mineral that was extensively used in construction and in the naval, railway, automobile, chemical, food, metal, plastic productions. The dramatic link between exposure to the fibrous mineral and certain diseases emerged in these years in many industrial sites all over the world. The long latency and cumulative effects, but delayed in time, deriving from the exposure to some pollutant agents were also discovered combining medicine, statistics and other important data on workers' health and safety collected by trade unions, social and health organizations both private and public.

An important role in the emergence of a new attitude towards technological hazards management was also played by the media that contributed to increased knowledge of the public opinion of these threats. But probably the most relevant contribution to this change was due to the complaints also of associations and organizations representing the interests of victims and workers and, especially since the end of the 1950s and the 1960s, to the emergence of the ecologist and environmentalist movements.

The effects on public health and environment of pollution caused by population growth, industrial expansion, and technological change were another relevant problem in main industrialized countries during the so called "golden age" of western capitalism.[16] On December 1952, for instance, a yellowish dense fog remained for 5 days all over the metropolitan area of London, semi-paralyzing private and public transports and causing serious damages to public health. During these same decades also emerged a growing concern over public health and safety associated with hazardous wastes.

Among the most impressive technological disasters registered in the second post-war years there is undoubtedly that which occurred in a coastal station of Japan: Minamata.[17] In this village, since the mid 1950s, some medical investigations suggested the possible link between exposure to residues of the production of fertilizer that the Chisso Corporation chemical industry usually discharged into the sea, and the very serious disease to the central nervous system and peripheral which affected many of the inhabitants. The so called "Minamata disease" was one of the first documented cases of mass poisoning with mercury. As revealed by the findings of a commission of inquiry appointed by central authorities, the inorganic mercury discharged into the sea from Chisso, was transformed as a result of the bacteria into methylmercury, entered the food chain, poisoning fish and consequently persons who ate. The effects of the poisoning were particularly serious on pregnant women and consequently on newborns. Around 3,000 persons were recognized as victims of the poisoning, although the estimates about the number of inhabitants contaminated over the years ranged from 10,000 to 30,000.

The fact that these news came from a country like Japan, that in the collective imagery was associated with the atomic bomb and its devastating effects on civilian populations, deeply impressed the international public opinion. Associations of the victims were able to create a movement that not only contributed to inform public opinion on the effects of the disaster and that promoted the determination of liabilities but also achieved positive results in terms of prevention and control. The strong opposition of the company involved in the case but also of some public authorities and even of some part of local population—who for instance feared that the investigation could halt the production and even cause the closure of the factory—were gradually overcome by a growing mobilization that reached its climax at the beginning of the 1970s, with a documentary-complaint, which was projected abroad and found wide echo in North America. As result of this campaign for civil rights of the victims of the disaster were then signed several compensation agreements with Chisso for damages caused and were created centers for the study and treatment of victims.

The Minamata disease, but also the Niigata disease (both syndromes caused by mercury), the Yokkaichi asthma (caused by the presence of sulfur dioxide and nitrogen dioxide) and Itai-Itai (caused by poisoning from cadmium) were among the first cases that were publicized by mass media, and that by consequence impressed international public opinion. Equally serious incidents of poisoning by heavy metals were also registered in other countries—notably in Iraq (in 1956 and 1971 in the so called Basra poison grain, a mass methylmercury poisoning incident

[16] Hobsbawm (1994).

[17] George (2002).

that involved a shipment of contaminated wheat) and in Guatemala—but, in spite of their severity, they received only a low media exposure.

Behind every industrial accident directly or indirectly caused by technological factors in some cases there is no fatality but negligence and underestimation of risk. These factors played a decisive role—though not exclusive—in the Kiev large-scale mudslide occurred near a brick factory in 1961 and in Italy, in the Vajont disaster. This accident was caused by the terrible effects of two giant waves caused by the collapse of a landslide over a dam built below Mount Toc on Longarone and other small villages near Belluno, in North-Eastern Italy. While in the Minamata disaster Chisso Corporation acted for a long time in complete secrecy, denying any involvement even after the discovery of contamination, the Vajont disaster was a tragedy foretold. In fact, before the collapse of the dam, the electrical company that was the owner of the site, the local and central authorities were informed about the risks of a possible structural failure of the dam by the inhabitants, by some journalists and above all by direct signs. Notwithstanding this, the village of Longarone and other small communities near the dam were completely swept away in the night of October 9, 1963 (1,900 victims). In the days that followed the disaster domestic public opinion was informed in detail by the newspapers, radio and television news about the apocalyptic effects caused by the "wave of death", high more than seventy meters, that hit the villages below the dam, about the blanket of rock and mud that had erased everything, about the rescuers and about the stories of those who, still in shock, were pure fatality escaped the disaster. The controversies that had accompanied the construction of the dam on an area geologically at risk did not find that a pale echo on the newspapers and other media. When the rituals and rhetoric of disasters ended, the Vajont disaster gradually vanished from the media.[18]

8.5 From 1970s to the Early 1980s

For many reasons, the early 1970s are an important turning point in the history of disasters. It was in fact during these years that the major international organizations began to work, for the first time, in this area in an organic way. In 1972, for example, the UN created a special office, the United Nations Disaster Relief Organization (UNDRO) with special competence in rescue and humanitarian aid and primarily operated in case of natural disasters.

Since the mid-1970s, industrial and technological disasters had a great increase throughout the world, especially in those countries the most affected by the process of modernization and development. This was particularly evident in Asia, where these accidents assumed in many cases the character of a real emergency. All this was partly the result of the development process involved in these areas as a result of the globalization process and of the gradual relocation of industrial plants from

[18] Reberschak and Mattozzi (2009) and Silei (2013).

the more developed countries as a consequence of the advent of post-industrial society. For instance, Vila Parisi, a *favela* (slums) in Cubatao (Brazil) was a site of one of major industrial accident on February, 1984, becoming a sort of case-study, almost an emblem ("the chemical dirtiest town in the world") of the "destructive powers of the developed risk industry".

The environmental damages caused by the process of industrialization and urbanization that were recorded during this period involved both ancient and more recent industrialization areas. There are many examples in this regard: from the heavily industrialized areas of certain neighborhoods and suburban areas of large cities (e.g. Greenpoint, Brooklyn, where from early 1800s had been built large refineries and where in the late 1970s it was discovered extensive contamination of soil) to the huge industrial and urban areas in Asia whose harmful emissions, along with the fires used to deforestation policies, caused the so-called Southeast Asian haze of 2006.

It should also be considered the enormous environmental damage and health of the population (in terms of incidence of cancer, deformities, chronic diseases and serious pathologies) recorded in many areas subjected to decades of indiscriminate mining or oil exploitation as the region of the Bolivar Coastal Field, the largest oil field in South America, or the region of the Niger Delta and of natural resources exploitation, as in the case of the Aral Sea, almost dried as a consequence of the massive exploitation of its waters planned since the second post-war years by central and local Soviet Union authorities to promote intensive cultivation and for industrial and civil purposes and whose shores were contaminated the systematic use of herbicides and pesticides.

The progressive globalization of technological hazards and therefore their nature and the relevant environmental impact of many technological disasters led the United Nations. In 1974, for instance, the Secretariat of the United Nations Scientific Committee on the Effects of Atomic Radiation (UNSCEAR) was moved from New York to Vienna and its functions were linked with the United Nations Environment Programme (UNEP).

As a result of many serious accidents that have occurred in the 1980s and confirming the need of a growing awareness on the potential risk to people and environment of technological hazards, the United Nations declared the 1990s, the "International Decade for Disaster Reduction" and especially to make a reorganization of the international organisms which until then had dealt with the management of emergencies. In 1992, the United Nations Disasters Relief Organization has been transformed into the Department of Humanitarian Affairs (DHA), based in Geneva and New York. The new body has new tasks to operate in a more specific and effective way in case of disasters. In order to operate in the field of prevention has been established a special secretariat, the International Strategy for Disaster Reduction (ISDR). In 1994, the ISDR has organized a conference in Yokohama, which led to the drafting of the Yokohama Strategy and Plan of Action for a Safer World. Although this work of prevention and response to disasters has been largely focused on natural disasters, the ISDR has been increasingly concerned, especially since the mid 1990s, to include the technological risks, paying particular attention to the developing countries. This broader approach to the problems of disasters has resulted in the report *Living with Risk: A global review of disaster reduction initiatives*. Published in 2004, this report has confirmed the importance, even in cases of technological disasters, of the creation of national and international early warnings systems and procedures to cope with possible emergencies.[19]

The occurrence of technological accidents was also the result of the inevitable and unpredictable character of technological systems themselves. Increasingly complex, these systems consist of multiple elements that because of to their structural features make it difficult if not impossible, effective security controls. With the result that accidents, more and more difficult to predict, have become "normal events". Moreover, according to some sociological interpretations, at this stage difficulties arise not only in the perception and risk assessment but also in risk management: the traditional external risks of the industrial society, that national institutions were able to anticipate and manage—for example through welfare policies or through safety legislation—in this stage have been multiplied. Compared with the old industrial risks this new typology of threats, as result of technological progress, the so-called manufactured risks, are much more difficult to predict and their negative impact can be multiplied by the fragmentation of powers and responsibilities in the prevention and management of security.[20]

According to some literature, the increasing impact of technological disasters has also been caused by specific decisions of main developed countries in the management of risks. Faced with difficulties in the regulation of technological risks, many governments adopted a neoconservative and neoliberal approach to these issues and gradually shifted the responsibility for protecting against risks from public agencies to individuals. This new prudential tendency is one of the traits that characterize the society of risk and uncertainty and postmodern ethics.

Technological accidents in the transport sector are an example of the new approach of the neo-liberal approach to risk management. Some technological accidents were an indirect consequence of marketing strategies harmful for the safety of consumers. An example this conduct is the case of the Ford *Pinto* motor vehicles models that were produced by the General Motors industries in 1970s. For a defect in design, these low-cost cars, that were conceived to a large segment of the market, had a high risk of fire and explode in case of collision. Despite being aware of a defect in design, and although this might conflict with the principle of acceptable risk, the manufacturer decided—it was said on a cold analysis of cost-benefit analysis—not to intervene. The model was finally withdrawn from the market only in 1978 after numerous accidents, many of them fatal, and after that a journalist had denounced executives of Ford motor to have deliberately produced a "firetrap". The case of the Ford *Pinto* became for some interpretations a symbol of criminal behavior of the great corporations and for other ones a classic example of bad decision-making processes.[21]

---

19 United Nations Office for Disaster Risk Reduction (UNISDR) (2004).

20 Kates et al. (1985).

21 Dowie (1977), Strobel (1980), Gioia (1996) and Lee (1998).

Apart from cynical assessments, in this case ethically condemnable, assumed as a result of simple cost-benefit analysis, this peculiar approach to hazards was characterized, also in the transport sector, by a gradual process of privatization and deregulation. In the railways and public transportation services this led some advantages in terms of better competition and cost of services but it had also some negative consequences, including a significant reduction, as a consequence of the need to reduce the operating costs, of some safety standards. An equal process of deterioration of safety standards for the lack of control caused by organizational weaknesses, but above all for the economic crisis even occurred, from the end of the 1980s, in many countries of Central and Eastern Europe and in the former Soviet Union. In fact, it was here that there have been two of the major accidents ever registered in the history of rail and subway lines: the disaster of Ufa in 1989, when a gas pipeline exploded at the passage of two trains (at least 600 victims), and the disaster of Baku, in Azerbaijan, in 1995, when a metro train caught fire and 337 people die.

Fire accidents in major tunnels, in the railway sector and in the transport sector in general, represent another type of technological disaster. Particularly significant were the Channel tunnel fire and the Ekeberg fire of 1996 and the Gotthard incident of 1997, that severely damaged equipment and tunnel infrastructures. But the most serious accidents can probably be considered those that occurred in 1999 in the Mont Blanc tunnel (39 victims) and in the Tauern tunnel in Austria (12 victims). In both cases, among the main causes of the disaster there were accidents caused by heavy vehicles. Another dramatic disaster was the fire that occurred in an ascending railway car in a tunnel in Kaprun, Austria, on November 2000 (155 victims). After these incidents national governments and European Union institutions have intervened to prevent the recurrence of similar catastrophic events that receive enormous media attention and consequently have a dramatic impact on the public.

Among the most serious technological accidents in transportations there are the so called Boiling Liquid Expanding Vapor Explosions (BLEVE): among them the Los Alfaques Disaster occurred in 1978 near Tarragona, in Spain when a road accident generated the explosion of a tanker truck (217 victims) and the isobutane and propane explosions from a train derailment registered at Murdock, Illinois in 1983. This kind of disaster are very frequent in railway transportation: in 1997, in Germany a regional passenger train collided with a freight train that carried petrol tankers causing an explosion. Another serious accident caused by the explosion of flammable substances carried by rail was to Mont-Saint-Hilaire, Canada, 1999. Always a BLEVE, finally, was the cause of the disaster happened in the railway station of Viareggio (near Lucca, Italy) on July 2009. In that case, the derailment of a freight train determined the rupture of a tank of gas that caused several explosion and a fire that killed 30 people. The seriousness of these incidents lies not only in the particular dangers of the substances but also in the possibility to occur the so-called domino effect, much feared event by those involved in technological and industrial disasters multiplying material damages but most of all the number of victims.

The most dramatic BLEVE disasters cases involved industrial plants. Some of them have been particularly severe: the explosion at the refinery in Feyzin, not far from Lyon, which took place in 1966 and cost the lives of 18 persons, for example, is considered the first industrial disaster of recent French history. Other serious incidents were recorded in the United States (respectively in Kingman, Arizona, 1974, Texas City, Texas in 1978, Murdock, Illinois, 1983). The most serious disaster, however, happened in 1984 in San Juan Ixhuatepec, a center near Mexico City. The chain of explosions of liquid petroleum gas (LPG) from the tanks of the facilities struck the nearby village, causing death (from 500 to 600 deaths) and destruction. Equally devastating explosions were recorded in 1989 in the Houston Ship Channel in Texas (23 victims), and in 2000 in the Nigerian oil pipelines (over 300 victims).

Shipping casualties have continued to be even at this stage a significant part of technological accidents related to transports. This kind of disaster has affected not only merchant ships but also ferries and passengers ships, often with dramatic consequences. In the latter case, the main causes of disasters are not so much been strictly technical in nature but have often been determined, especially in disasters occurred in developing countries, by lack of maintenance of the ferries, overcrowding, severe negligence and errors for maneuver. In this type of disaster, the increase of traffic by sea and especially the need for continuous supply of crude oil by the most developed countries has added more advanced ones caused by the sinking of the supertankers. These technological accidents are particularly feared for their environmental consequences, immediate and in the long term.

The sinking of the *Exxon Valdez* in March of 1989 off the coast of Alaska, is considered one of the most serious environmental disaster in history until BP's *Deepwater Horizon* oil rig accident in the Gulf of Mexico of 2010. The tanker spilled about 40 million liters of crude oil into the sea, and the oil eventually polluted large parts of the coasts. Quite similar to the *Exxon Valdez* disaster were the fire and oil spill of the Norwegian tanker *Mega Borg*, off the coast of Texas (1990), the *Haven* oil tanker disaster in the Gulf of Genoa, Italy (1991), the *Aegeum Sea* disaster, off the Spanish port of La Coruna (1992), the sinking of the Liberian tanker *Braer* off Shetland Islands (1993), the *Sea Express* oil spill, off the coast of Wales (1996), and the *Prestige* sank in face of the Spanish coast (2002), that polluted the coasts of Galicia, as also as those of Portugal and France.[22]

Technological progress, the lowering in airfares and the process of globalization have also determined the rise in the trafficking of goods and passengers by air. Among the various consequences, this process, which began at the end of the 1970s, called into question the air safety standards. According to some interpretations, policies for reducing costs of management adopted by airlines during these years had as a consequence a reduction of the inspections and maintenance procedures and therefore a reduction of safety standards, especially in the low cost airlines or in those in greater financial difficulties. All this would have accordingly led to an increased risk of accidents. The disaster of *ValuJet* flight 592, crashed in the Everglades in southern Florida in May of 1996 is, according to some

[22] Silei (2011).

interpretations, a direct consequence of this situation. The disaster occurred by a fire caused by a load of flammable material placed in an aircraft of the *ValuJet* airlines, a discount operator that offered services at very low prices also saving on security procedures for boarding and maintenance of aircrafts. Trying to solve these problems, that emerged in many former Soviet bloc countries after the collapse of the USSR, the international inspection agencies and national and international air control authorities have responded in various ways: for example, the European Union, has decided to periodically publish lists of blacklisted airlines, that is unsafe companies, whose flights are subject to prohibition of operation in the European Union.

Among the technological disasters linked to transport could also be included those that occurred during aerospace missions. Since mid 1970s, with the success of the lunar missions the space race has suffered a drastic downsizing. Although reduced in numbers, the missions continued in the following years using different technologies and spacecrafts. Since the launch by the NASA of the Space Shuttle program on two occasions, these missions have had a disastrous outcome: in 1986, when the Space Shuttle *Challenger* exploded in flight because of the malfunction of a component of Solid-Fuel Rocket Booster, and in 2003, when the Space Shuttle *Columbia* disintegrates during return for a breach opened in a wing. In both cases, crew, consisting of seven astronauts died. Decision-making and risk assessment procedures adopted by the NASA were severely criticized.[23]

The increase in the complexity of technological hazards was the basis for new approaches to the study and the management of these issues. One of this is the kindunology, a definition introduced in France during the 1980s. This "hazard science" aims to analyze the natural disasters and those produced by both technological and economic-financial factors, by starting with a proper risk assessment and combining approaches from the natural sciences, social sciences and humanities.

This new, broader approach to the issue of disaster was also the result of some incidents, mainly industrial, that took place between the 1970s and 1980s, and that caused serious environmental contamination by chemicals, with equally dramatic consequences for health populations.[24]

The disaster of the Flixborough chemical plant (28 deaths and 40 serious injuries), an English village near Scunthorpe, which occurred in 1974, was one of the first examples of this type of accident. It was however the serious contamination occurred in 1976 in the small town of Seveso, in northern Italy, and in its surrounding localities to represent a genuine case school. A failure in a reactor of the ICMESA, a chemical plant of the Swiss Hoffmann La Roche company, officially used for the preparation of basic products for the cosmetics industry, released into the environment highly polluting substances, including the dioxin. These substances contaminate an area of over 1,800 ha in the town of Seveso, Meda, Desio, Cesano Maderno and other centers of the province of Milan. In addition to the serious

---

[23] Vaughan (1996).

[24] Brickman et al. (1985) and Mitchell (1996).

environmental disaster, the population suffered from short-term effects (almost 450 persons were found affected by skin lesions or chloracne) and long term, since it was recorded and proved the increase of serious diseases caused by exposure to pollutants.

The Seveso accident highlighted the lack of adequate inspections in industries with high risk but the need for a legislation on industrial risks prevention at a European level.[25] In fact, accidents in the production of trichlorophenol had occurred repeatedly in the past (in Germany, the Netherlands, France) and factories such as ICMESA were present in all countries of the European Community. After Seveso, the European Community put the problem of dealing with various aspects of these types of hazards and accidents: safety standards, inspections, monitoring tools, the procedures for the determination of penal liability and of civil penalties. The result was the adoption in 1982 of the first Seveso Directive, a set of rules which applied specifically to the risks and consequences that would have been registered from major industrial accidents.

Meanwhile, the growing energy needs and the oil crisis caused by the rising cost of crude oil, led many countries to step up the exploitation of energy produced through nuclear power. In early 1970s, the European Nuclear Agency, renamed to Nuclear Energy Agency (NEA), began a new phase, by paying greater attention to coordination of national programs but also to the issues of safety and respect of environmental legislations that was meanwhile adopted by national governments under the pressure of the environmental organizations. Also the United States operated a reorganization of their structures for the control of atomic energy industry, creating in 1974 a special government agency, the Nuclear Regulatory Commission (NRC). The old Atomic Energy Commission (AEC) was later replaced by the Energy Research and Development Administration, which in 1977 became the United States Department of Energy.

The question of the security of nuclear facilities became an urgent necessity when some serious accidents occurred in some major industrial sites for the production of atomic energy. In 1973 the nuclear power plants in the British Windscale were again the scene of an accident. The best known accident, however, happened a few years after in the United States in the nuclear power plant of Three Mile Island, located close to Harrisburg, the capital of Pennsylvania. In March 1979, a major reactor accident at the plant caused the release of significant amount of radiations. Even if there were not victims and were avoided more serious consequences, such as the total meltdown of the reactor core, the Three Mile Island accident was the most severe ever recorded in U.S. history, brought about great alarm throughout the country and all over the world. As for the Windscale site, despite requests from the environmental groups to proceed with its decommissioning, it remained active and, in 1981, for reasons of image it was decided to change its name to Sellafield.[26]

[25] Centemeri (1996).

[26] Sells et al. (1982) and Walker (2004).

Although not linked to the exploitation of nuclear energy, in 1979 there was another serious technological disaster, the so called Sverdlovsk anthrax leak. The accident occurred when there was an accidental release of anthrax spores from a military facility for treatment of biological weapons located in the city of Sverdlovsk in the Soviet Union. The final death toll was about a hundred victims. Although the Soviet Union authorities denied, because of military secrecy, the Sverdlovsk incident was one of the first documented cases of contamination by biological agents.

8.6 Technological Disasters During the World Risk Society: From Bhopal to Fukushima

The accident in Bhopal, the worst technological disaster ever happened instead was the first example of industrial disaster in the age of globalization. On the night of 2 December 1984 a deadly cloud of methyl isocyanate, leaked from the Bhopal chemical plant in India, owned by American multinational Union Carbide, and resulted in the following days about 20,000 deaths and a number of intoxicated that was calculated between 200,000 and the half-million. That of Bhopal has been considered a component failure accident, a disaster caused by a series of deliberate omissions, disorganization and of sloppy management. Despite some earlier episodes, in fact, the plant continued in producing and stocking large quantities of highly toxic chemicals that were later the cause of the disaster.[27]

The case of Bhopal has thus helped to develop a different sensibility towards this type of disasters that lead to the adoption of some legislation that impose an assessment of the possible harmful consequences of the processing of toxic chemicals, taking into account the so-called worst-case scenario. Moreover, despite the resistance of Union Carbide, the incident has raised the issue of the proper taking of responsibility by the large multinational corporations for accidents with catastrophic consequences for the population and the environment occurred in their industrial facilities.

Apart from being one of the most serious technological accidents ever occurred in history, the accident at Chernobyl, in its dynamics, was also a paradigmatic event not only from the point of view of the emergency management. The incident happened April 26 1986 in the nuclear power station at Chernobyl, in Ukraine, then in the Soviet Union and was caused by overheating of a reactor during a test. The heat and the excessive pressure caused an explosion then a fire. In the same time, a cloud of highly radioactive material was released in the environment.[28]

Although Soviet authorities were aware of the seriousness of the accident, they organized operations of fire containment and control the leakage of radiation from

---

27 Shrivastava (1987).

28 Mould (2000).

the central without informing and evacuating the civilian population living in cities close to facilities, so that thousand people were exposed to the effects of the radioactive cloud. It was only after the reporting of an excessive level of radiation in the atmosphere by Swedish nuclear central and international diplomatic pressures that the Soviet government officially admitted the accident. The inhabitants of the town of Pripyat and areas contaminated (about 350,000 people) were then evacuated, but much later than it would have been necessary. The leakage of radioactive particles caused an increase in the levels of radiation in Scandinavia, Eastern Europe and Central Europe but also in the West, forcing the authorities to carry out stringent checks on the level of radioactive substances in the air, in the environment and especially in food.

Apart from the heavy economic loss resulted from the explosion of the plant and especially by the complete evacuation of entire highly populated residential areas within a radius of tens of kilometers, the death toll caused by the Chernobyl accident cannot be quantified precisely. The official reports drawn up by the main organizations involved in the assessment of radiation exposures and health effects, about 65 victims, in large part the so called "liquidators", people who participated in containment operations near the reactor and at least another 4,000 presumed dead. But it should be considered that the estimated amount of people directly contaminated is about 600,000. Given the long-term effects of radiation on human health, it is impossible to know how many people have been (or will be) affected by serious illnesses (malformations, tumors, leukemia) for exposure to radioactive materials released to the environment at the disaster.

The Chernobyl accident had a dramatic impact from many points of view. First, it put in evidence in the eyes of world public opinion of risks arising not only from nuclear weapons but also from the exploitation of nuclear energy.[29]

In many countries, environmental and anti-nuclear movements made Chernobyl as a symbol of the catastrophic consequences from the exploiting of this source of energy. In Italy, after a referendum held in 1987 it was decides to abandon the use of nuclear energy for civilian use and proceed with the decommissioning of all nuclear site that were built up to that time. The accident also influenced a systematic review of all procedures so far adopted to deal with nuclear emergencies. The first consequence was to put the need to avoid in the future omissions or reticent behavior by the authorities and to inform the public and the authorities of the neighboring countries of possible threats. Chernobyl pose perhaps for the first time the problem of the adoption of international measures for monitoring and for early warnings to prevent and facing major technological disasters.

Moreover, in order to assess the severity of accidents in nuclear facilities and radiation sources and transport, the International Atomic Energy Agency (IAEA) adopted a classification system by considering three different areas of impact (people and the environment to exposition radiation; radiological barriers and control, defense-in-depth) and developed an International Nuclear Event Scale.

---

[29] Cameron et al. (1988) and Beck (1992).

According to this scale each event is classified at seven levels: level 0 indicates anomalies without safety significance, also called "deviations", levels 1–3 are defined "incidents"; levels 4–7 "accidents." Every increased level on the scale is ten times greater of the previous one. In this scale, the Chernobyl disaster has been considered a "major accident" (level 7), the Kyshtym disaster of 1957 is defined a "serious accident" (level 6), while Windscale (1957) and Three Mile Island events has been valuated as "accidents with wider consequences" (level 5).

The medical research conducted by national and international (e.g. activities of the United Nations Scientific Committee on the Effects of Atomic Radiation) also promoted specific medical and scientific knowledge to face the consequences of the accident. The biggest consequence of Chernobyl was also in law and of international security. In fact, the disaster exposed the serious deficiencies of international legal safeguards. The result was the signature, under the auspices of the IAEA, of two international Conventions on the subject of nuclear safety and a more general change in the provisions concerning the safety of power to exploit atomic energy. Moreover, even in the wake of the dissolution of the Soviet Bloc, the main international organizations on nuclear energy began to concentrate their efforts in monitoring the safety levels of plants to produce energy as well as deposits of waste storage in the world. In spite of these efforts the occurrence of new accidents was not avoided. Among these events, could be mentioned the incident occurred in 1999 in the central Japanese Toikamura, the shutdown of a nuclear reactor in Sweden in 2006 and the dramatic Fukushima Daiichi Nuclear Power Plant disaster (classified as level 7 accident according to the International Nuclear Event Scale) occurred on March 2011 as a consequence of the tsunami of the Tōhoku earthquake.

Besides this type of disasters it should be added those events occurred in other areas that previously were shown to be particularly at risk of serious accidents. The mining sector, especially in China, Soviet Union and later in Russia and other countries of the former Soviet Bloc, but also in many developing countries, continued to record incidents cost the lives of dozens and sometimes hundreds of miners. Even the dams and ponds, despite the improvement of construction techniques and safety procedures, have maintained a high level of risk. The collapse of Banquiao and Shimantan reservoir dams in China, which took place in 1975 (but that Chinese authorities confirmed only after many years), with its 26,000 dead is probably one of the worst disasters in history. Serious incidents of this type have been registered also in western countries. In Italy, despite the serious Vajont disaster experience, in July 1985, the collapse of two retention basins and the resulting landslide of mud, sand and water was the cause of the disaster of Stava (268 victims). The disaster of Stava, as previous accidents, such as the Aberfan disaster of 1966 in Wales, which claimed 144 lives or the Buffalo Creek disaster happened in the United States in 1972 (125 victims), placed in evidence weaknesses in control procedures and the bad management by the authorities responsible for management of invaded that led to talk of the inevitability of the disaster to institutional and even cultural reasons.[30] In this respect,

[30] Mclean and Johnes (2000) and Stern (2008).

the proposal to prevent the recurrence of incidents of this kind has been to promote the creation of special authorities for the regulation, monitoring and control of technological risks for occupational safety, health and the environment. In the United States, for example, Congress has entrusted this task to special federal agencies like the Environmental Protection Agency (EPA), the National Institute for Occupational Safety and Health (NIOSH) and in particular the Occupational Safety and Health Administration (OSHA) who operates on the principles of the so-called whistle-blower protection, or rigid capillaries and safety standards.

The major technological accidents occurred in the early twenty-first raised again the issue of increasing environmental impact of such events. A prime example of these accidents was the disaster occurred in Baia Mare in January 2000 with the release in the Danube river of cyanide and heavy metals from a gold processing plant in Romania and the subsequent poisoning of the water, flora and fauna in vast areas of Romania, Serbia, Hungary and Bulgaria. Another case of a serious industrial accident was the Enschede fireworks disaster, which occurred on the same year in a warehouse in the suburbs of the Dutch city. The explosion killed 22 people and injured about 950. Equally relevant was the explosion occurred in a fertilizer factory near Toulouse, in France, in 2001. The explosion, according to some reconstructions accidental, according to other sources result of an attack of terrorists, caused 29 deaths and 2,500 serious injuries, significant material damages and serious environmental consequences.

These disasters have called into question the EU legislative framework on environmental and industrial risks. The European Parliament has begun a review of the Seveso II Directive, which in the meantime had been adopted in 1996. It has thus come to the Directive 2003/105/EC, which has broadened the scope of previous directives to other potentially hazardous industrial facilities and tightened the procedures to be taken in the event of an accident. To better monitor the possible industrial accidents, the European Commission has therefore established a special system of reporting and complaint, the Major Accident Reporting System (MARS) managed by a special body, the Major Accident Hazards Bureau (MAHB). The debate on these rules is then taken up with the progressive enlargement of the EU Member States. The discipline of industrial risks in the enlarged Europe is one of the central issues in the debate on technological risks at Community level.

The release of poisonous chemicals continued to be cause of recurrent emergencies: major accidents occurred in these years has been the Camelford water pollution incident, occurred in 1988 in Cornwall; the release of huge quantities of sulfur dioxide into the environment from the Al-Mishraq sulfur plant near Mosul, Iraq (2003); the Jilin City and the Formosa Plastics chemical plant explosions both of them occurred in 2005 respectively in China and in Texas.

Apart from potentially catastrophic environmental impact not only of man-made disasters, but more generally the model of global development (above all the issue of global warming caused by emissions of CO_2) and the increasingly tenuous boundary between natural and technological disasters, the actual debate over these topics is focused on whether to consider technological disasters also events such as the extensive black outs and power outages or other technological failures in communications.

The discussion is still open; what seems certain is that the constant technological development, the globalization of economies and societies will have as a consequence a further increase of the technological risks, making it necessary timely and effective answers by national and above all international authorities in order to ensure collective and environmental security.

References

Beck U (1992) The risk society: on the way to an alternative modernity. Sage, Newbury Park

Brickman R, Jasanoff S, Ilgen T (1985) Controlling chemicals: the politics of regulation in Europe and the U.S. Cornell University Press, Ithaca

Cameron PD, Hancher L, Kühn W (eds) (1988) Nuclear energy law after Chernobyl. Graham & Trotman, London

Centemeri L (2006) Ritorno a Seveso. Il danno ambientale, il suo riconoscimento, la sua riparazione. Bruno Mondadori, Milano

Dowie M (1977) Pinto madness. Mother Jones, vol 2, pp 18–32

Dynes RR (1970) Organized behaviour in disaster. Disaster Research Center, Ohio State University, Columbus

Fritz CE (1961) Disaster. In: Merton RK, Nisbet RA (eds) Contemporary social problems. Harcourt, Brace and World Inc., New York, pp 651–669

George TS (2002) Minamata: pollution and the struggle for democracy in postwar Japan. Harvard University Press, Cambridge

Gerhard WP (1899) The safety of theatre audiences and the stage personnel against danger from fire and panic. British Fire Prevention Committee, London

Gioia DA (1996) Why i didn't recognize pinto fire hazards. In: Ermann MD, Lundman RJ (eds) Corporate and governmental deviance. Oxford University Press, New York, pp 139–157

Hobsbawm EJ (1994) Age of extremes. The short twentieth century, 1914-1991. Michael Joseph, London

Karmis M (2001) Mine health and safety management. Society for Mining, Metallurgy, and Exploration, Littleton

Kates RW (1971) Natural hazards in human ecological perspective: hypothesis and model. Econ Geogr 47:438–451

Kates RW, Hohenemser C, Kasperson JX (eds) (1985) Perilous progress: managing the hazards of technology. Westview, Boulder

Krejsa P (1997) Report on early warning for technological hazards. UN-IDNDR Secretariat, Geneva

LaPiere RT (1938) Collective behavior. McGraw-Hill, London

Lee MT (1998) The Ford Pinto case and the development of auto safety regulation, 1893-1978. Bus Econ Hist 27(2):390–401

Mclean I, Johnes M (2000) Aberfan. Government and disasters. Welsh Academic Press, Cardiff

Merewether ERA, Price CW (1930) Report on effects of asbestos dust on the lungs and dust suppression in the asbestos industry. HMSO, London

Mitchell JK (1996) The long road to recovery: community responses to industrial disasters. UN University Press, New York

Mould RF (2000) Chernobyl record: the definitive history of the Chernobyl catastrophe. Institute of Physics Publishing, Bristol

Nash JR (1976) Darkest hours: a narrative encyclopedia of worldwide disasters from ancient times to the present. Nelson-Hall, Chicago

Perrow C (1984) Normal accidents: living with high risk technologies. Basic Books, New York

Prince SA (1920) Catastrophe and social change. Based upon a sociological study of the Halifax disaster. Columbia University Press, New York

Quarantelli EL (1966) Organization under stress. In: Brictson RC (ed) Symposium on emergency operations. System Development Corporation, Santa Monica, pp 3–19

Reberschak M, Mattozzi I (eds) (2009) Il Vajont dopo il Vajont 1963-2000. Marsilio, Venezia

Sagan SD (1993) The limits of safety: organizations, accidents, and nuclear weapons. Princeton University Press, Princeton

Schlager N (1994) When technology fails. Significant technological disasters, accidents, and failures of the twentieth century. Gale Research, Detroit

Schlager N (1995) Breakdown. Deadly technological disasters. Visible Ink Press, Detroit

Sells DL, Wolf CP, Shelanski VB (eds) (1982) Accident at Three Mile Island: the human dimension. Westview Press, Boulder

Shrivastava P (1987) Bhopal. Anatomy of a crisis. Ballinger Publishing Company, Cambridge

Silei G (2011) Imparare dalle catastrofi: disastri navali e incidenti petroliferi. Storia e Futuro 27:1–32

Silei G (2013) Una lezione dai disastri? Il Vajont e l'alluvione di Firenze. Storia e Futuro 33:1–15

Smith RA (1872) Air and rain. The beginnings of a chemical climatology. Longmans, Green, and Co, London

Smith K, Petley DN (1990) Environmental hazards. Assessing risk and reducing disasters. Routledge, London

Stern GM (2008) The Buffalo Creek disaster: how the survivors of one of the worst disasters in coal-mining history brought suit against the coal company… an won. Vintage Books, New York

Strobel LP (1980) Reckless homicide: Ford Pinto's trial. And Books, South Bend

United Nations Office for Disaster Risk Reduction (UNISDR) (2004) Living with risk: a global review of disaster reduction initiatives. UN Publications, Geneva

Vaughan D (1996) The challenger launch decision. Risky technology, culture, and deviance at NASA. University of Chicago Press, Chicago

Walker JS (2004) Three Mile Island: a nuclear crisis in historical perspective. University of California Press, Berkeley

Walter F (2008) Catastrophes: une histoire culturelle. Seuil, Paris

MIX
Papier aus verantwortungsvollen Quellen
Paper from responsible sources
FSC® C105338

If you have any concerns about our products,
you can contact us on
ProductSafety@springernature.com

In case Publisher is established outside the EU,
the EU authorized representative is:
Springer Nature Customer Service Center GmbH
Europaplatz 3, 69115 Heidelberg, Germany

Printed by Libri Plureos GmbH
in Hamburg, Germany